9/0·2023
FSC HE

Tourism, Recreation and Climate Change

ASPECTS OF TOURISM
Series Editors: Professor Chris Cooper, *University of Queensland, Australia*
Dr C. Michael Hall, *University of Otago, Dunedin, New Zealand*
Dr Dallen Timothy, *Arizona State University, Tempe, USA*

Aspects of Tourism is an innovative, multifaceted series which will comprise authoritative reference handbooks on global tourism regions, research volumes, texts and monographs. It is designed to provide readers with the latest thinking on tourism world-wide and in so doing will push back the frontiers of tourism knowledge. The series will also introduce a new generation of international tourism authors, writing on leading edge topics. The volumes will be readable and user-friendly, providing accessible sources for further research. The list will be underpinned by an annual authoritative tourism research volume. Books in the series will be commissioned that probe the relationship between tourism and cognate subject areas such as strategy, development, retailing, sport and environmental studies. The publisher and series editors welcome proposals from writers with projects on these topics.

Other Books in the Series
Classic Reviews in Tourism
Chris Cooper (ed.)
Progressing Tourism Research
Bill Faulkner, edited by Liz Fredline, Leo Jago and Chris Cooper
Managing Educational Tourism
Brent W. Ritchie
Recreational Tourism: Demand and Impacts
Chris Ryan
Coastal Mass Tourism: Diversification and Sustainable Development in Southern Europe
Bill Bramwell (ed.)
Sport Tourism Development
Thomas Hinch and James Higham
Sport Tourism: Interrelationships, Impact and Issues
Brent Ritchie and Daryl Adair (eds)
Tourism, Mobility and Second Homes
C. Michael Hall and Dieter Müller
Strategic Management for Tourism Communities: Bridging the Gaps
Peter E. Murphy and Ann E. Murphy
Oceania: A Tourism Handbook
Chris Cooper and C. Michael Hall (eds)
Tourism Marketing: A Collaborative Approach
Alan Fyall and Brian Garrod
Music and Tourism: On the Road Again
Chris Gibson and John Connell
Tourism Development: Issues for a Vulnerable Industry
Julio Aramberri and Richard Butler (eds)
Nature-based Tourism in Peripheral Areas: Development or Disaster?
C. Michael Hall and Stephen Boyd (eds)

For more details of these or any other of our publications, please contact:
Channel View Publications, Frankfurt Lodge, Clevedon Hall,
Victoria Road, Clevedon, BS21 7HH, England
http://www.channelviewpublications.com

ASPECTS OF TOURISM 22
Series Editors: Chris Cooper (*University of Queensland, Australia*),
C. Michael Hall (*University of Otago, New Zealand*)
and Dallen Timothy (*Arizona State University, USA*)

Tourism, Recreation and Climate Change

Edited by
C. Michael Hall and James Higham

CHANNEL VIEW PUBLICATIONS
Clevedon • Buffalo • Toronto

This book is dedicated to:

Al Gore, in recognition of his political foresight and initiatives with respect to global change;

Geoff McBoyle and Geoff Wall, for their pioneering work in the field of tourism and climate change research and education; and

Jody Cowper and Linda Buxton, for their untiring support

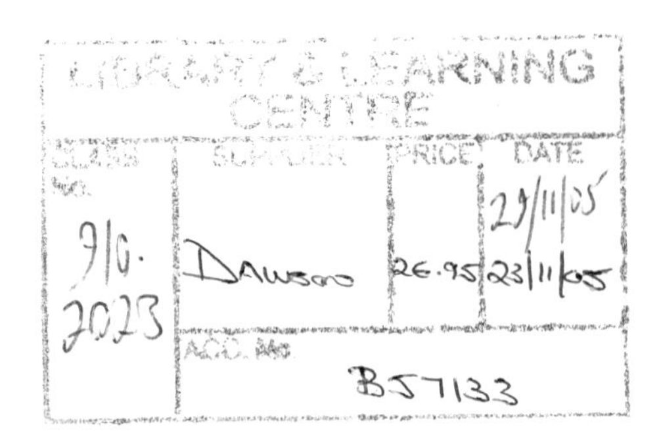

Library of Congress Cataloging in Publication Data
Tourism, Recreation and Climate Change/Edited by C. Michael Hall and James Higham.
Aspects of Tourism: 22
Includes bibliographical references and index.
1. Tourism–Environmental aspects. 2. Outdoor recreation–Environmental aspects.
3. Climatic changes–Environmental aspects.
I. Hall, Colin Michael. II. Higham, James E.S. III. Series.
G155.A1T592435 2005
363.738'741–dc22 2004016906
A catalog record for this book is available from the Library of Congress.

British Library Cataloguing in Publication Data
A catalogue entry for this book is available from the British Library.

ISBN 1-84541-004-1 (hbk)
ISBN 1-84541-003-3 (pbk)

Channel View Publications
An imprint of Multilingual Matters Ltd

UK: Frankfurt Lodge, Clevedon Hall, Victoria Road, Clevedon BS21 7SJ.
USA: 2250 Military Road, Tonawanda, NY 14150, USA.
Canada: 5201 Dufferin Street, North York, Ontario, Canada M3H 5T8.

Typeset by Florence Production Ltd.
Printed and bound in Great Britain by the Cromwell Press.

Contents

Contributors

Carlo Aall, Western Norway Research Institute, PO Box 163, N-6851 Sogndal, Norway (caa@vestforsk.no).

Bruno Abegg, Economic Geography, University of Zurich, Winterthurerstrasse 190, CH8057 Zurich, Switzerland (brunoabegg@yahoo.com).

Susanne Becken, Landcare Research, Canterbury Agriculture & Science Centre, Gerald Street, Lincoln, New Zealand (beckens@landcareresearch.co.nz).

Rolf Bürki, Economic Geography, University of Zurich, Winterthurerstrasse 190, CH8057 Zurich, Switzerland (rbuerki@bluewin.ch).

Jean-Paul Ceron, Centre de recherche interdisciplinaire en droit de l'environnement, de l'aménagement et de l'urbanisme (CRIDEAU: Université de limoges, CNRS, INRA), 32 rue Turgot, 87000 Limoges, France (ceron@chello.fr).

Stephen Craig-Smith, The School of Tourism and Leisure Management, The University of Queensland, Queensland, Australia (s.craigsmith@mailbox.uq.edu.au).

Keith Dewar, Faculty of Business, University of New Brunswick – Saint John, PO Box 5050, Canada, E2L 4L5 (kdewar@unbsj.ca).

Ghislain Dubois, Tourisme Environnement Consultants (TEC), 89 rue de la République 13002, Marseille, France (Ghislain.Dubois@tec-conseil.com / www.tec-conseil.com).

Hans Elsasser, Economic Geography, University of Zurich, Winterthurerstrasse 190, CH8057 Zurich, Switzerland (elsasser@geo.unizh.ch).

Lotta Frändberg, Department of Human and Economic Geography, Göteborg University, Sweden (lotta.frandberg@geography.gu.se).

C.R. de Freitas, School of Geography and Environmental Science, The University of Auckland, Private Bag 92019, Auckland, New Zealand (c.defreitas@auckland.ac.nz).

Stefan Gössling, Department of Service Management, Lund University, Helsingborg, Sweden (stefan.gossling@humecol.lu.se).

C. Michael Hall, Department of Tourism, University of Otago, Dunedin, New Zealand; Honorary Professor, Department of Marketing, Stirling University, Scotland (cmhall@business.otago.ac.nz).

S.J. Harrison, Honorary Research Fellow, Department of Geography, University of Dundee; St John's Vicarage, 129 Main Street, Spittal, Berwick upon Tweed, Northumberland, YD15 1RP, UK (Johnandaveril@aol.com).

James Higham, Department of Tourism, University of Otago, Dunedin, New Zealand (jhigham@business.otago.ac.nz).

Karl G. Høyer, Western Norway Research Institute, PO Box 163, N-6851 Sogndal, Norway (kgh@vestforsk.no).

R.C. Johnson, Mountain Environments, Callander, Scotland (info@mountain-environments.co.uk).

Urs Koenig, Economic Geography, University of Zurich, Winterthurerstrasse 190, CH8057 Zurich, Switzerland (urs@koenigcoaching.com).

John B. Loomis, Department of Agricultural and Resource Economics, B-320 Clark Building, Colorado State University, Fort Collins, CO 80523–1172, USA (John.Loomis@colostate.edu).

Geoff McBoyle, Faculty of Environmental Studies, University of Waterloo, Waterloo, Ontario, Canada (gmcboyle@fes.uwaterloo.ca).

Sue Mather, Travel Research International, Beaconsfield, Buckinghamshire, UK (skm@travelresearch.co.uk).

Paul Peeters, NHTV Centre for Sustainable Tourism and Transport, Breda University of Professional Education, PO Box 3917, 4800 DX Breda, The Netherlands (peeters.p@nhtv.nl).

Allen Perry, Department of Geography, University of Wales – Swansea, Singleton Park, Swansea, SA2 8PP, Wales (A.H.Perry@Swansea.ac.uk).

R.A. Preston-Whyte, School of Life and Environmental Sciences, University of Natal, Durban, 4041, South Africa (preston@nu.ac.za).

Robert B. Richardson, School of Field Studies, Centre for Rainforest Studies, PO Box 141, Yungaburra, Queensland 4872, Australia (rrichardson@qldnet.com.au).

Lisa Ruhanen, The School of Tourism and Leisure Management, The University of Queensland, Queensland, Australia (uqlruhan@staff.uqi.uq.edu.au).

Daniel Scott, Faculty of Environmental Studies, University of Waterloo, Waterloo, Ontario, Canada (dj2scott@fes.uwaterloo.ca).

David G. Simmons, Environment, Society and Design Division, Lincoln University, New Zealand (dsimmons@lincoln.ac.nz).

Graham Todd, Travel Research International, Beaconsfield, Buckinghamshire, UK (gjt@travelresearch.co.uk).

L. Michael Trapasso, Director of the College Heights Weather Station, Department of Geography and Geology, Western Kentucky University, Bowling Green, KY 42101, USA (michael.trapasso@wku.edu).

David Viner, Climatic Research Unit, University of East Anglia, Norwich, NR4 7TJ, UK (d.viner@uea.ac.uk).

Geoff Wall, Faculty of Environmental Studies, University of Waterloo, Waterloo, Ontario, Canada (gwall@watserv1.uwaterloo.ca).

H.K. Watson, School of Life and Environmental Sciences, University of Natal, Durban, 4041, South Africa (watsonh@nu.ac.za).

S.J. Winterbottom, School of Biological and Environmental Sciences, University of Stirling, Stirling, Scotland (sj.winterbottom@stir.ac.uk).

Preface

Climate change has emerged as a major issue in tourism and recreation management and planning in recent years with a numbers of papers being published in journals such as *Climate Change*, *Journal of Sustainable Tourism* and *Tourism Analysis*. It has also been the subject of meetings of the Tourism and Climate Commission of the International Society of Biometerology and a special meeting of the World Tourism Organisation. Yet despite such interest in the subject, the present volume presents the first edited book published explicitly on the subject of the relationships between tourism, recreation and climate change. It therefore aims to present the work of some of the leading researchers in the field in what is undoubtedly one of the major challenges facing the tourism industry in the 21st century.

The book is divided into the three main sections. The first three chapters provide an introduction to the issues dealt with in the remainder of the book through an examination of the context of the tourism/recreation and climate change relationship. The second section provides a number of chapters that examine the effects of climate change on tourist flows and recreation patterns for specific geographical areas and locations with some of the policy implications also being noted. The third section examines tourist, recreationist and industry responses and adaptation to the issues of global climate change, although there is considerable interplay between the chapters in the second and third sections. The last three chapters in particular highlight the complexities involved in the tourism–climate change relationship by highlighting the difficulties that are inherent in making tourism sustainable.

Given the significance of global climate change for tourism, we trust that this volume will be of interest not only to students of tourism in the narrow sense, but also to industry and policy makers. Climate change is a politically charged subject, as the editors found in attempting to bring the book together, even the act of editing such a book seemed to raise issues of politics, turfdom and petty resentment. Yet, as a number of contributors to this book note, dealing with the massive implications and complexities of climate change requires not only greater cooperation and understanding between government, industry and researchers, but also between physical and social scientists, those who understand climate and those who understand the tourism and recreation phenomena. Without the development of such relationships our understanding and response to climate change and tourism relationships are bound to stay piecemeal.

For the editors this book also continues a long-standing interest in sustainable tourism, for James from a base in ecotourism, for Michael from the framework of contemporary mobility. In completing this book we would also like to acknowledge our colleagues in the Department of Tourism, School of Business, University of Otago, and particularly the assistance of Mel Elliott and Monica Gilmour. Michael would also like to acknowledge his hosting by the Department of Social and Economic Geography, University of Umeå in autumn 2004, and Dieter Müller in particular, which greatly contributed to the editing and completion of this book. We wish to thank all at Channel View Publications for their support for this project and particularly Sami Grover for his enthusiastic response when we originally approached him with the idea for such a book. In that vein we must also thank all of the contributors with chapters in this book for their enthusiastic and timely support. Similarly, the editing process was greatly assisted by Billy Bragg, Geoff Buckley, Nick Cave, Paul Kelly, Ed Kuepper, Lucinda Williams and Chris Wilson. Finally, we wish to thank our family, friends and colleagues for their ongoing support in our research and teaching endeavours.

C. Michael Hall and James Higham
Dunedin, New Zealand

Part 1: Context

Chapter 1
Introduction: Tourism, Recreation and Climate Change

C. MICHAEL HALL AND JAMES HIGHAM

In my view, climate change is the most severe problem we are facing today, more serious even than the threat of terrorism – David King, Chief Scientific Advisor to the UK Government. (King, 2004: 176)

Human-induced changes in the global climate system and in stratospheric ozone pose a range of severe health risks and potentially threaten economic development and social and political stability – Declaration of the Third Ministerial Conference on Environment and Health, 1999. (WHO, Regional Office for Europe, Global Change and Health, http://www.euro.who.int/globalchange)

It is one of the great truisms that everybody talks about the weather. However, in recent years, interest in the weather has grown as high magnitude storm events, floods, droughts, snowstorms and record high temperatures have become associated with potential changes in the world's climate. For example, the record high temperatures experienced in Europe in the northern hemisphere summer of 2003 focused enormous attention on climate-related issues. Paris experienced the highest night-time temperatures ever recorded on 11 and 12 August (25.5C), and several countries, including Belgium, Finland, Germany, Switzerland and the United Kingdom, also experienced record temperatures. The heatwave was unusual in that it affected several countries and persisted for at least ten days; in fact the whole northern summer (June, July, August) was much hotter than usual (Schär *et al.*, 2004; see also Perry, Chapter 5, this volume).

In France the Minister for the Elderly admitted that 10,000 people had most likely died because of the heatwave. In the last week of August, President Jacques Chirac addressed the nation saying that weaknesses in the French health system had contributed to these heat-related deaths. Despite similar heatwave conditions in the United Kingdom, with temperatures peaking on 10 August in Bogdale, near Faversham, Kent, where 38.5C (101.3F) was recorded, the British government response was much more low key. However, in October 2003 official figures released by the Office of National Statistics (ONS) suggested that the death toll in England and Wales as a result of the ten-day heatwave in August 2003 may have been around 2000 people, which was much higher than those admitted at the time. According to the ONS between 4 and 13 August there were 15,187 deaths in England and Wales, 2045 above the average for the previous five years. In commenting on the ONS figures Carvel (2003: 10) said: 'Although the statisticians were not yet able to provide an analysis of the ages of the victims and causes of death, it seemed almost certain that extreme heat was the reason for the higher mortality.' Indeed, more recent preliminary estimates of the impact of the European heatwave on mortality suggest that in England and Wales 2045 or 16% excess deaths occurred, in France 14,802 deaths (60%), Italy 3134 deaths (15%) and Portugal 2099 deaths (26%) (Kovats *et al.*, 2004). While reaction to the heatwave from European governments raised substantial issues regarding government preparedness for such extreme climate events in relation to public health, the forest and scrub fires in Portugal, Spain and France also created a powerful image in the media of the impacts of such heatwaves on the landscape. As Jose Manuel Durao Barroso, the Portuguese Prime Minister, stated, 'We are standing before a tragedy which is unprecedented in Portugal in terms of fires . . . We are facing an exceptional situation. It's been brought about by absolutely exceptional weather conditions, so we have to respond with exceptional measures' (BBC News, 2003). Arguably, the impacts of the 2003 heatwave on European perceptions of climate were even more stark because of the comparisons that could be made with the floods that affected central Europe the previous year (BBC News, 2002).

Given this kind of context it should therefore be of no great surprise that prospects of climate change have become the focus of media attention as well as substantial scientific debate (e.g. O'Riordan and Jäger, 1996; Houghton, 1997; Jepma & Mohan, 1998; Mendlesohn, 1999; Drake, 2000; Harvey, 2000; Sarewitz & Pielke Jr, 2000; Claussen, 2002; King, 2003). The extent of media coverage of global climate change issues is illustrated in Table 1.1, which shows major stories on climate change reported in the *Guardian* and *Observer* newspapers at the end of 2003/ beginning of 2004. While undoubtedly highlighting the range of issues associated with global climate change and some of the policy debates that surround them, the newspaper reports also begin to indicate the potential role of the media in influencing the public's perception of places and activities. Not only do the news stories indicate the potential direct impact of climate change on tourism, e.g. the sale of Scottish ski resorts (Seenan, 2004), but also indirect impacts because of changes to resources that are part of tourism product offerings, e.g. coral reef bleaching (Radford, 2004b), species extinction (Brown, 2004a), and changes to

Table 1.1 Key climate change-related stories in the *Guardian* and *Observer* newspapers in late 2003/early 2004

Date	*Story heading*	*Outline*
22 February	Now the Pentagon tells Bush: Climate change will destroy us	Secret security report to President Bush warns of rioting and nuclear war as a result of climate change, describes threat as greater than terrorism. According to the report climate change over the next 20 years could result in a global catastrophe costing millions of lives in wars and natural disasters and warns that Britain will be 'Siberian' in less than 20 years (Townsend & Harris, 2004).
19 February	Careful with that planet, Mr President	Diana Liverman was a senior climate adviser in the US. Back in the UK, she argues for American scientists to be freed from their fear of speaking out on global warming (Liverman, 2004).
19 February	Bonfire of the promises	The Global Climate Coalition, a powerful alliance of car makers, oil drillers and electricity generators, believes that the White House under President Bush shares their view that global warming is a hoax (Goldenberg, 2004).
19 February	The White Death	Refers to the bleaching of coral reefs as a result of increased ocean temperatures (Radford, 2004b).
14 February	Meltdown	Alaska is a wealthy huge oil producer but has suffered the consequences of global warming, faster and more terrifyingly than anyone could have predicted (Lynas, 2004).
14 February	Global warming forces sale of Scottish winter sports resorts	The future of skiing and snowboarding in Scotland appeared bleak after two of the country's five ski resorts were put up for sale after large financial losses (Seenan, 2004).
3 February	Summer heatwave matches predictions	A climate scientist warned that the heatwave in the northern summer of 2003 that killed thousands across Europe and saw temperatures in Britain pass 100F (38C) is a sign of things to come (Adam, 2004).

Table 1.1 continued

Date	*Story heading*	*Outline*
20 January	CO_2 limits suicidal for competitiveness, says industry	British industry urged ministers to undertake a drastic revision of their plans for cuts of up to 20% in carbon dioxide emissions and warned they could be suicidal for manufacturing's competitiveness (Gow, 2004b).
17 January	CO_2 cuts will raise prices, says industry	Britain's heavy industry, including power stations, will have to cut carbon dioxide emissions by 16% over the next few years under strict new national guidelines for implementing EU regulations (Gow, 2004a).
12 January	Freak summers 'will happen regularly'	Study suggests costly extremes of weather will become the norm (Radford, 2004a).
11 January	Giant space shield plan to save planet	Key talks involving the UK government's most senior climate experts produced proposals to site a massive shield on the edge of space, deflecting the sun's rays and stabilising the climate (Townsend, 2004).
9 January	Top scientist attacks US over global warming	Climate change is a more serious threat to the world than terrorism, David King, the UK government's chief scientist, writes in an article in *Science* magazine, attacking governments for doing too little to combat global warming (Brown & Oliver, 2004).
9 January	Midwinter spring is the new season	Climate change confusing wildlife (Brown, 2004b).
8 January	An unnatural disaster	Climate change over the next 50 years is expected to drive a quarter of land animals and plants into extinction, according to the first comprehensive study into the effect of higher temperatures on the natural world, killing off one million species, and one in ten animals and plants extinct by 2050 (Brown, 2004a).

Table 1.1 continued

Date	*Story heading*	*Outline*
8 January	Action now could still save some threatened species	Estimates of extinction risk associated with climate change should compel people to start thinking about the consequences of massive species loss (Thomas & Cameron, 2004).
21 December	Britain can start dreaming of a green Christmas with swallows	As temperatures rise in the UK, spring is earlier and snow will become only a memory (Jowit, 2003).
18 December	Earth is 20% darker, say experts	Scientists report that human activity is making the planet darker as well as warmer, with levels of sunlight reaching earth's surface having declined by up to 20% in recent years (Adam, 2003).
13 December	Drowning islands halt effort to postpone climate change talks	A coalition of 40 small islands, some of which are in imminent danger of disappearing beneath the ocean, blocked attempts by major states to delay climate talks for 18 months (Brown, 2004e).
12 December	Global warming kills 150,000 a year: Disease and malnutrition the biggest threats, UN organisations warn at talks on Kyoto	At least 150,000 people die needlessly each year as a direct result of global warming, three major UN organisations warned. The belief that the effects of climate change would become apparent in 10, 20 or 50 years time was misplaced, they said in a report. The changes had already brought about a noticeable increase in malnutrition, and outbreaks of diarrhoea and malaria (Brown, 2003d).
11 December	Extreme weather of climate change gives insurers a costly headache	Economic losses in the European agricultural sector because of the 2003 summer drought exceeded £7bn because of loss of crops and livestock, the insurance industry announced (Brown, 2003c).

Table 1.1 continued

Date	*Story heading*	*Outline*
11 December	Global warming is killing us too, say Inuit	The Inuit of Canada and Alaska are launching a human rights case against the Bush administration claiming they face extinction because of global warming. The Inuit claim that by repudiating the Kyoto Protocol and refusing to cut US carbon dioxide emissions, which make up 25% of the world's total, Washington is violating their human rights (Brown, 2003b).
10 December	Climate change doubles Britain's stormy weather	Britain has become twice as stormy in the past 50 years as climate change has forced the deep depressions that used to hit Iceland further south. The Hadley Centre for Climate Prediction and Research also reported that pressure changes in the atmosphere had caused storms to become more intense. Low pressure areas which bring high wind and rain are getting deeper, and the high pressure areas which bring calm, settled periods are getting stronger. The increased gradients between the two means more dramatic weather and expensive insurance claims (Brown, 2003a).
15 November	Shrinking ice in Antarctic sea 'exposes global warming'	Australian scientists report that frozen oceans that affect currents such as the Gulf Stream have decreased dramatically, suggesting that sea ice around Antarctica had shrunk 20% in the past 50 years (Fickling, 2003).
9 September	Warming warning for Antarctica	Researchers warned that the appearance of Antarctica will dramatically change in the next 100 years as ice melts, glaciers retreat, penguins move south and green plants begin to colonise bare rocks of the Antarctic Peninsula (Radford, 2003).

seasonal weather patterns (Brown, 2004b; Radford, 2004a). Undoubtedly, tourism will not be the only industry to feel the effects of climate change. However, as one of the main world industries in terms of employment and economic returns, and of particularly importance in a number of developing countries and small island states, any impacts on tourism will have substantial economic, social and political repercussions. Moreover, tourism is arguably even more susceptible to global climate change because of the reliance on the environment in many destinations for their attractiveness, especially in coastal and mountain regions, while outdoor recreation activities are also susceptible to climatic extremes. Tables 1.2 and 1.3 detail the ideal climatic requirements for some outdoor recreation activities.

The size of the international tourism industry is substantial. Preliminary estimates of full year results for 2003, published by the World Tourism Organization (WTO) (2004) indicate that even though the number of international arrivals fell by 1.2% (8.5 million) from the previous year, there were still some 694 million arrivals. With respect to 2004 the World Travel and Tourism Council (WTTC) (2004) forecast that the combined direct and indirect impact of the travel and tourism economy is expected to total 10.4% of the world economy and total 73.7 million jobs or 2.8% of total world employment. The direct and indirect impact of the travel and tourism economy is expected to lead to a total of 214.7 million jobs being dependent on travel and tourism or 8.1% of total employment. An international overview of the economic contribution of tourism undertaken by the WTTC (World Travel and Tourism Council, 2003) is shown in Table 1.4.

However, while the numbers provided by the WTO and the WTTC are impressive, they only give a partial perspective of the impacts of tourism. They do not account for the overall direct and indirect costs and benefits, particularly with respect to social and environmental impacts, nor do they provide any assessment of the opportunity costs that might be associated with tourism development. Nevertheless, they do provide an indication of why some governments and industry bodies are now starting to take the relationships between tourism and climate change seriously.

The Interrelationships between Tourism and Climate Change

Given the long awareness of the relationship between climate and tourism (see Paul, 1972; Adams, 1973; Mieczkowski, 1985; Harrison *et al.*, 1986; de Freitas, 1990, 2003, this volume; Smith 1990, 1993; Ewert, 1991; Harlfinger, 1991; Perry, 1997), it is perhaps surprising that an overt focus on the implications of climate change has occurred so recently (see

Table 1.2 Ideal climate-related requirements for summer water-based and dry terrain recreation activities

Water-based activities	*Motor boating*	*Water skiing*	*Sailing*	*Fishing*	*Swimming/ sunbathing*
Air temperature (0C)	15–35	18–35	10–35	15–30	15–30
Wind (km/h)	<50	<15	15–50	<15	<15
Water temperature (0C)	2–20	10–20	10–18	<18	15–20
Precipitation	Nil	Nil	Nil	Nil	Nil
Lake size:					
– Minimum (ha)	>80	>100	>30–>100	20–80	20–40
– Maximum (ha)	400	800	800	400	800
Lake depth (m)	1.5–2.5	>2.0	1.5–2.0	0.5–1.0	0.5–2.0
Carrying capacity	1 ha/boat	5 ha/boat	10 ha/boat	–	–
Dry terrain activities	***Camping***	***Picnicking***	***Golf***		
Air temperature (0C)	>10	10–25	10–30		
Wind (km/h)	<10	<20	<20		
Precipitation	Nil to light	Nil	Nil		

Source: After More (1988)

Table 1.3 Ideal climatic requirements for winter recreation activities

Environmental condition	*Nordic skiing*	*Alpine skiing*	*Snow shoeing*	*Snow-mobiling*
Snow depth (cm) – Minimum – Optimum	 20–30 60	 20–30 60	 20–30 60	 30 60
Snow density (g/cm^3)	<0.6	<0.6	0.2–0.6	0.4–0.1
Air temperature (0C)	–2 to –15	5 to –20	10 to –40	10 to –30
Snow making (0C)[1]	–3 to –15	–3 to –15	Not applicable	–3 to –15
Wind (km/h)	<20	<15	<45	<45
Wind chill (watts/m^2)	700	700	1600	1400

Source: After More (1998)
Note: 1 The lower the humidity, the more snow a system can make at a given temperature. This is because evaporation furnishes a significant part of the cooling in the snow making process. The lower the humidity, the more evaporation per unit of water, the more snow can be made.

Harrison *et al.*, 1986; Wall *et al.*, 1986; McBoyle & Wall, 1987; Ewert, 1991; Lipski & McBoyle, 1991; Wall, 1993; Wall & Badke, 1994; Agnew & Viner, 2001; Scott *et al.*, Chapter 3, this volume). According to Scott *et al.* (Chapter 3, this volume), the overall development of tourism and climate research can be categorised into four stages: an initial period of activity in the 1960s and 1970s, which has been labelled the 'formative stage', followed by a decade when very little research was published ('period of stagnation'). The first peer-reviewed journal publications on the implications of climate change for tourism or recreation appeared in the mid-1980s (Harrison *et al.*, 1986; Wall *et al.*, 1986). These publications are regarded as signifying the start of a third phase ('emergence of climate change'), which pre-dated the formation of the United Nations Intergovernmental Panel on Climate Change (IPCC) by two years. The final era is labelled 'maturation' reflecting the tremendous growth in the volume of research since the end of the 1990s and which is detailed in many of the chapters in the present volume. Arguably, the relative lack of interaction between climate and tourism research, until recently, may relate to the time period in which governments have started to pay attention to climate change issues. However, several other factors may be considered including:

- the overall perception of tourism and recreation as a 'serious' area of study in the physical and social sciences;

Table 1.4 WTTC Tourism Satellite Account global tourism estimates and forecasts

	2004 World US$ Billion	*2004 % of Total*	*2004–13 % Growth*[1]
Personal travel and tourism	2295	10.1	3.7
Business travel	525	–	3.7
Government expenditures	237	3.9	3.0
Capital investment	731	9.6	4.3
Visitor exports	605	6.0	7.1
Other exports	535	5.3	7.2
Travel and tourism demand	4927	–	4.6
Travel and tourism industry GDP	1375	3.7	3.6
Travel and tourism economy GDP	3787	10.3	3.9
Travel and tourism industry employment	69,738[2]	2.7	2.2
Travel and tourism economy employment	200,967[2]	7.7	2.4

Source: After WTTC (2003: 4). For most up to date figures, visit WTTC website: www.wttc.org

Notes:

1. Annualised real growth adjusted for inflation (%).
2. '000 of jobs.

- the relative lack of interaction between physical and social sciences with respect to impact analysis;
- lack of institutional support in the financing of such research;
- relative lack of baseline data;
- methodological difficulties in undertaking relevant analysis; and
- diversion into other, more attractive, areas of research.

Scott (2003; see also Scott *et al.*, Chapter 3, this volume) also suggested that one possible explanation for the lack of research on the tourism–climate relationship is that a resource management perspective is almost absent from tourism discourse. However, in contrast Hall and Page (2002) identified such a resource management approach as a significant theme in tourism and recreation research, at least with respect to geography and geographers. Pigram and Jenkins (1999) have similarly recognised a resource management approach in the context of outdoor recreation. Arguably, of far more significance is the scale at which research on tourism impacts has occurred and the means by which tourism has been conceptualised.

Most research on tourism impacts has tended to occur at the destination. This has meant that research has often focused on local factors rather than trying to incorporate all stages of the travel process (the tourism generating region, travel to and from the destination, and the destination itself) and their occurrence over time and space. Conceptualising tourism in terms of wider aspects of human mobility therefore has considerable importance with respect to assessing the complete impacts of tourism (Høyer, 2000; Gössling, 2002, Chapter 20, this volume; Gössling *et al.*, 2002; Hall & Williams, 2002; Frändberg & Vilhelmson, 2003; Hall 2003a, 2004, 2005; Coles *et al.*, 2004; Hall & Müller, 2004). Therefore, the study of tourism must be willing to formulate a coherent approach to understanding the meaning behind the range of mobilities undertaken by individuals, not just 'tourists' (Coles *et al.*, 2004). People's travel time budgets have not changed substantially over time, but the ability to travel further at a lower per unit cost within a given time budget (Schafer, 2000) has led to a new series of social interactions and patterns of leisure production and consumption which are often discussed under the heading of tourism but, arguably, are perhaps better dealt with under the overall rubric of mobility. Assessment of the relationship between tourism and climate change therefore needs to be undertaken over the totality of the tourism consumption and production system, particularly with respect to transport impacts, rather than just specific elements of the tourism system, as significant as they might be.

Figure 1.1 illustrates some of the interrelationships between the tourism system and climate change. All demand and supply facets of tourism are regarded as being affected by global climate change, but just as importantly tourism has direct and indirect affects on climate change itself. In addition, tourism is impacted by other factors that influence global and local change, such as processes of political, economic and cultural globalisation, technology, especially information technology, mega-urbanisation, and environmental change (Johnston *et al.*, 1995), processes which themselves interact with global climate change in ways that we are only beginning to identify. Within the tourism generating regions climate change affects the tourism decision-making process in a number of different ways. Climate change can directly impact tourist behaviour because of changed perceptions not only of the climatic appeal and image of certain destinations but also the activities that can be engaged in. Moreover, climate change may impact patterns of seasonal attraction and associated visitor flows. At a broad scale of analysis climate

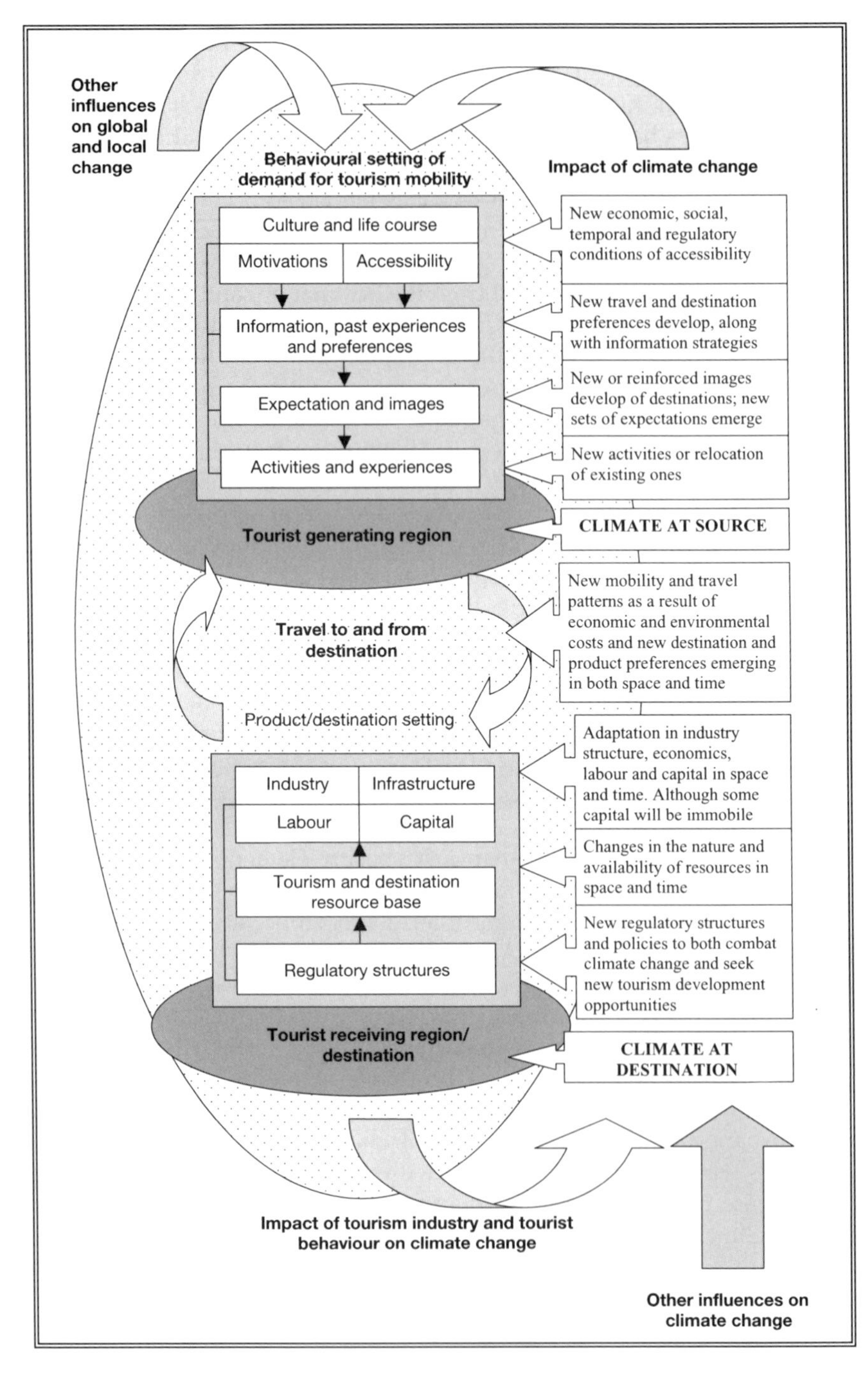

Figure 1.1 The tourism system and climate change (after Hall, 2005)

change will likely mean not that people will stop travelling but that they will change their travel preferences in both space and time. Nevertheless, such shifts assume that economic, social, regulatory and environmental conditions at the generating region stay relatively constant, unfortunately, it is likely that this will not be the case in many areas, particularly with respect to impacts on agricultural and manufacturing production as well as the location of human settlements (IPCC, 1990, 1996, 2001).

Under conditions of rapid human-induced global climate change, tourist flows as well as other human mobility patterns will undoubtedly shift in both space and time. Indeed, the question of will the Mediterranean become too hot for summer holidays has already been posed (Whitfield, 2003; Mather *et al.*, Chapter 4, this volume; Perry, Chapter 5, this volume). Changes in overall mobility patterns will be a result of new destinations and activity locations developing, while at the same time some existing destinations will cease to be as attractive (in some extreme cases, some coastal and island destinations may even disappear), with some transport routes potentially bypassing current destinations as new time and economic budgets develop in the tourist market. The energy demands of tourism transport will undoubtedly be a focal point for new regulatory structures that may act to restrict long-distance travel in particular. Gössling *et al.*'s (2002) study of the ecological footprint of travellers to the Seychelles revealed that the major environmental impact of travel was a result of transportation to and from the destination: more than 97% of the energy footprint was a result of air travel. This suggests that efforts to make destinations more sustainable through local energy initiatives

> can only contribute to marginal savings in view of the large amounts of energy used for air travel. Any strategy towards sustainable tourism must thus seek to reduce transport distances, and, vice versa, any tourism based on air traffic needs per se to be seen as unsustainable. Obviously, these insights also apply to ecotourism based on long-distance travel. (Gössling *et al.*, 2002: 208) (Also see Gössling, Chapter 20, this volume; Høyer & Aall, Chapter 18, this volume)

Nevertheless, this is not to suggest that destination based environmental initiatives are not without value.

New regulatory structures to manage the impacts of global climate change will undoubtedly develop at the national and sub-national level and many of these will impact tourism at the destination level. Natural resource attractions at the species, community and ecosystem level will be particularly affected by climate change. As Root *et al.* (2003: 57) argue:

> The synergism of rapid temperature rise and other stresses, in particular habitat destruction, could easily disrupt the connectedness among

> species and lead to a reformulation of species communities, reflecting differential changes in species, and to numerous extirpations and possibly extinctions.

Species loss and ecosystem change will have enormous impacts on nature-based tourism especially in peripheral regions. Ironically, these are the very areas that often most need tourism as a tool for regional economic development as well as a justification for conservation measures (Hall & Boyd, 2005), and, by virtue of their peripheral locations, will also be affected by any new regulatory regimes on long-distance travel. The potential for species and ecosystem loss is severe (Pounds *et al.*, 1999; IPCC, 2001; Parmesan & Yohe, 2003; Pounds & Puschendorf, 2004). Thomas *et al.* (2004) used projections of the future distributions of 1103 animal and plant species to provide 'first-guess' estimates of extinction probabilities associated with climate change scenarios for 2050. On the basis of mid-range climate-warming scenarios for 2050, they predict that 15–37% of species in their sample of regions and taxa will be 'committed to extinction'. Indeed, the Australian tropical rainforest, which is a major nature-based tourism resource, much of which has World Heritage status, is recognised as being under severe threat because of climate change (Williams *et al.*, 2003). As Hall (2003b: 300) observed:

> The rainforests of Queensland, New South Wales and Tasmania are now an integral part of those states' marketing strategies, and the long-term development of tourism in those areas may well depend on the conservation of those resources for both direct tourist experiences and as a setting for other activities.

In addition to changes to the nature and use of tourism resources and the development of new regulatory regimes, climate change will also affect tourism capital in its various forms. Some destinations and resorts will have to find adaptive reuses for much tourism-related fixed capital and infrastructure because of its immobility if they are negatively affected by climate change. Indeed, climate change is already having an impact on insurance and investment strategies for tourism development (see Bürki *et al.*, Chapter 10, this volume; Craig-Smith & Ruhanen, Chapter 12, this volume; Harrison *et al.*, Chapter 9, this volume). The changed accessibility and attractiveness of tourist destinations under conditions of climate change will also have implications not just for tourist mobility but also labour force mobility. Indeed, the movement of the human capital of tourism will be a flexible response to climate change. However, such movement will obviously have implications for community change and the associated social and economic effects of such change.

One of the key lessons that we draw from the tourism–climate relationship is that feedback processes within the tourism and climate systems will undoubtedly have unforeseen impacts on numerous locations around the world. Nevertheless, as the various chapters in this book demonstrate, for many locations the implications do appear to set numerous policy and management challenges. The parallels between human and natural system dynamics are, not surprisingly, extremely close. The comment of Root *et al.* (2003: 57) with respect to wild animals and plants that 'many of the most severe impacts of climate-change are likely to stem from interactions between threats, factors not taken into account in our calculations, rather than from climate acting in isolation' is likely to hold true for the impacts of climate change on tourism. Indeed, the observations of Root *et al.* (2003) apply to the human species as much as to plant and animal species: 'The ability of species to reach new climatically suitable areas will be hampered by habitat loss and fragmentation, and their ability to persist in appropriate climates is likely to be affected by new invasive species' (2003: 57).

Climate Change as the New Security Threat

The opening of this introductory chapter quotes David King, Chief Scientific Advisor to the UK government (King, 2004: 176), arguing that climate change is a greater security threat than that of terrorism. Although David King was reportedly substantially criticised from within the British government for his comments (Connor, 2004; Connor & Grice, 2004), and particularly his targeting of the United States' failure to sign the Kyoto Protocol and general inaction with respect to climate change, it is salutory to note that his comments have been echoed in a leaked US security assessment of climate change. According to a report in the *Observer* (Townsend & Harris, 2004):

> A secret report, suppressed by US defence chiefs and obtained by *The Observer*, warns that major European cities will be sunk beneath rising seas as Britain is plunged into a 'Siberian' climate by 2020. Nuclear conflict, mega-droughts, famine and widespread rioting will erupt across the world.

The report commissioned by Pentagon defence adviser Andrew Marshall, and reportedly suppressed by the White House for four months was authored by Peter Schwartz, CIA consultant and former head of planning at Royal Dutch/Shell Group, and Doug Randall of the California-based Global Business Network. According to Townsend and Harris (2004) the report states that climate change 'should be elevated beyond a scientific debate to a US national security concern', and concludes that an imminent scenario of catastrophic climate change is 'plausible and would challenge United States national security in ways that should be considered immediately' (see the *Observer*, 2004 for further details from the report).

The identification of climate change as a security threat already has echoes in the security literature and, more recently, in the tourism literature. In a special issue of the *Journal of Travel and Tourism Marketing* on security and safety in tourism, Hall *et al.* (2004) identify environmental security, including climate change, as a major issue facing tourism policy and development.

Environmental issues such as global climate change, as well as the system of international governance by which such concerns are governed, now clearly lie within contemporary understandings of security (Boulding, 1991). The notion of environmental security arguably developed as an offshoot of concerns over defence security in the 1970s and 1980s (Myers, 1989, 1993; Käkönen, 1994; Stern, 1995). Concerns over the potential impact of a nuclear winter following a nuclear war provided a basis for international environmental problems, such as global climate change, to contribute to an appreciation of the security dimensions of ecological interdependence (Mische, 1989; Gore, 1992; Käkönen, 1994; Levy, 1995a, 1995b). At least four sets of relationships between more traditional notions of security and the concept of environmental security may be identified. First, the relationship between environmental security and the potential for resource wars fought over increasingly scarce resources such as water (Käkönen, 1988; Homer-Dixon, 1991, 1994; Homer-Dixon *et al.*, 1993; Falkenmark & Lundqvist, 1998; Haftendorn, 2000; Maxwell & Reuveny, 2000). Second, direct threats to environmental health because of environmental change and biosecurity threats (Alexander, 2000). Third, the mass migration of ecological refugees who are abandoning resource-poor or damaged areas. Refugee flows often generate new resource, political, economic and cultural tensions in those nations that receive them (Jamieson, 1999). Fourth, there is the environmentally destructive capability of the military itself (Käkönen, 1994; Kirchner, 2000). Arguably, all of the above security issues fit in with concerns over the potential impacts of global climate change.

In recent years substantial attention has been given by government and industry to the security implications of terrorism. Yet, objectively, the potential impacts of climate change on tourism and on the wider economy, society and environment are far greater than that of terrorism, however terrible such terrorist events may have been. Fortunately, international bodies such as the World Tourism Organization are starting to par attention to the threat of climate change. However, while many national science organisations are paying attention to climate change, most national tourism authorities are not. Government and industry response is piecemeal and, arguably, the one area of the tourism industry that does seem to be paying attention to climate change is some of the winter resorts whose traditional economic base is under threat even though they seek to adapt through the development of new products and the use of snow-making equipment (Bürki *et al.*, Chapter 10, this volume; Harrison *et al.*, Chapter 9, this volume). Nevertheless, there are numerous island and coastal resorts, as well as second-home locations that are also under threat, in which no institutional or industry response has been forthcoming. The reasons for such a lack of response given the potential magnitude of the problem are complex. The short-sightedness of the tourism industry may be one reason (Whitfield, 2003). However, such short-sightedness is often based on, what is to industry and government, a lack of definite information on which to make investment and other decisions. Another factor may well be not wishing to acknowledge the potential for negative images or stories in an industry that is built on the foundation of positive images (see Trapasso, Chapter 15, this volume, with respect to attitudes towards the ozone hole; Bürki *et al.*, Chapter 10, this volume; Dewar, Chapter 16, this volume). Regardless of the reason, it is becoming increasingly apparent that there are potentially enormous issues facing tourism, as well as other industries, that require urgent attention if climate change affects are to be dealt with. The costs of not doing so become increasingly costlier than the mitigation and regulatory initiatives that are required (Bretteville & Aaheim, 2003; King, 2003), particularly given that the majority of international tourism generation occurs in the same countries that are themselves the greatest contributors to greenhouse gases.

Table 1.5 shows various regions' contribution to climate change in 2000, 2050 and 2100, respectively, based on greenhouse gas (GHG) emissions from 1890 to 2000. Setting the evaluation year farther in the future increases the share of climate change that can be attributed to OECD countries, proportionally decreasing the share that can be attributed to Asia. This is because emissions from OECD countries have a greater share of long-lived gases than emissions from Asia. Table 1.6 shoes the situation of contribution to climate change when land-use change is not taken into account. In this case, the contribution of the OECD countries is significantly larger. The contribution from developing countries drops significantly in later evaluation years because without land-use change, a larger share of their emissions is made up of short-lived gases (Romstad *et al.*, 2003).

The results of efforts to reduce GHG emissions are mixed. For example, the European Union reduced its GHG emissions by 2.3% between 1990 and 2001. However, this is only just over one quarter of the way towards achieving the 8% emissions reduction from base-year levels required by 2008–12 under the Kyoto Protocol. Ten member states (Austria, Belgium, Denmark, Finland, Greece, Ireland, Italy, the Netherlands, Portugal and Spain) are not on course to meet their national targets with domestic policies and measures. Significantly for tourism, although EU GHG emissions decreased in most sectors (energy supply,

Table 1.5 Contribution to climate change in 2000, 2050 and 2100, respectively, based on emissions from 1890 to 2000

Region	*2000*	*2050*	*2100*
OECD	40%	46.5%	49.5%
Africa/Latin America/ Middle East	19.5%	19.5%	19.5%
Asia	25.5%	19.5%	16.5%
Eastern Europe and former Soviet Union	15%	14.5%	14.5%

Source: After Romstad *et al.* (2003)

Table 1.6 Contribution to climate change when land-use change is not taken into account

Region	*2000*	*2050*	*2100*
OECD	46%	57%	65%
Africa/Latin America/ Middle East	15%	13%	11.5%
Asia	12.5%	13%	6.5%
Eastern Europe and former Soviet Union	16.5%	17%	17%

Source: After Romstad *et al.* (2003)

industry, agriculture, waste management) between 1990 to 2001, emissions from transport increased by nearly 21% in the same period, with passenger transport by road being a major contributor. Carbon dioxide emissions from transport account for 20% of total EU emissions:

> The increase in carbon dioxide emissions from international aviation and navigation was even higher (an 82 %increase from 1990 to 2001 of emissions from international aviation), but these are currently not addressed in the Kyoto Protocol or in EU policies and measures. (European Environmental Agency, 2003: 13)

In 1994 Kaplan referred to 'the coming anarchy' with respect to the prospect of greater international conflict in relation to resource and environmental security. Hopefully, this will not be the future in which we find ourselves regardless of the effects of climate change. The political stability of collective security is essential for tourism to continue to thrive and human well-being enhanced. Whether for self-interest or out of social and environmental responsibility, the tourism industry, as well as government, has a responsibility to do all it can to mitigate the effects of climate change. In this, the research community has an enormously important role to play. The responsibility of academia and science is not just to undertake research but also to communicate it. It is therefore in this spirit that this book has been written. However, we would note that in the current political climate within which global climate change research finds itself, such communication will not be without criticism, and scientific debate will often turn into an overtly political debate. In conveying some aspects of tourism and climate change issues in the present volume, we are therefore conscious that we will be contributing to such debates. However, that is one of the responsibilities of researchers in a civil society. We believe that in terms of the future of tourism, as well as the societies within which we live, there are probably few policy and development concerns as significant as global climate change. We hope that this book makes at least some small contribution to action and understanding of the issues that face us.

Further Reading: Website Sources on Global Climate Change Policies, Research and Resources

- Australian Greenhouse Office: http://www.greenhouse.gov.au/ (as of 2004 Australia had not signed the Kyoto Protocol)
- BBC Science: http://www.bbc.co.uk/science/hottopics/climatechange/ (provides a useful basic introduction to issues surrounding climate change)
- Canada's National Climate Change Process: http://www.nccp.ca/NCCP/index_e.html
- Cato Institute, global warming: http://www.cato.org/current/global-warming/index.html (a private US policy research foundation focusing on 'limited government, individual liberty, free markets and peace')
- Center for International Climate and Environmental Research (Oslo): http://www.cicero.uio.no/index_e.asp
- Changingclimate.org (Oxford University): http://www.changingclimate.org/index.asp
- Climate Action Network: http://www.climatenetwork.org/ (a network of NGOs focusing on climate change issues)
- Climate Change Research Unit (University of East Anglia): http://www.cru.uea.ac.uk/tourism/
- Cooler Heads Coalition: http://www.globalwarming.org (the Cooler Heads Coalition is a sub-group of the National Consumer Coalition. It was formed on 6 May 1997 'to dispel the myths of

global warming by exposing flawed economic, scientific, and risk analysis'

- Emissions Database for Global Atmospheric Research: http://www.rivm.nl/en/milieu/
- European Environmental Agency, climate change theme: http://themes.eea.eu.int/Environmental_issues/climate
- European Union, Environment Commission, climate change: http://europa.eu.int/comm/environment/climat/home_en.htm
- German Federal Ministry for the Environment, nature conservation and nuclear safety: http://www.bmu.de/english/fset800.php (details German climate change policies and research)
- Global Change Coalition: http://www.globalclimate.org/ ('A voice for business in the global warming debate' – a business coalition that sought to influence President Bush's climate change policy, 'deactivated' as of early 2004)
- Greenpeace Climate Campaign: http://www.greenpeace.org/international_en/archive: http://archive.greenpeace.org/climate/
- Greenpeace Climate Countdown campaign: http://archive.greenpeace.org/climate/climatecountdown/
- Indicators of Climate Change in the UK: http://www.nbu.ac.uk/iccuk/
- Intergovernmental Panel on Climate Change: http://www.ipcc.ch/
- International Energy Agency, Greenhouse Gas Programme: http://www.ieagreen.org.uk
- International Federation of Chemical, Energy, Mine and General Workers' Unions (ICEM) Labour and Climate Change report: http://www.icem.org/climaen.html
- Japanese Ministry for the Environment, climate change topics: http://www.env.go.jp/en/topic/cc.html
- Netherlands National Research Programme on Global Air Pollution and Climate Change: http://www.nop.nl/
- New Zealand climate change position: http://www.climatechange.govt.nz/
- Pew Centre on Global Climate Change: http://www.pewclimate.org/
- Tyndall Centre for Climate Change Research http://www.tyndall.ac.uk/
- United Nations Environment Programme: http://grida.no/climate/
- United Nations Framework Convention on Climate Change: http://unfccc.int/
- United Nations Framework Convention on Climate Change (Kyoto Protocol): http://unfccc.int/resource/docs/convkp/kpeng.html

- United States Climate Change Science Program: http://www.climatescience.gov/
- United States Department of State: http://www.state.gov/g/oes/climate (as of 2004 the USA had not signed the Kyoto Protocol)
- United States Environmental Protection Agency: http://www.epa.gov/globalwarming/
- United States Global Change Research Program (USGCRP): http://www.usgcrp.gov
- United States National Oceanic and Atmospheric Administration: http://www.noaa.gov/climate.html
- World Climate Research Programme: http://www.wmo.ch/web/wcrp/wcrp-home.html
- World Health Organization, global change and health: http://www.euro.who.int/globalchange
- World Meterological Organization: http://www.wmo.ch/

References

Adam, D. (2003) Earth is 20% darker, say experts. *Guardian* 18 December.

Adam, D. (2004) Summer heatwave matches predictions. *Guardian* 3 February.

Adams, D.L. (1973) Uncertainty in nature . . . weather forecasts and New England beach trip decision. *Economic Geographer* 49, 287–97.

Agnew, M. and Viner, D. (2001) Potential impact of climate change on international tourism. *Tourism and Hospitality Research* 3, 37–60.

Alexander, G.A. (2000) Ecoterrorism and nontraditional military threats. *Military Medicine* 165 (1), 1–5.

BBC News (2002) Europe counts cost of flood damage. BBC News, Thursday 15 August, 12.12 GMT. Available at: http://news.bbc.co.uk/1/hi/business/194038.stm) (accessed 25 January 2003).

BBC News (2003) Portugal declares fire calamity, BBC News World Edition, Monday 4 August, 14.10 GMT. Available at: http://news.bbc.co.uk/2/hi/europe/3121967.stm (accessed 5 August 2003).

Boulding, E. (1991) States, boundaries and environmental security in global and regional conflicts. *Interdisciplinary Peace Research* 3 (2), 78–93.

Bretteville, C. and Aaheim, H.A. (2003) *Option Values and the Timing of Climate Policy*. Working Paper 2003:04. Oslo: Center for International Climate and Environmental Research.

Brown, P. (2003a) Climate change doubles Britain's stormy weather. *Guardian* 10 December.

Brown, P. (2003b) Global warming is killing us too, say Inuit. *Guardian* 11 December.

Brown, P. (2003c) Extreme weather of climate change gives insurers a costly headache. *The Guardian* 11 December.

Brown, P. (2003d) Global warming kills 150,000 a year: Disease and malnutrition the biggest threats, UN organisations warn at talks on Kyoto. *Guardian* 12 December.

Brown, P. (2003e) Drowning islands halt effort to postpone climate change talks. *Guardian* 13 December.

Brown, P. (2004a) An unnatural disaster. *Guardian* 8 January.

Brown, P. (2004b) Midwinter spring is the new season. *Guardian* 9 January.
Brown, P. and Oliver, M. (2004) Top scientist attacks US over global warming. *Guardian* 9 January.
Carvel, J. (2003) Heatwave may have claimed 2,000 lives. *Guardian* 4 October p. 10.
Claussen, E. (2002) *Emissions Reductions: Main Street to Wall Street 'The Climate in North America'*. Pew Center on Global Climate Change, New York 17 July. Available at: http://www.pewclimate.org/press_room/speech_transcripts/transcript_swiss_re.cfm (accessed 25 January 2003).
Coles, T., Duval, D. and Hall, C.M. (2004) Tourism, mobility and global communities: New approaches to theorising tourism and tourist spaces. In W. Theobold (ed.) *Global Tourism: The Next Decade* (3rd edn). Oxford: Butterworth Heinemann.
Connor, S. (2004) How chief scientist attracted heat from No. 10. *Independent* 8 March.
Connor, S. and Grice, A. (2004) Scientist 'gagged' by No. 10 after warning of global warming threat. *Independent* 8 March.
de Freitas, C.R. (1990) Recreation climate assessment. *International Journal of Climatology* 10, 89–103.
de Freitas, C.R. (2003) Tourism climatology: Evaluating environmental information for decision making and business planning in the recreation and tourism sector. *International Journal of Biometeorology* 47 (4), 190–208.
Drake, F. (2000) *Global Warming: The Science of Climate Change*. New York: Oxford University Press.
European Environmental Agency (2003) *Greenhouse Gas Emission Trends and Projections in Europe 2003 Summary: Tracking Progress by the EU and Acceding and Candidate Countries Towards Achieving their Kyoto Protocol Targets*. Luxembourg: Office for Official Publications of the European Communities.
Ewert, A. (1991) Outdoor recreation and global climate change: Resource management implications for behaviors, planning and management. *Society and Natural Resources* 77 (4), 365–77.
Falkenmark, M. and Lundqvist, J. (1998) Towards water security: Political determination and human adaptation crucial. *Natural Resources Forum* 22 (1), 37–50.
Fickling, D. (2003) Shrinking ice in Antarctic sea 'exposes global warming'. *Guardian* 15 November.
Frändberg, L. and Vilhelmson, B. (2003) Personal mobility – a corporeal dimension of transnationalisation. The case of long-distance travel from Sweden. *Environment and Planning A* 35 (10), 1751–68.
Goldenberg, S. (2004) Bonfire of the promises. *Guardian* 19 February.
Gore, A. (1992) *Earth in the Balance: Ecology and the Human Spirit*. New York: Houghton Mifflin.
Gössling, S. (2002). Global environmental consequences of tourism. *Global Environmental Change* 12 (4), 283–302.
Gössling, S., Borgström-Hansson, C., Hörstmeier, O. and Saggel, S. (2002) Ecological footprint analysis as a tool to assess tourism sustainability. *Ecological Economics* 43 (2–3), 199–211.
Gow, D. (2004a) CO_2 cuts will raise prices, says industry. *Guardian* 17 January.
Gow, D. (2004b) CO_2 limits suicidal for competitiveness, says industry. *Guardian* 20 January.
Haftendorn, H. (2000) Water and international conflict. *Third World Quarterly* 21 (1), 51–68.

Hall, C.M. (2003a) Tourism and temporary mobility: Circulation, diaspora, migration, nomadism, sojourning, travel, transport and home. Paper presented at International Academy for the Study of Tourism Conference, 30 June–5 July 2003, Savonlinna, Finland.

Hall, C.M. (2003b) *Introduction to Tourism: Dimensions and Issues*. Melbourne: Pearson Education.

Hall, C.M. (2004) Space-time accessibility and the tourist area cycle of evolution: The role of geographies of spatial interaction and mobility in contributing to an improved understanding of tourism. In R. Butler (ed.) *The Tourism Area Life-Cycle*. Clevedon: Channel View.

Hall, C.M. (2005) *Tourism*. Harlow: Prentice-Hall.

Hall, C.M. and Boyd, S. (eds) (2005) *Nature-based Tourism in Peripheral Areas: Development or Disaster*. Clevedon: Channel View.

Hall, C.M. and Müller, D. (eds) (2004) *Tourism, Mobility and Second Homes: Between Elite Landscape and Common Ground*. Clevedon: Channel View.

Hall, C.M. and Page, S.J. (2002) *The Geography of Tourism and Recreation: Environment, Place and Space* (2nd edn). London: Routledge.

Hall, C.M. and Williams, A. (eds) (2002) *Tourism and Migration: New Relationships between Production and Consumption*. Dordrecht: Kluwer.

Hall, C.M., Timothy, D.J. and Duval, D. (2004) Security and tourism: Towards a new understanding? *Journal of Travel and Tourism Marketing*, in press.

Harlfinger, O. (1991) Holiday bioclimatology: A study of Palma de Majorca, Spain. *GeoJournal* 25 (4), 377–81.

Harrison, R., Kinnaird, V., McBoyle, G., Quinlan, C., and Wall, G. (1986) Climate change and downhill skiing in Ontario. *Ontario Geographer* 28, 51–68.

Harvey, D. (2000) *Global Warming: The Hard Science*. Harlow: Prentice Hall.

Homer-Dixon, T.F. (1991) On the threshold: Environmental changes as causes of acute conflict. *International Security* 16, 76–116.

Homer-Dixon, T.F. (1994) Environmental scarcities and violent conflict: Evidence from Cases. *International Security* 19, 5–40.

Homer-Dixon, T.F., Boutwell, J.H. and Rathjens, G.W. (1993) Environmental change and violent conflict. *Scientific American* February, 38–45.

Houghton, J. (1997) *Global Warming: The Complete Briefing*. New York: Cambridge University Press.

Høyer, K.G. (2000) Sustainable tourism or sustainable mobility? The Norwegian case. *Journal of Sustainable Tourism* 8 (2), 147–61.

IPCC (Intergovernmental Panel on Climate Change) (1990) *Impacts Assessment of Climate Change – Report of Working Group II*. Geneva: United Nations Intergovernmental Panel on Climate Change.

IPCC (1996) *Climate Change 1995: Impacts, Adaptations and Mitigation of Climate Change: Scientific-Technical Analyses*. Geneva: United Nations Intergovernmental Panel on Climate Change.

IPCC (2001) *Climate Change 2001: Impacts, Adaptation and Vulnerability*. Geneva: United Nations Intergovernmental Panel on Climate Change.

Jamieson, J.W. (1999) Migration as an economic and political weapon. *Journal of Social, Political and Economic Studies* 24 (3), 339–48.

Jepma, C. and Mohan, M. (1998) *Climate Change Policy: Facts, Issues, and Analysis*. New York: Cambridge University Press.

Johnston, R.J., Taylor, P.J. and Watts, M.J. (eds) (1995) *Geographies of Global Change: Remapping the World in the Late Twentieth Century*. Oxford: Blackwell.

Jowit, J. (2003) Britain can start dreaming of a green Christmas with swallows. *Observer* 21 December.

Käkönen, J. (1988) *Natural Resources and Conflicts in the Changing International System: Three Studies on Imperialism*. Brookfield: Avebury.
Käkönen, J. (ed.) (1994) *Green Security or Militarized Environment*. Aldershot: Dartmouth Publishing Company.
Kaplan, R. (1994) The coming anarchy. *Atlantic Monthly* February, 45–76.
King, D.A. (2003) *The Science of Climate Change: Adapt, Mitigate, or Ignore? Ninth Zuckerman Lecture*. London: Foundation for Science and Technology/Office of Science and Technology.
King, D.A. (2004) Climate change science: Adapt, mitigate, or ignore? *Science* 303 (9 January), 176–7.
Kirchner, A. (2000) Environmental protection in time of armed conflict. *European Environmental Law Review* 9 (10), 266–71.
Kovats, S., Wolf, T. and Menne, B. (2004) Heatwave of August 2003 in Europe: Provisional estimates of the impact on mortality, Eurosurveillance Weekly Archives 8 (11), 11 March. Available at: http://www.eurosurveillance.org/ew/2004/040311.asp#7 (accessed 12 March 2004).
Levy, M.A. (1995a) Time for a third wave of environment and security scholarship? *Environmental Change and Security Project Report* 1, 44–6.
Levy, M.A. (1995b) Is the environment a national security issue? *International Security* 20 (2), 35–62.
Lipski, S. and McBoyle, G. (1991) The impact of global warming on downhill skiing in Michigan. *East Lakes Geographer* 26, 37–51.
Liverman, D. (2004) Careful with that planet, Mr President. *Guardian* 19 February.
Lynas, M. (2004) Meltdown. *Guardian* 14 February.
Maxwell, J.W. and Reuveny, R. (2000) Resource scarcity and conflict in developing countries. *Journal of Peace Research* 37 (3), 301–22.
McBoyle, G. and Wall, G. (1987) The impact of CO_2-induced warming on downhill skiing in the Laurentians. *Cahiers de Géographie de Québec* 31, 39–50.
Mendlesohn, R. (1999) *The Impact of Climate Change on the United States*. New York: Cambridge University Press.
Mieczkowski, Z. (1985) The tourism climatic index: A method of evaluating world climates for tourism. *Canadian Geographer* 29 (3), 220–33.
Mische, P. (1989) Ecological security and the need to reconceptualize sovereignty. *Alternatives* 14, 389–427.
More, G. (1988) Impact of climate change and variability on recreation in the Prairie Provinces. In B.L. Magill and F. Geddes (eds) *The Impact of Climate Variability and Change on the Canadian Prairies: Proceedings of the Symposium/Workshop*, Edmonton, Alberta 9–11 September 1987. Edmonton: Alberta Environment.
Myers, N. (1989) Environment and Security. *Foreign Policy* 74 (Spring): 23–41.
Myers, N. (1993) *Ultimate Security: The Environmental Basis of Political Stability*. New York: W.W. Norton & Co.
Observer (2004) Key findings of the Pentagon. *Observer* 22 February.
O'Riordan, T. and Jäger, J. (1996) *Politics of Climate Change: A European Perspective*. London: Routledge.
Parmesan, C. and Yohe, G. (2003) A globally coherent fingerprint of climate change impacts across natural systems. *Nature* 421, 37–42.
Paul, A.H. (1972) Weather and the daily use of outdoor recreation areas in Canada. In J.A. Taylor (ed.) *Weather Forecasting for Agriculture and Industry* (pp. 132–46). Newton Abbot: David and Charles.

Perry, A. (1997) Recreation and tourism. In R. Thompson and A. Perry (eds) *Applied Climatology* (pp. 240–8). London: Routledge.
Pigram, J.J. and Jenkins, J. (1999) *Outdoor Recreation Management*. London: Routledge.
Pounds, J.A., Fogden, M.L.P. and Campbell, J.H. (1999) Biological response to climate change on a tropical mountain. *Nature* 398, 611–5.
Pounds, J.A. and Puschendorf, R. (2004) Ecology: Clouded futures. *Nature* 427, 107–9.
Radford, T. (2003) Warming warning for Antarctica. *Guardian* 9 September.
Radford, T. (2004a) Freak summers 'will happen regularly'. *Guardian* 12 January.
Radford, T. (2004b) The white death. *Guardian* 19 February.
Romstad, B., Fuglestvedt, J.S. and Berntsen, T. (2003) Hvem har skylden for klimaendringene? *Cicerone* 1, 8–10.
Root, T.L., Price, J.T., Hall, K.R., Schneider, S.H., Rosenzweig, C. *et al.* (2003) Fingerprints of global warming on wild animals and plants. *Nature* 421 (2 January), 57–60.
Sarewitz, D. and Pielke Jr, R. (2000) Breaking the global-warming gridlock. *Atlantic Monthly* 286 (1), July, 54–64.
Schafer, A. (2000) Regularities in travel demand: An international perspective. *Journal of Transportation and Statistics* 3 (3), 1–31.
Schär, C., Vidale, P.L., Lüthi, D., Frei, C., Häberli, C. *et al.* (2004) The role of increasing temperature variability in European summer heatwaves. *Nature* 427 (2 January), 332–6.
Scott, D. (2003) Climate change and sustainable tourism in the 21st century. In J. Cukier (ed.) *Tourism Research: Policy, Planning, and Prospects*. Waterloo: Department of Geography Publication Series, University of Waterloo.
Seenan, G. (2004) Global warming forces sale of Scottish winter sports resorts. *Guardian* 14 February.
Smith, K. (1990) Tourism and climate change. *Land Use Policy* April, 176–80.
Smith, K. (1993) The influence of weather and climate on recreation and tourism. *Weather* 48, 398–404.
Stern, E.K. (1995) Bringing the environment in: The case for comprehensive security. *Cooperation and Conflict* 30 (3), 211–37.
Thomas, C. and Cameron, A. (2004) Action now could still save some threatened species. *Guardian* 8 January.
Thomas, C.D., Cameron, A., Green, R.E., Bakkenes, M., Beaumont, L.J. *et al.* (2004) Extinction risk from climate change. *Nature* 427 (8 January), 145–8.
Townsend, M. (2004) Giant space shield plan to save planet. *Observer* 11 January.
Townsend, M. and Harris, P. (2004) Now the Pentagon tells Bush: Climate change will destroy us. *Observer* 22 February.
Wall, G. (1993) *Impacts of Climate Change for Recreation and Tourism in North America*. Washington, DC: Office of Technology Assessment, US Congress.
Wall, G. and Badke, C. (1994) Tourism and climate change: An international perspective. *Journal of Sustainable Tourism* 2 (4), 193–203.
Wall, G., Harrison, R., Kinnaird, V., McBoyle, G. and Quinlan, C. (1986) The implications of climatic change for camping in Ontario. *Recreation Research Review* 13 (1), 50–60.
Whitfield, J. (2003) Researchers make holiday plans in Milan: Tourist industry must prepare now for climate change, warn scientists. *Nature* 5 June.

Williams, S.E., Bolitho, E.E. and Fox, S. (2003) Climate change in Australian tropical rainforests: An impending environmental catastrophe. *Proceedings of the Royal Society of London B* 270, 1887–92.
World Tourism Organization (WTO) (2004) *Global troubles took toll on tourism in 2003, growth to resume in 2004, Press Release,* 27 January. Madrid: WTO.
World Travel and Tourism Council (WTTC) (2003) *Blueprint for New Tourism.* London: WTTC.
World Travel and Tourism Council (WTTC) (2004) *Global Travel & Tourism Poised for Robust Growth in 2004, Press Release,* 10 March. London: WTTC.

Chapter 2

The Climate–Tourism Relationship and its Relevance to Climate Change Impact Assessment

C.R. DE FREITAS

Introduction

For many regions of the world, climate is the main impetus for attracting visitors and, in this way, forms an important part of the region's natural resource base for tourism. Clearly, therefore, any change in climate will affect the resource. The problem is that comparatively little is known, other than in very general terms, about the relative contribution of climate to this resource base: in particular, the particular role climate plays and its effects on tourism. Even less is known about the economic impacts of climate or climate change on commercial prospects for tourism. There has been some progress in assessing which climate-related criteria people use to make decisions related to tourism, but it is true to say that in the published research on tourism and climate generally, the nature of the relationship between the two is usually based on assumptions rather than empirical data. With so many uncertainties surrounding both what future climate condition to expect and the nature relationship of climate to tourism, there is no way of knowing what impact climate change could have.

In light of the above, the aim of this chapter is to explore approaches to research in tourism from a climatological perspective and identify important concepts and theoretical frameworks. It is presented as a first step towards developing a coherent set of research methods and models that might bridge observations with theory and help build a coherent basis for understanding, explanation and prediction.

Climate as a Resource

According to Hibbs (1966), the concept of tourism climate recognises a climatically controlled resource, which along with weather at various times and locations, may be classified along a favourable-to-unfavourable spectrum. Thus, climate is a resource exploited by tourism. It can be treated as an economic asset for tourism, which is capable of being measured and assessed. But there are numerous problems. For example, what exactly are the criteria for *ideal, suitable, acceptable* or *unacceptable* conditions? When is the best time to visit? What clothing or equipment is needed? What are the weather hazards or climate extremes likely to be? Only after appropriate climatological criteria have been clearly identified can key questions be answered.

In defining the resource, the type of climatic data and manner in which they are presented depend on the intended use of the information. Information can be used by (1) the tourist planner; (2) the tourist operator; and (3) the individual tourist. For example, a ski facility planner needs information on the length of snow season, whereas the skier wants seasonal distribution of probabilities that a skiable depth of snow will exist at a particular location and time. A tourist operator for a tropical island resort needs to know the length of period in which the climate is acceptable to tourists. Prospective tourists need to know when and where conditions will be optimal, acceptable, tolerable or unacceptable.

Climate information must be presented in a form that relates to the response of an individual or business to weather or climate, that is *events* (or real conditions) rather than *averages*. Averages have no physiological or psychological meaning. Since tourists respond to the integrated effects of the atmospheric environment, it is generally accepted that statistical weather data or even secondary climatic variables are not always reliable indicators of the significance of atmospheric conditions. For example, at any given air temperature, the thermal conditions experienced will vary depending on the relative influence and often offsetting effects of wind, humidity, solar radiation and level of a person's activity. The information provided should give an impression of the likelihood of the occurrence of the climate or weather conditions (or events). Data should also reflect the fact that individuals respond to various *facets of climate* (see section below). Equal importance should be given to the nature and form of output data, which should be presented in a form that can be readily interpreted and understood by the user. Often we have to rely on standard meteorological or climate station data, which may not be representative of the recreational area, e.g. valleys, peaks, hills, coast or a beach. Climate station data are intended to be representative of the bottom of the atmospheric column rather than a particular microclimate or location such as a beach, park or ski slope.

Facets of Tourism Climate

The nature of the relationship between the atmospheric environment and the enjoyable pursuit of outdoor recreational activity can be taken to be a function of facets or attributes of on-site climatic conditions. A conceptual framework for this is shown schematically in Figure 2.1. The *facets of tourism climate* given at the top of Figure 2.1 are: (1) *thermal*; (2) *physical*; and (3) *aesthetic* (de Freitas, 1990).

Assessment of the *thermal* attribute of climate involves four steps:

(1) Integration the physical factors influencing the body's thermal state, which must include both the attributes of those exposed and the functional attributes of the environment, as well as the complete range of atmospheric variables. For the environment these include air temperature, humidity, wind, solar and longwave radiation and nature of the physical surroundings, and for the body, metabolic rate, posture and clothing.
(2) Provide a rational index with sound physiological basis that adequately describes the net thermal effect on the human body.
(3) Identify relationships between the thermal states of the body and the condition of mind that expresses the thermal sensation associated with these states.
(4) Provide a rating of the perceived thermal sensation and corresponding thermal index according to the level of satisfaction experienced. This means identifying subjective reaction classified on a favourable-to-unfavourable spectrum as a measure of desirability of climate conditions.

The *physical* category shown in Figure 2.1 recognises the existence of specific meteorological elements such as rain, snow and high wind that directly or indirectly affect participant satisfaction other than in a thermal sense. For example, the occurrence of high wind can have either a direct mechanical effect on the vacationer, causing inconvenience (personal belongings having to be secured or weighted down), or an indirect effect such as blowing sand or dust causing annoyance. Others things that fall into the physical category are rain (duration), rain days (frequency), ice, snow, severe weather, air quality and ultraviolet radiation.

The *aesthetic* aspects relate to the climatically controlled resource attributes of the environment, which Crowe *et al.* (1973) have termed the atmospheric component of the 'aesthetic natural milieu'. Included within this category are 'weather' factors such as visibility, sunshine or cloud associated with the prevailing synoptic condition (for example, 'a nice, clear, sunny day'), day length and visibility.

Using this conceptual framework, de Freitas (1990) examined methods capable of giving information that could be used to appraise and rate tourism climates in terms of user sensitivity and satisfaction. To identify and describe the experience of on-site atmospheric conditions, the study used two separate forms of user response, shown in Figure 2.1: (1) sensory perception of the immediate atmospheric surrounds expressed

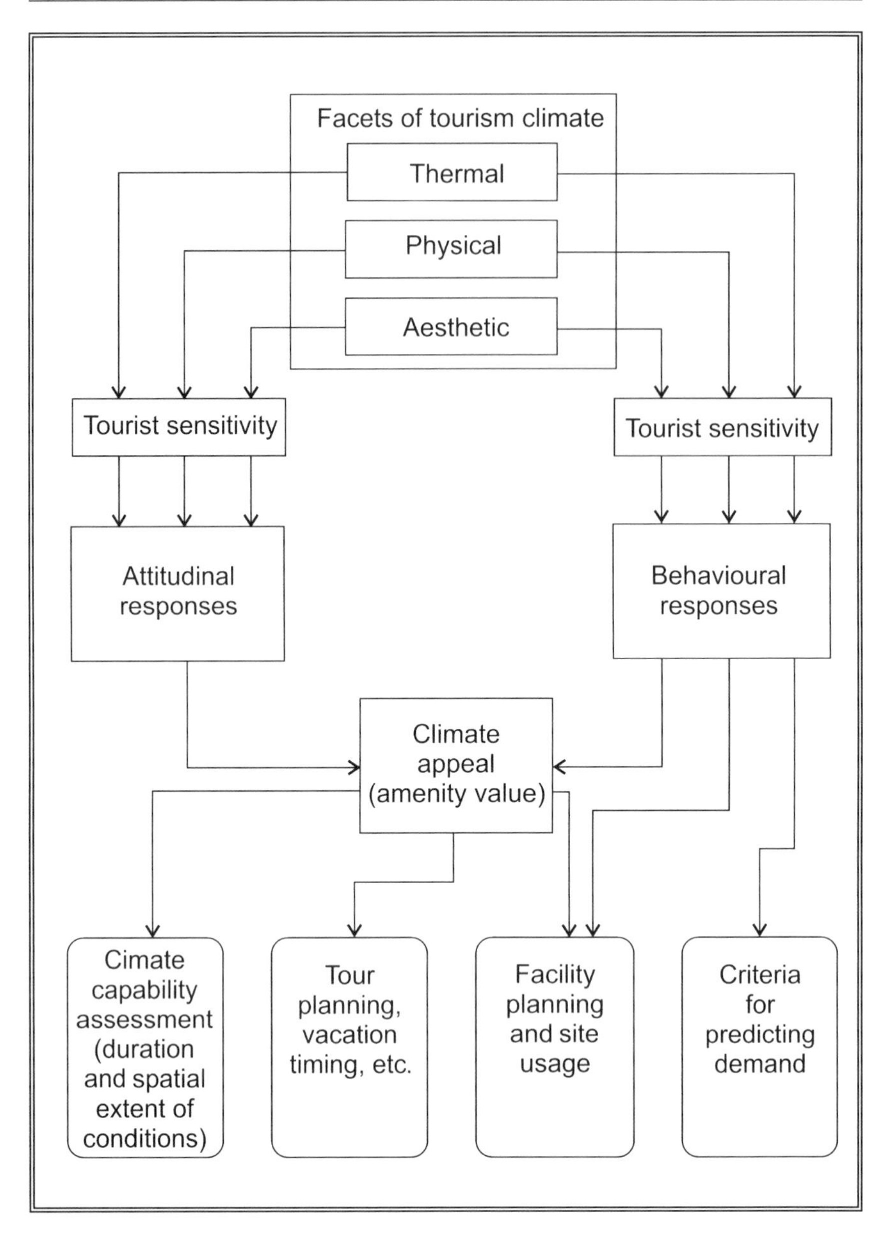

Figure 2.1 Conceptual framework for the study of tourism climate showing the facets or attributes of climate and two independent methods for assessing human response. These can be used for rating the amenity value of climate in terms participant sensitivity and satisfaction to conditions, predicting demand (e.g. participation, attendance) and a range of planning decisions

verbally; (2) behavioural responses, including those that modify or enhance effects of the atmosphere. By employing, independently, separate indicators of the on-site experience, the reliability of each was examined and interpreted by comparison, and apparent threshold conditions verified.

The above research showed that a body-atmosphere heat balance model can be used to integrate the effect thermal components of climate as well as thermophysiological variables such as type of activity (i.e. metabolic rate) and produce a unitary index that can be used to derive the levels of thermal comfort or discomfort experienced by tourists. Also, it provides a means for isolating the thermal component of climate, enabling the identification of important non-thermal recreational resource attributes of the atmospheric environment.

The results showed that optimum thermal conditions are located in the zone of slight heat stress, interpreted as warm, rather than precisely at the point of minimum stress or thermal neutrality. Sensitivity to thermal conditions appeared to be greatest in the zone of moderate environmental heat stress. Also, the immediate thermal environment of the tourist is the main contributing factor to assessments of the overall desirability of on-site climate conditions, followed by cloud cover and wind. Rainfall events of half an hour duration or longer have an overriding effect on the perceived level of attractiveness of climatic conditions, resulting in ratings dropping to their lowest levels. Cloud cover/sunshine is the main aesthetic variable. High wind at speeds in excess of $6m/s^{-1}$ has an important direct physical or mechanical effect on the beach user (personal belongings blown around) as well as an indirect effect stemming from the annoyance caused by blowing sand. Generally, ideal atmospheric conditions are those producing 'slightly warm' conditions in the presence of scattered cloud (0.3 cover) and with wind speeds of less than $6m/s^{-1}$. In the absence of rainfall, the relative percentage weighting of the various facets of tourism climate are 60% thermal (i.e. heat balance of the body including activity level, air temperature solar and longwave radiation, wind), 30% aesthetic (e.g. cloud/sunshine) and 10% physical (e.g. wind).

The Ideal Tourism Climate Index

De Freitas (2003) has reviewed in some detail the various ways climatic information has been portrayed for tourism. Considerable effort has gone into devising numerical indices that summarise the significance of climate for tourism. This is because of the multivariable nature of climate and the complex way these variables come together to give meaning to a particular weather or climate condition in terms of recreation or tourism. These indices aim to facilitate interpretation of the integrated effects of various atmospheric elements and permit places to be compared. For example, Mieczkowski (1985) devised a broadly based Tourism Climate Index (TCI) for evaluating world climates. However, with the exception of the work by Harlfinger (1991), meaning attached to these measures has been secondarily derived and interpreted without field investigation.

The popularity of the TCI shows that there is a demand for this type of unitary indicator of climate. What is now required is research that tests the accuracy of such an index, or work that devises a similar index using systematic surveys to interpret it rather than relying on arbitrary and subjective value judgements of the researcher, as in the case of TCI. Using these criteria, a Climate Amenity Index (CAI) is proposed. Ideally a CAI would:

- Rely only on standard climate data.
- Minimise reliance on averaged climate data and maximise reliance on actual (real) observations.
- Use as input all attributes of the atmospheric environment.
- Use an integrated body-atmosphere energy balance assessment of the thermal component of climate.
- Includes all three attributes of tourism climate: thermal, aesthetic and physical/mechanical.
- Recognise the notion of climate as a limiting factor, or climate limits to tourism, with a focus on thresholds.

In all of this, the aim would be to adopt standard methods and indices as far as possible. There is also a need to provide potential tourists with probabilistic information on the climate to be expected at various destinations.

Measuring the Climate Component in Tourism Demand

Given that people freely engage in tourism and recreation activities for personal satisfaction or pleasure, the behaviour is voluntary proceeding from one's own free choice. As a result, participation will only occur if the potential participant perceives the climate condition to be suitable. The voluntary and discretionary nature of tourism means that participation will decrease as discomfort and dissatisfaction increase. Thus satisfaction affects participation, which can be taken as a measure of demand for the climatic resource, the so-called 'demand factor'. Examples of indicators of demand in this context are visitation or attendance numbers (Paul, 1972; de Freitas, 1990) and hotel/motel occupancy or hotel 'tourist nights' (Rense, 1974).

The climate or weather circumstances to which the tourist may react (i.e. those that affect decisions) are (1) conditions anticipated by the tourist (say, gleaned from weather or climate forecasts, and/or travel brochures and advertising); and (2) on-site weather. These are collectively referred to as human responses to weather and climate. They can be identified and assessed using 'demand indicators'.

There are two categories of methods that have been used for assembling data on human response to climate and thus the demand for the climate resource: (1) by assessing conditional behaviour using, say, questionnaires and images (e.g. Adams, 1973) to determine how people react or think, which includes assessing the influence or role of weather or climate forecasts; or (2) by examining on-site experience, as described above. Since individuals are experiencing conditions first hand, the latter is a more reliable method than questionnaires and interviews. Ideally, the approach must be activity-specific. And it is best not to lump all tourism together, but deal with specific categories of activities, either (1) *active* or (2) *passive*.

Should climate change, so will be the tourism demand. The work of de Freitas and Fowler (1989) and Fowler (1999) deals with this in four steps for evaluating climate change impacts on the tourism: (1) projection of future climate; (2) transformation of this into changes to the tourism climate resource base; (3) transformation of changes to the tourism climate resource into societal impacts; (4) identification of possible societal responses to these impacts, that is, adaptation and mitigation. The last of these can be assessed in terms of: (1) exposure – what sort and amount of stress is the tourism sector exposed to? (2) sensitivity – how big will the response be to a given stress? and (3) adaptive capacity – to what extent is there potential for adaptation, or of a society having resources, institutional capability etc. to adjust?

Approaches to Climate Change Impact Assessment

Assessment of impact of climate change on tourism requires both knowledge of future climate as well as methods capable of transforming this knowledge into likely societal effects. There are two ways of approaching this: (1) the *scenario approach* or (2) the *sensitivity assessment method*. The scenario approach is by far the most common and is driven by our inability to adequately forecast future climate. Scenarios are effectively 'what if' statements that represent plausible future climate states from a range of possibilities based on the current state of knowledge of how the global climate system works using hypothetical general circulation models (GCMs) of global climate along with estimates of future rates GHG emissions and atmospheric concentrations. They are emphatically not forecasts. However, they are frequently treated this way.

In the scenario approach, a future climate state is identified and impacts evaluated. But this method is hampered by the unreliability of GCMs, especially at the regional level. Moreover, we do not know which climate change scenario to use. Clearly, the consequences of the scenarios being 'wrong' could have serious planning implications. Sometimes there is an implicit assumption that a specific changed climate condition is predicted, reinforced by the fact that a GCM is limited to calculating an equilibrium response condition, presented in terms of a small array of climatic variables with temporal detail limited to mean monthly data at best. But planners require more information about the altered climatic environment than simply changes expressed as time averaged, secondary climatic variables that might exist at some given time in the future. Climate data must be presented in a form that relates to the response of a tourist or tourism business to weather or climate: that is, events or real conditions as experienced rather than averages.

In the alternative approach using sensitivity assessment, many of these problems are circumnavigated. First, sensitivity of a tourism activity to climate is assessed, then the question asked: What is the net effect of change on the tourism activity or related social exposure unit? By identifying the sensitivity to climate and evaluating it in terms of the adaptive capacity of the exposure unit, *vulnerability* of tourism to change may be determined and assessed. With this information, planning decisions would be possible without knowing precisely the magnitude of climate change that will occur. Research is needed to develop and test sensitivity assessment methodologies to cope with this, but some methods are currently in use.

Assessment of Sensitivity

The aim of climate change impact assessment is to determine how the availability of climatic resources will change and which regions will lose or gain from these changes. The nature and magnitude of impacts caused by change are the joint products of interactions between climate and society, bearing in mind that similar climatic variations may result in different impacts under different sets of social or economic conditions. However, the impact potential of a given change in climate is related to overall sensitivity of a particular tourism activity to those aspects of climate that do change. Or it may be related to the particular climate type or climate regimes in which change occurs. For example, an average 1°C temperature rise and 10% increase in rainfall may be of little consequence in an equatorial climate region where high temperatures are commonplace and there are already extended periods of high rainfall throughout the year. On the other hand, sub-temperate environment may be highly sensitive and respond dramatically to even the smallest decrease in temperature or increase in precipitation in an already short summer beach recreation season.

There are a variety of ways of identifying sensitivity. In theory, sensitivity of a region to changes in climate does vary depending on climate type or regime. Climatic types can be characterised and assessed on the basis of this sensitivity since a given change will perturb some climatic regimes more than others (de Freitas & Fowler, 1989). But the net effect is not always intuitive. For example, in coastal and maritime climates the occurrence of higher average annual air temperatures due to GHG induced warming could be moderated by the time lag of ocean temperature. However, this could be misleading since, for any given increase in annual average air temperature, the sensitivity to that change could be quite different. An example of the relationship is shown in Figure 2.2 where a given increase in mean monthly temperature results in a greater increase in the length of the tourist season (A–A increased to B–B) at a site with a maritime climate (Figure 2.2a) than at a site with a continental climate (Figure 2.2b).

Climate change impact assessment of the type described above relies on a greatly simplified picture of the role of climate, mainly because it deals with change in terms of single, secondary climatic variables that allow for only elementary statistical connections to be made with impacts. This approach is of limited use since the significance of the impact will depend on the net effect of several changed climatic variables. For example, thermal state of the climatic environment in terms human

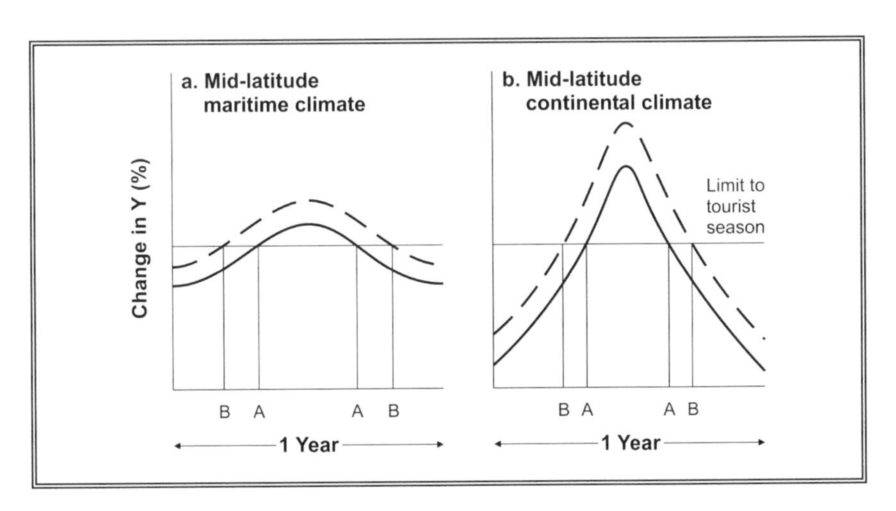

Figure 2.2 The effect of a small change in mean annual temperature on the length of the tourist season for a mid-latitude maritime climate (a) and a mid-latitude continental climate (b). Counter-intuitively, the effect (period B–A) will be greater in the mid-latitude maritime climate. Adapted from Ford (1982)

comfort is a function of the combined effect of air temperature, humidity, solar radiation and wind. Impact will also depend on the timing as well as the magnitude of change. For example, increases in the number of rain days in winter may have no consequences for mid-latitude locations geared to summer beach recreation while increases summer may destroy the amenity value of a place with an already a marginal beach recreational climate. A response surface analysis allows the decision maker to these into account simultaneously.

Spatial Dimensions of Sensitivity and Vulnerability

The observation that, in some areas, tourism conforms to climate regions that are preferred or are optimal for a particular type of tourist activity has given rise to labels and climate connotations, for example, the 'Sun Belt' in the United States, the Costa del Sol and the Riviera in Europe and the Gold Coast in Australia. Building on this conceptually, imagine how these 'zones' may evolve spatially. In theory, neglecting non-climatic constraints, if there is a significant change in climate, the size and appeal of the zone will not necessarily change. Rather, the geographical location of the zone will shift. As climate changes, there is a shift of the margins or transition zones at the boundaries. The change in location of the zone is a spatial manifestation of response to changed climatic conditions. Figure 2.3 shows how the boundaries are affected. Taking a southern hemisphere example and using a very simple case of air temperature as an index of climate, an increase in temperature will result in a southerly shift of a hypothetical holiday climate 'zone'. For 'Continent A', the zone to the north is most vulnerable to change since climatic conditions are no longer suitable. The central region is unaffected, in that there is no change in appeal or suitability. It is therefore labelled a zone largely 'insensitive' to the specified change in climate. To the south, there is a zone in which conditions for tourism improve, assuming that tourists and the tourist industry respond accordingly and exploit the changed opportunities. For 'Island B' in Figure 2.3, there is no zone of vulnerability, and generally conditions improve.

Thermal time indices such as degree days, or others such as a rain free days, can be used as measures of the changing appeal of tourism climate. These indices may be employed along with climate-change scenarios to approximate possible spatial shifts in boundaries to identify zones of high risk or vulnerability to change. In analogous examples from agricultural climatology, Newman (1980) and Blasing and Solomon (1983) found that climatic warming would displace the United States Corn Belt approximately 170km per degree of warming in a roughly northerly direction. Williams and Oakes (1978) describe a similar northward expansion of the Canadian Small Grain Belt, neglecting all environmental barriers other

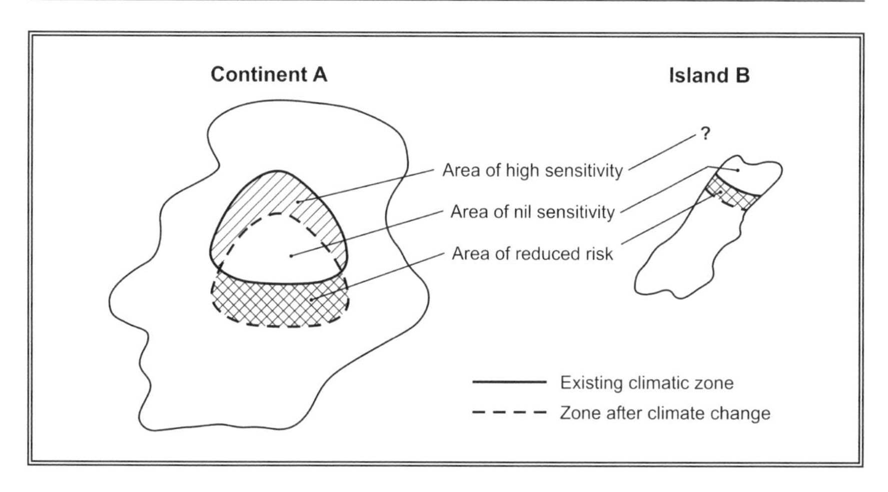

Figure 2.3 Simulated geographical shift in a climatically determined holiday region (e.g. 'Sun Belts') in the southern hemisphere resulting from climatic warming showing zones of relative sensitivity

than climate. Using data for New Zealand from Salinger (1988), there would be a 220km shift southward of the geographic range for most crops for every 1°C increase in mean annual air temperature.

Sensitivity Assessment Using Response Surfaces

A response surface is a two-dimensional representation of the sensitivity of a specific response variable (Q in Figure 2.4) to change in the two controlling features of climate (X and Y in Figure 2.4). It is presented as a plot of the response variable against the values of two driving climate variables, say number of rain days and thermal comfort, on the graph axes. The relationship between the response variable and climate are determined from a pre-tested set of relationships, usually in the form of an empirical model, called a transfer function (de Freitas & Fowler, 1989; Fowler & de Freitas, 1990). Changes might be simply percentage adjustments to the each of the driving variables. In some cases, however, expressing the response in percentage change terms in not meaningful. The response variable is represented in the body of the graph as isolines. The three variables can be plotted using absolute values, or as values relative to the unamended baseline data representing no climate change (Figure 2.4). The latter representation is a step removed from the input and output but does have the advantage of providing a direct measure of sensitivity. For example, a 20% response to a 10% change in a controlling climate variable is clearly an example of impact amplification.

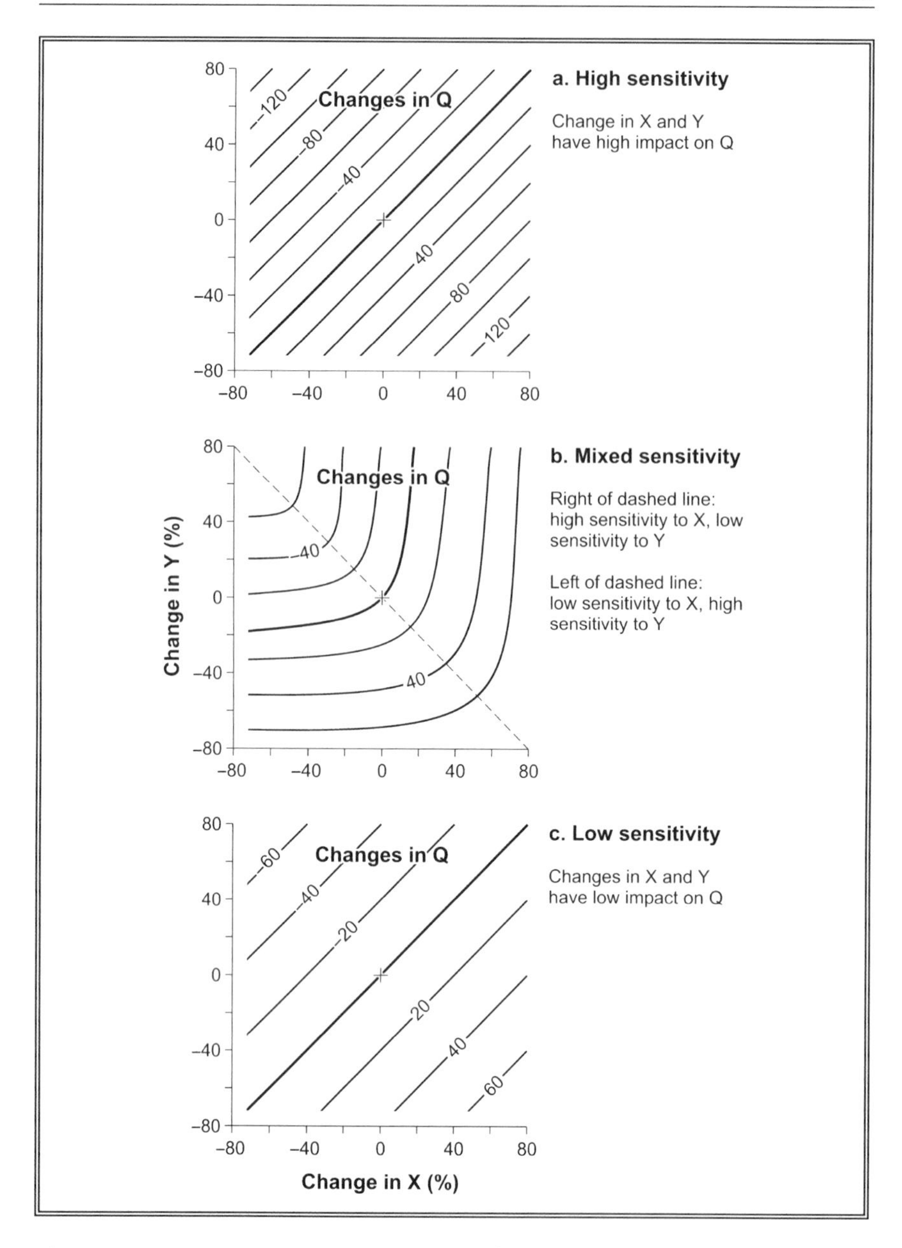

Figure 2.4 Response surface diagrams showing the effects of changes in tourism climate elements X and Y on response element Q. Patterns and spacing of lines are a measure of the sensitivity of Q to a variety of relative changes in X and Y. The 'no-change' point is shown as cross, which refers to 'current' relationship of X, Y and Q. The dashed line in case b denotes a threshold where there is a discontinuity in response

Response surface isolines are a summary of a matrix of response points associated with various combinations of changes to the two driving climate variables (Figure 2.4). The required data are derived from repeated runs of the transfer function with the prescribed changes to the input. The slope and closeness of the isolines are an indicator of sensitivity, and discontinuities an indicator of change in response (Figure 2.4).

Plotting climate change scenarios on the response surface enables it to be used for impact analysis. A scenario of say a 10% increase in rain days and a 20% increase in the thermal comfort index, for example, can be plotted on the response surface to assess the anticipated impact on the response variable, say change in the climate amenity index (Figure 2.5), or visitor numbers. An advantage of the response surface method is that it is less likely to obscure inherent sensitivities to change that can occur in scenario approach. Another is its flexibility. A wide range of new or changed scenarios can be easily handled by plotting them on the response surface. This avoids the need to rerun the transfer function, thus facilitating use by non-climate specialists such as planners and policy makers wanting to reassess impacts.

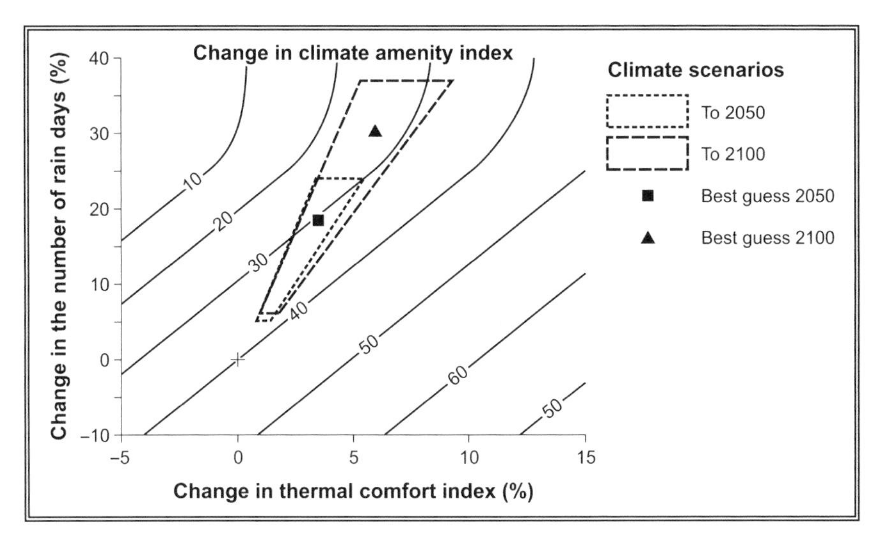

Figure 2.5 Response surface showing the sensitivity of tourism climate, expressed as change in a CAI, to climate change expressed as change in number of rain days (%) and change in thermal conditions expressed as an integrated comfort index. Climate change envelopes show incremental change based on scenarios to 2050 and 2100 (modified from Semadeni-Davies, 2003). Envelopes are defined by maximum and minimum possible changes and show the location of 'best-guess' estimates for 2050 and 2100, respectively

In the scenario approach the impression is given that a future climate state will occur at a particular time. This may not be particularly useful since a variety of planning time frames may be required. In contrast, the response surface method has an additional advantage of allowing, through interpolation, both longer- and shorter-term impacts to be assessed by way of response envelopes (Figure 2.5).

Conclusions

Given that, for many regions, climate is the main impetus for attracting visitors, it forms an important part of the natural resource base for tourism. Any change in climate will affect not only the resource but also demand for the resource. The capacity of society to respond to climate change or changed climate variability will depend on the role climate plays in this and its effects on tourism. Assessing changed opportunities will require an understanding of which climate-related criteria people use to make decisions related to tourism. Given that the nature of the relationship between the two is usually based on assumptions rather than empirical data, and with so many uncertainties surrounding what future climate condition to expect, there is no way of knowing what impact climate change could have. In view of this, a coherent set of research methods and models is required that might bridge observations with theory and help build a coherent basis for understanding, explanation and prediction.

Future research directions will depend to a large extent on what climatic information tourism planners, members of the tourism industry and tourists themselves require. Determining what exactly these requirements are should be highest on the list of research priorities.

References

Adams, D.L. (1973) Uncertainty in nature . . . weather forecasts and New England beach trip decision. *Economic Geographer* 49, 287–97.

Blasing, T.J. and Solomon, A.M. (1983) Response of North American corn belt to climatic warming. Washington, DC: US Department of Energy, DOE/NBB-004.

Crowe, R.B., McKay, G.A. and Baker, W.M. (1973) *The Tourist and Outdoor Recreation Climate of Ontario. Volume I: Objectives and Definitions of Seasons*. Publications in Applied Meteorology, REC-1–73. Toronto: Meteorological Applications Branch, Atmospheric Environment Service, Canada, Department of the Environment.

de Freitas, C.R. (1990) Recreation climate assessment. *International Journal of Climatology* 10, 89–103.

de Freitas, C.R. (2003) Tourism climatology: Evaluating environmental information for decision making and business planning in the recreation and tourism sector. *International Journal of Biometeorology* 47 (4), 190–208.

de Freitas, C.R. and Fowler, A.M. (1989) Identifying sensitivity to climatic change at the regional scale: The New Zealand example. In R. Welch (ed.) *Proceedings of 15th Conference New Zealand Geographical Society* (pp. 254–61). New Zealand Geographical Society Conference Series, No. 15. Dunedin: New Zealand Geographical Society.

Ford, M.J. (1982) *The Changing Climate: Response of the Natural Flora and Fauna.* London: George Allen and Unwin.

Fowler, A. (1999) Potential climate change impacts on water resources in the Auckland Region (New Zealand). *Climate Research* 11, 221–45.

Fowler, A.M. and de Freitas, C.R. (1990) Climate impact studies from scenarios: Help or hindrance? *Weather and Climate* 10, 3–10.

Harlfinger, O. (1991) Holiday bioclimatology: A study of Palma de Majorca, Spain. *GeoJournal* 25 (4), 377–81.

Hibbs, J.R. (1966) Evaluation of weather and climate by socio-economic sensitivity indices. In W.R.D. Sewell (ed.) *Human Dimensions of Weather Modification* (pp. 91–110). Research Paper No. 105. Chicago: University of Chicago, Department of Geography.

Mieczkowski, Z. (1985) The tourism climatic index: A method of evaluating world climates for tourism. *Canadian Geographer* 29 (3), 220–33.

Newman, J.E. (1980) Climate change impacts on the growing season of the North American Corn Belt. *Biometeorology* 7 (2), 128–42.

Paul, A.H. (1972) Weather and the daily use of outdoor recreation areas in Canada. In J.A. Taylor (ed.) *Weather Forecasting for Agriculture and Industry* (pp. 132–46). Newton Abbot: David and Charles.

Rense, W.C. (1974) Weather as an influencing factor in the use of Oregon's coastal recreation areas. Ph.D. thesis, Department of Geography, Oregon State University.

Salinger, M.J. (1988) Climatic warming: Impact on New Zealand growing season and implications for temperate Australia. In G.I. Pearman (ed.) *Greenhouse: Planning for Climate Change* (pp. 564–75). Leiden: E.J. Brill.

Semadeni-Davies, A. (2003) Urban water management vs. climate change: Impacts on cold region waste water inflows. *Climatic Change* 59 (4) (in press).

Williams, G.D.V. and Oakes, W.T. (1978) Climatic resources for maturing barley and wheat in Canada. In K.D. Hage and E.R. Reinhelt (eds) *Essays on Meteorology and Climatology, in Honor of Richard W. Longley* (pp. 367–85). Studies in Geography 3. Edmonton: University of Alberta.

Chapter 3
The Evolution of the Climate Change Issue in the Tourism Sector

DANIEL SCOTT, GEOFF WALL AND GEOFF MCBOYLE

Introduction

Although tourism and recreation are considered to be a climate-sensitive sector of the economy, a number of researchers have lamented that few investigations have examined the relationships between climate and tourism and, as a consequence, the vulnerability of individual tourism industries and destinations to climate variability remains largely unknown (de Freitas, 1990, 2003; Smith, 1990; Perry, 1997). Similarly, several authors (Wall, 1992; Wall & Badke, 1994; Abegg *et al.*, 1998; Perry, 2000; Agnew & Viner, 2001; Scott, 2003) have expressed concern that our understanding of the potentially profound consequences of global climate change for the tourism sector remains equally limited and that research in this field is still in its infancy. Nonetheless, many that are new to the field, are not aware that there is 30 years of relevant climate and tourism research to draw on and that the earliest work to consider the implications of climate change and tourism is almost 20 years old.

The purpose of this chapter is to review the chronological development of scientific inquiry into the implications of global climate change for tourism. A comprehensive bibliography of the English language literature was developed in order to examine the amount and nature of scientific inquiry related to climate-weather and tourism-recreation over time. The methods used to develop this bibliography are presented in the following section. Through analysis of this bibliography and the authors' combined experiences in this field, four distinct phases are identified in the evolution of this field of inquiry. The remainder of this chapter describes each of these eras and provides an overview of some of the major initiatives that have contributed to the development of research on global climate change and tourism. The intent was not to conduct a thorough meta-analysis and therefore the chapter does not critically assess and synthesize the available body of literature. Such an analysis was beyond the scope of this chapter, but would be a very useful contribution.

Evolution of the Literature on Climate-Weather and Tourism-Recreation

In order to provide a quantitative measure of the level of research activity in the field of weather and climate, and tourism-recreation over time, a detailed literature search was undertaken and a bibliography containing over 250 items was developed. The criteria for inclusion in the bibliography were for the term 'tourism' or 'recreation' and either 'weather', 'climate', 'climate change', or 'global warming' to appear in the title, keywords or abstract of the publication. The bibliography contains several types of publications, including peer-reviewed journal articles, reports (by government, industry, non-governmental organizations, academic institutions), book chapters and papers in conference proceedings. The choice of publications deemed relevant for the bibliography is somewhat subjective. In some instances publications that were obviously suitable for the bibliography based on their titles, but which were missing one of the terms in the above criteria, were included. One example would be a publication on climate change and skiing that did not specifically use the terms recreation or tourism in its title, keywords or abstract. The bibliography is considered reasonably comprehensive for the English language literature, particularly for the peer-reviewed journal category, where all major tourism, recreation and climatology journals were searched from the period 1970 to the present. The bibliography is available to researchers through the University of Waterloo (www.fes.uwaterloo.ca/u/dj2scott) and because some omissions have undoubtedly occurred, the authors invite readers, especially those who have published in other languages, to submit additional publications to the authors in order to make this bibliographical resource as comprehensive as possible.

The bibliographical data in Figures 3.1 and 3.2 illustrate the overall pattern of published research in the climate-weather and tourism-recreation literature over the past 40 years. The general pattern revealed is an initial period of activity in the 1960s and 1970s, which has been labelled the 'formative stage,' followed by a decade when very little research was published ('period of stagnation'). The first peer-reviewed journal publications on the implications of climate change for tourism or recreation appeared in 1986 (Harrison *et al.*, 1986; Wall *et al.*, 1986). These publications signify the start of a third phase ('emergence of climate change') and pre-dated the formation of the United Nations Intergovernmental Panel on Climate Change (IPCC) by two years. A decade later, the number of peer-reviewed journal publications with a climate change focus exceeded those with a climate-weather focus by approximately a 3:1 ratio (Figure 3.1). The final era is labelled 'maturation' reflecting the tremendous growth in the volume of publications in

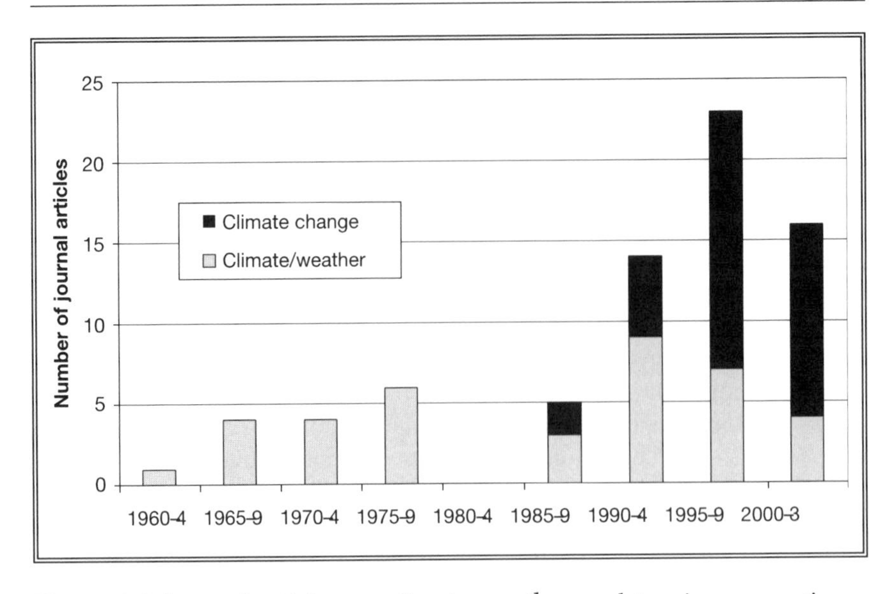

Figure 3.1 Journal articles on climate-weather and tourism-recreation

Note: only articles published as of 30 Sept. 2003 are recorded in the 2000–3 period.

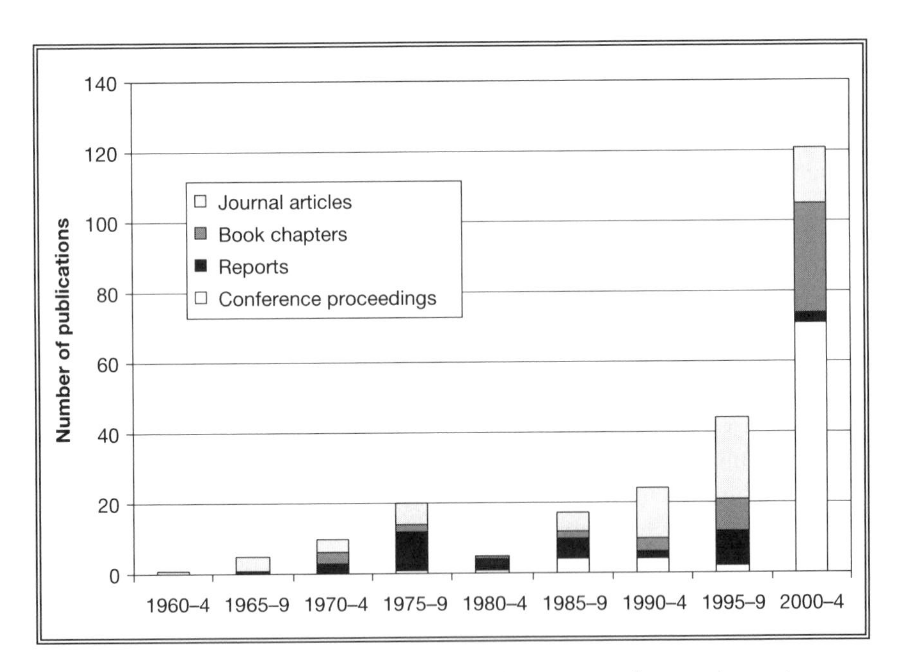

Figure 3.2 Volume of publications on climate-weather and tourism-recreation

Figure 3.2, the increasing level of organization among researchers in this field (specialized conferences and establishment of collaborative research networks), and the increasing interest and support from national and international agencies for research in this field. The following sections will discuss each of these phases in sequence, although the majority of the discussion will focus on the latter two eras that pertain specifically to climate change.

Formative Phase (1960–79)

The earliest research on the relationship between climate-weather and tourism-recreation began during what Lamb (2002) calls the 'climate revolution' during the 1960s and 1970s. Lamb (2002: 4) indicates that, 'climatology was neither respected nor valued in the 1950s and early 1960s'. Technological advances, such as improved radar, satellites, communications and computers, significantly improved climate modelling, forecasting and archiving. Extensive publicity of several extreme climate events (e.g. Sahel drought and famine, Soviet crop failures, and very cold North American winters) prompted substantive government investment in climate research programmes.

During this period, applied climatologists expanded their research to a wide range of socio-economic sectors, including tourism and recreation. The 15 journal articles and larger number of government reports during this formative period were polarized into two areas with distinct spatial, temporal and substantive concentrations. The influence of weather on recreation activities at the local scale was one focus (i.e. the effect of short-term weather on specific recreation activities like skiing, ballooning, beach visits). The second focus was the interaction of climate and tourism at the global scale, most often developing approaches to assess the suitability of climate for tourism and recreation. While the notion of variability was acknowledge in the former, climate was considered a static resource and notions of climatic 'change' were not incorporated in either set of studies. Virtually all of this work was conducted by climate scientists or geographers and, with two exceptions, published in climate or geography journals. As de Freitas (1990: 89) noted, 'much of the research in recreation climatology appears to be motivated by the potential usefulness of climatological information within planning processes for tourism and recreation'.

At the end of this formative phase, Masterton (1980) noted several factors that limited the number of detailed studies on climate-weather and tourism-recreation: (1) insufficient awareness of the significance of weather and climate for tourism; (2) studies on weather and climate are complicated; and (3) weather and climate data are not available in sufficient spatial and temporal resolution for detailed study. Interestingly, the availability of tourism and recreation data at appropriate temporal and spatial scales was not identified as a research barrier. This can be attributed, in part, to the fact that several of the studies completed during this period did not use tourism or recreation data to establish relationships with climate and weather. This lack of field investigation was a central criticism of de Freitas (1990). Abegg *et al.* (1998) levelled similar criticisms at some of the research from this period for its 'climate-deterministic viewpoint' and over simplification of the complexities of the climate-weather and tourism-recreation relationship.

Period of Stagnation (1980s)

Following growth in the level of research activity on climate-weather and tourism-recreation throughout the 1970s, there was a notable decline in research during the following decade. Publication of research in this field almost stopped during the early 1980s (Figures 3.1 and 3.2) and did not regain the level of activity of the late 1970s until the 1990s. Interestingly, of the five Ph.D. dissertations from North America that were identified as falling within this field during the formative phase, none of the researchers went on to continue research in the field or published beyond their dissertation research.

A possible explanation for the lack of continued development in the 1980s was that climate scientists, who predominately did the early research in this field, were deflected into new, salient and better funded atmospheric science issues, such as acid rain, ozone depletion and the re-emergence of air pollution issues as attempts were made to weaken environmental legislation for atmospheric protection (e.g. the *US Clean Air Act*). In addition, the concept of anthropocentric global warming was not yet widely accepted in the early 1980s and applied climatologists were therefore wary of suggesting implications for economic sectors.

Emergence of Climate Change (1990s)

Although the first publications to examine the implications of climate change for tourism and recreation appeared in 1986, global climate change only emerged as a salient international political issue in the late 1980s. The United Nations IPCC was founded in 1988 and published its First Assessment Report (FAR) in 1990 (IPCC, 1990). Tourism and recreation were not mentioned in this report. As Wall (1998: 65) indicated, 'Since [the FAR] was based on a compilation and evaluation of existing knowledge ... this was a reflection of the paucity of attention given by tourism researchers to climate and, similarly, by climatologists to tourism.' Figure 3.1 further illustrates this point, for, in the decade preceding the FAR, only five peer-reviewed journal articles on climate and tourism-recreation were available for the IPCC to draw on, with only two addressing the possible implications of climate change specifically.

An important contribution to the literature was made by Wall and Badke (1994). In late 1989, they conducted a survey with national tourism and meteorological organizations on climate change and tourism. The survey was completed before the release of the IPCC FAR and therefore provides a baseline of the international awareness and concern regarding the implications of climate change for tourism. The vast majority (81%) of respondents felt climate and weather were major determinants of tourism and recreation in their nation. An almost equal proportion (75%) believed climate change would have significant future implications for tourism in their country. Notably, 70% of the respondents from tourism organizations did not answer this question, suggesting that at the time, most were not in a position to form a scientific opinion about the possible implications of climate change for tourism. Importantly, less than 20% of the respondents were aware of any ongoing research on climate change and tourism. Based on the findings of this survey and review of the extant literature, Wall (1992: 215) argued that:

> Although the implications (of climate change) are likely to be profound, very few researchers have begun to formulate relevant questions, let alone develop methodologies which will further understanding of the nature and magnitude of the challenges that lie ahead.

After the publication of the IPCC FAR, a number of what Wall and Badke (1994) call 'somewhat speculative overviews' of the possible implications of climate change and related environmental change for tourism were published (for example, Smith, 1990; Ewert, 1991; Wall, 1992; Guoyu, 1996). These 'speculative overviews' were useful for raising awareness of the possible implications of climate change for tourism, but too rarely have been followed by more rigorous studies to substantiate the impacts on the tourism and recreation sector that were projected. Perhaps an unfortunate reflection on this field is that such 'speculative overviews' continue to be published and presented at conferences more than a decade after the first such publications appeared. The 'speculative overviews' and other more empirical assessments of potential climate change impacts to specific tourism and recreation activities in specific locations were the basis for a broader consideration of tourism and recreation in the IPCC Second Assessment Report (SAR) (IPCC, 1996). Wall's (1998) review of the SAR found that it gave much greater attention to tourism and recreation than the FAR. The implications of climate change for tourism and recreation in four types of environments (mountains, oceans and coasts, small islands and aquatic ecosystems) were discussed. The anticipated adverse impact of climate change on skiing received considerable discussion, however there was no scientific basis for the projected loss of US$1.7 billion annually in the US ski industry. Even today there is little research to even speculate on the potential magnitude of economic losses in the US ski industry. Tourism in coastal areas and small island nations was identified as particularly vulnerable to sea level rises associated with climate change. The impact of the potential loss of economically important beaches was seen as a key threat to economic development in some developing nations where tourism is a vital component of the economy. Aquatic ecosystems and the recreation and tourism they support were recognized as vulnerable to changes in hydrological regimes. Once again, estimates of economic impacts related to changes in fish populations and the associated loss of recreational benefits were highly speculative. The contribution of national and international tourism travel to global greenhouse gas emissions and the implications of mitigation policies for travel patterns were not raised in the SAR.

Abegg *et al.*'s (1998) critique of the climate change and tourism research provided another important contribution to the growing body of literature in the 1990s. Abegg *et al.* (1998) argued that in very few cases did climate change and tourism research adopt a holistic climate impact assessment approach advocated in the IPCC's 'Technical Guidelines for Assessing Climate Change Impacts and Adaptations' (Carter *et al.*, 1994). It was noted that much of the climate change and tourism research focused largely on first-order impacts (i.e. biophysical effects of climate change, such as changes in water levels, ecosystems and snow cover) and, apart from identifying these biophysical changes as detrimental to tourism, rarely investigated the implications for tourism in a meaningful manner. The concentration on first-order impacts can be largely attributed to the fundamental lack of knowledge related to the relationship between climate and tourism-recreation and the observation by Wall and Badke (1994) and Abegg *et al.* (1998), supported by this bibliographical analysis, that the preponderance of climate change and tourism-recreation research was being conducted by climate scientists and others whose primary expertise was not tourism and recreation.

Abegg *et al.* (1998) also identified three core methodological limitations of climate change assessments of tourism and recreation. These criticisms are still largely applicable to more recent climate change and tourism-recreation research and are therefore worth reiterating. The first methodological limitation was related to the uncertainties and limited spatial and temporal resolution of climate change models. The inability to project how climate variability and extremes may be altered in the future was seen as a critical limitation. However, as Wall (1993: 27) pointed out, even if better climate change scenarios were available, '... it is doubtful if the (tourism) industry has, at present, sufficient understanding of its sensitivity to climate variability to plan rationally for future conditions'.

The universal application of the simplifying assumption of a static tourism sector (i.e. 'all else will remain equal') in climate change assessments of tourism and recreation was the second methodological limitation identified. As Scott (2003) noted, this assumption is common to the vast majority of climate change impact assessments and is not a shortcoming specific to tourism and recreation research. Knowing that 'all will not remain equal' over the next 20 to 80 years in any economic sector, let alone the rapidly evolving tourism sector, researchers must also consider how climate change may interact with other major influencing variables in the tourism sector (globalization and economic fluctuations, fuel prices, ageing populations in industrialized countries, increasing travel safety and health concerns, increased environmental and cultural awareness, advances in information and transportation technology, environmental limitations – water supply and pollution). Abegg *et al.* (1998) acknowledge that there is no easy way to overcome this important uncertainty. A number of authors have argued that the climate change impacts and adaptation research community need to develop methodologies to better integrate socio-economic changes into future assessments (Berkhout & Hertin, 2000; Lorenzoni *et al.*, 2000; US National Assessment Team, 2000).

The final methodological limitation identified by Abegg *et al.* (1998) was the inadequate consideration of adaptation. Assessing the potential of climate change adaptations to reduce the risks posed by climate change and to maximize the benefit of new opportunities is critical to understanding the vulnerability of tourism industry to climate change. Because the tourism industry is so dynamic, the opportunities for supply- (by recreation-tourism operators) and demand-side (by recreationists-tourists) adaptation to climate change will be numerous (Scott, 2003). Although some progress has been made, adaptation still remains an underdeveloped theme in climate change assessments of tourism and recreation.

The volume of journal articles related to climate change and tourism-recreation increased threefold between 1990–4 and 1995–9. These publications were available for consideration in the IPCC Third Assessment Report (TAR), which was eventually published in 2001. A review of the TAR showed that progress had been made since the SAR. Each of the regional chapters of volume two (Impacts, Adaptations and Vulnerabilities) mentioned implications for tourism in several contexts and three of the regional chapters dedicated a section to tourism (Chapter 13 – Europe, Chapter 15 – North America, and Chapter 17 – Small Island States). As in the SAR, a range of potential implications of climate change and related biophysical impacts for tourism were identified (e.g. water supply and water levels for recreational boating, wildlife habitat and ecotourism, snow cover and winter tourism, and sea level rise risks to infrastructure and recreational beaches).

Several other notable advances in the discussion of the tourism sector occurred in the TAR. For the first time, the European chapter (IPCC, 2001: 643) placed a confidence level on the impact of climate change on tourism:

> Recreational preferences are likely to change with higher temperatures. Outdoor activities will be stimulated in northern Europe, but heat waves are likely to reduce the traditional peak summer demand at Mediterranean holiday destinations, and less reliable snow conditions could impact adversely on winter tourism [*medium confidence*].

Climate analogues were used for the first time in the Asia chapter (IPCC, 2001: 573) to illustrate potential impacts of climate change on tourism: 'The increased frequency of forest fires because of drier conditions in Indonesia during the 1997 El Niño resulted in haze that affected the tourism industries of Indonesia, Singapore, and Malaysia.' The first discussion of adaptation strategies for the tourism sector in the IPCC reports occurred in the Small Island States chapter (IPCC, 2001: 862):

> To ensure the sustainability of the tourist industry in Cyprus, it has been recommended that a strategy of protection of infrastructure combined with planned retreat would be effective and appropriate to local circumstances. The overall goal would be to maintain the limited beach area to sustain the vital tourist industry, specifically by erecting hard structures, enforcing building set-backs, and use of artificial nourishment, although the latter measure may require external sources of sand. Although not all these strategies may be applicable to the atoll states, many other island nations – such as Barbados, Jamaica, Grenada, St. Lucia, and Singapore – already have begun to implement similar approaches as part of the Integrated Coastal Management process.

The North American chapter raised the possibility that climate change will bring new opportunities to the tourism and recreation sector and addressed the difficult question of the net impact of climate change on tourism (IPCC, 2001: 770), stating that:

> The net economic impact of altered competitive relationships within the tourism and recreation sector is highly uncertain. Studies by Mendelsohn and Markowski (1999) and Loomis and Crespi (1999) attempt to put an economic value on climate change impacts in the United States. Although these were pioneering efforts, the assumptions and methods employed limit the confidence that can be placed in the findings. Until systematic national-level analyses of economically important recreation industries and integrated sectoral

> assessments for major tourism regions have been completed, there will be insufficient confidence in the magnitude of potential economic impacts to report a range (based on disparate climate, social, technical, and economic assumptions) of possible implications for this sector.

Finally, the Small Island States chapter (IPCC, 2001: 862) also discussed the potential for integrated effects of climate change and greenhouse gas emission mitigation policies to impact tourism:

> A high proportion of tourism in small island states is motivated by the desire of visitors from developed countries of the north (their largest market) to escape cold winters. Small island states are becoming increasingly concerned that projected milder winters in these markets could reduce the appeal of these islands as tourist destinations. It is projected that tourism could be further harmed by increased airline fares if GHG mitigation measures (e.g., levies and emission charges) were to result in higher costs to airlines servicing routes between the main markets and small island states.

Maturation (2000–present)

Several trends were observed in the bibliographical analysis in the late 1990s and early 2000s that portend a positive future for the field of climate change and tourism and recreation research. Although the volume of journal publications appears of have declined slightly from 1995–9 to 2000–3 (Figure 3.1), the latter period only covers three and a half years. With the strong growth in the number of other types of publications (Figure 3.2), especially conference papers, it is anticipated that the number of journal papers published in 2000–4 will eventually exceed that of the 1995–9 period and continue the growth trend that began in the early 1990s. Two other observable trends support this supposition. At the same time that the volume of publications has increased, so too has the number of researchers involved in this field. Examining the authorship of journal papers in each decade, the number of different authors publishing in the field has increased from 11 in the 1980s, to 47 in the 1990s and 42 in only the first three and a half years of the 2000s. A similar analysis of the other types of publications included in Figure 3.2 would show a much greater increase in the number of researchers involved in climate change and tourism-recreation research. While the number of researchers involved in this field has increased substantially over the past five years, so too has the diversity of research approaches and academic disciplines involved. This infusion of new ideas and research techniques offers exciting prospects for multidisciplinary research in the future.

One notable concern remains – the low level of involvement from tourism and recreation experts. Of all the journal publications included in this analysis (Figure 3.1), 40% have appeared in climate-meteorological journals, 42% in a range of geography-environmental management-planning focused journals, and only 18% in tourism-recreation journals. While tourism-related articles have appeared in the top climate change journals, some of tourism journals have yet to publish a paper related to climate change (e.g. *Annals of Tourism Research*). Scott (2003) suggested that one possible explanation can be found in the contention by Carter *et al.* (2001: 266) that a resource management perspective is almost absent from tourism discourse and that:

> While the importance of ecological sustainability and resource management is recognized in the tourism literature, it appears to remain poorly integrated or, at least, it appears to fit uncomfortably within tourism thought that is driven by disciplines associated with sociology and business.

Advances in understanding of the vulnerability of tourism sectors and destinations to climate change require that these disciplinary barriers be overcome and the level of collaboration between tourism and climate change experts increases.

Another important trend in the field of climate-weather and tourism-recreation research has been the increasing organization of its contributors. Since 2000 two research organizations have been established and three conferences have taken place to advance the state of knowledge in this field and foster collaboration among its researchers. The first of these organizations was the International Society of Biometeorology's (ISB) Commission on Climate, Tourism and Recreation (CCTR), which was founded during the 14th Congress of the ISB (Ljubljana, Slovenia) in September 1996. The ISB CCTR organized the first international conference dedicated to climate and tourism-recreation in October 2001 (Porto Carras, Greece). This meeting brought together researchers and tourism officials from 12 nations and a wide range of disciplines (meteorology-climatology, tourism, geography, architecture, planning, economics, landscape architecture) to review the current state of knowledge of tourism climatology, identify research needs, and establish a network of scientists and tourism stakeholders with climate interests. The 270-page proceedings produced from this conference make up the largest collection of research papers in this field to date.

The World Tourism Organization (WTO), together with several other United Nations agencies, hosted the first international conference on climate change and tourism in April 2003 (Djerba, Tunisia). One of the central aims of this meeting was to develop awareness among tourism administrations, companies and other tourism stakeholders about the climate change issue. This initiative to involve the tourism community was particularly valuable because Butler and Jones (2001: 300), in their concluding summary of the *International Tourism and Hospitality in the 21st Century* conference, forthrightly stated that climate change

> could have greater effect on tomorrow's world and tourism and hospitality in particular than anything else we've discussed ... The most worrying aspect is that ... to all intents and purposes the tourism and hospitality industries ... seem intent on ignoring what could be *the* major problem of the century. [original emphasis]

Delegates from 45 nations signified the salience of climate change for the sustainability of the global tourism industry in the *Djerba Declaration on Climate Change and Tourism* (see Appendix). The Djerba declaration recognized that climate change impacts are already occurring in some tourism destinations and that the impacts of climate change are anticipated to be more pronounced over the course of the 21st century. The two-way relationship between tourism and climate change was also recognized and the obligation of the tourism industry to reduce its greenhouse gas emissions highlighted. Finally, like other authors who have emphasized the need for greater collaboration among climate and tourism researchers (Smith, 1993; Scott, 2003), the conference stressed the need for further studies in which the tourism sector itself would take a more proactive position to ensure its interests (e.g. investment decisions, marketing programmes, physical infrastructure) are adequately addressed.

The European Science Foundation (ESF) convened an interdisciplinary workshop of climate change and tourism experts in June 2003 (Milan, Italy), with the key aims of formulating a future research agenda on the vulnerability of tourism to climate change and developing an international network of scientists and stakeholders to address collectively the identified research needs. A virtual international network for the study of the interactions between climate and tourism (e-CLAT) was founded at this meeting and is now based in the Climate Research Unit at the University of East Anglia (www.e-clat.org). A detailed science plan was developed collaboratively to represent the research themes and priorities identified by the delegates attending the workshop. The delegates reiterated the sentiments of previous authors (Wall, 1998; Viner & Agnew, 2001; Scott, 2003) that limited research on climate change and tourism by the climate change research community is not justified considering the current and growing significance of the tourism industry to the global economy. As was the case at the WTO conference in Djerba, there was agreement among the delegates of the need to raise the profile of tourism within the IPCC Fourth Assessment Report (AR4). The interactions of climate change, the environment and tourism provide a cross-cutting research area that is highly relevant to the economies of many nations (Viner & Amelung, 2003) and the delegates called for the IPCC to commission a special report on tourism. Other important recommendations of the ESF workshop included, the adoption and implementation of the Djerba declaration (see Appendix), the need for greater international collaboration in the field, the importance for the tourism community to get involved in assessing the vulnerability of their own industry and new opportunities that might emerge from climatic change, and the need to prioritize research to focus on the components of the tourism sector and destinations that are anticipated to be the most vulnerable (i.e. developing nations where tourism is a vital aspect of the economy, coastal areas, areas with low water supplies, nature-based tourism industries).

Conclusion

This chapter has, in a broad manner, traced the chronological development of the climate change and tourism-recreation field, from the formative climate-weather and tourism-recreation assessments in the 1970s, to its place in the first three IPCC assessments and the recent initiatives that have led to notable growth in the field over the past five years. It is hoped that this overview provides a valuable context for the remainder of the chapters in this volume.

Appendix: *The Djerba Declaration on Climate Change and Tourism*

The participants gathered at the First International Conference on Climate Change and Tourism, held in Djerba, Tunisia, from 9 to 11 April 2003, convened by the World Tourism Organization, upon an invitation of the Government of Tunisia,

Having listened to the presentations by the representatives of the:

- Tunisian Government
- Intergovernmental Oceanographic Commission (IOC) – UNESCO
- Intergovernmental Panel on Climate Change (IPCC)
- United Nations Convention to Combat Desertification (UNCCD)
- United Nations Environment Programme (UNEP)
- United Nations Framework Convention on Climate Change (UNFCCC)
- World Meteorological Organization (WMO)
- World Tourism Organization (WTO)

and by representatives from the private and public sectors, as well as the points of view of a number of national governments, tourism companies, academic institutions, NGOs and experts;

Acknowledging that the objectives of this Conference are fully in line with the concerns, pursuits and activities of the United Nations system in the field of climate change, and more generally, in that of sustainable development;

Recognizing the key role of the Kyoto Protocol as a first step in the control of greenhouse gas emissions;

Taking into consideration that in convening this Conference WTO did not intend a purely science-based debate, neither to cover all the well-known social and environmental implications that climate change can have on societies, but rather to put emphasis on the relationships between climate change and tourism, given the economic importance that this sector of activity is having on many countries, especially small island and developing states, and with a view to raising awareness of these relationships and strengthening cooperation between the different actors involved;

Having carefully considered the complex relationships between tourism and climate change, and particularly the impacts that the latter are producing upon different types of tourism destinations, while not ignoring that some transport used for tourist movements and other components of the tourism industry, contribute in return to climate change;

Aware of the importance of water resources in the tourism industry and of its links with climate change;

Recognizing the existing and potentially worsening impact of climate change, combined with other anthropogenic factors on tourism development in sensitive ecosystems, such as the drylands, coastal and mountain areas as well as islands; and
Taking into consideration that the right to travel and the right to leisure are recognized by the international community, that tourism is now fully integrated in the consumption patterns of many countries, and that WTO forecasts indicate that it will continue to grow in the foreseeable future,

***Agree* the following:**

1. *To urge* all governments concerned with the contribution of tourism to sustainable development, to subscribe to all relevant intergovernmental

and multilateral agreements, especially the Kyoto Protocol, and other conventions and similar declarations concerning climate change and related resolutions that prevent the impacts of this phenomenon from spreading further or accelerating;

2. *To encourage* international organizations to further the study and research of the reciprocal implications between tourism and climate change, including in the case of cultural and archaeological sites, in cooperation with public authorities, academic institutions, NGOs, and local people; in particular, *to encourage* the Intergovernmental Panel on Climate Change to pay special attention to tourism in cooperation with WTO and to include tourism specifically in its Fourth Assessment Report;

3. *To call upon* UN, international, financial and bilateral agencies to support the governments of developing, and in particular of least developed countries, for which tourism represents a key economic sector, in their efforts to address and to adapt to the adverse effects of climate change and to formulate appropriate action plans;

4. *To request* international organizations, governments, NGOs and academic institutions to support local governments and destination management organizations in implementing adaptation and mitigation measures that respond to the specific climate change impacts at local destinations;

5. *To encourage* the tourism industry, including transport companies, hoteliers, tour operators, travel agents and tourist guides, to adjust their activities, using more energy efficient and cleaner technologies and logistics, in order to minimize as much as possible their contribution to climate change;

6. *To call upon* governments, bilateral and multilateral institutions to conceive and implement sustainable management policies for water resources, and for the conservation of wetlands and other freshwater ecosystems;

7. *To call upon* governments to encourage the use of renewable energy sources in tourism and transport companies and activities, by facilitating technical assistance and using fiscal and other incentives;

8. *To encourage* consumer associations, tourism companies and the media to raise consumers' awareness at destinations and in generating markets, in order to change consumption behaviour and make more climate friendly tourism choices;

9. *To invite* public, private and non-governmental stakeholders and other institutions to inform WTO about the results of any research study relevant to climate change and tourism, in order for WTO to act as a clearing house and to create a database on the subject and disseminate know-how internationally; and

10. *To consider* this Declaration as a framework for international, regional and governmental agencies for the monitoring of their activities and of the above mentioned action plans in this field.

Source: World Tourism Organization (2003)

References

Abegg, B., Konig, U., Bürki, R. and Elsasser, H. (1998) Climate impact assessment in tourism. *Applied Geography and Development* 51, 81–93.

Agnew, M. and Viner, D. (2001) Potential impact of climate change on international tourism. *Tourism and Hospitality Research* 3, 37–60.

Berkhout, F. and Hertin, J. (2000) Socio-economic scenarios for climate impact assessment. *Global Environmental Change* 10, 165–8.

Butler, R. and Jones, P. (2001) Conclusions – problems, challenges and solutions. In A. Lockwood and S. Medlik (eds) *Tourism and Hospitality in the 21st Century* (pp. 296–309). Oxford: Butterworth-Heinemann.

Carter, T., Parry, M., Harasawa, N. and Nishioka, S. (1994) *IPCC Technical Guidelines for Assessing Climate Change Impacts and Adaptations*. Intergovernmental Panel on Climate Change. Department of Geography, University College London, UK and the Centre for Global Environmental Research, Tsukuba, Japan.

Carter, R., Baxter, G. and Hockings, M. (2001) Resource management in tourism: A new direction? *Journal of Sustainable Tourism* 9, 265–80.

de Freitas, C. (1990) Recreation climate assessment. *International Journal of Climatology* 10, 89–103.

de Freitas, C. (2003) Tourism climatology: Evaluating environmental information for decision making and business planning in the recreation and tourism sector. *International Journal of Biometeorology* 4, 45–54.

Ewert, A. (1991) Outdoor recreation and global climate change: Resource management implications for behaviors, planning and management. *Society and Natural Resources* 77 (4), 365–77.

Guoyu, R. (1996) Global climate changes and the tourism of China. *The Journal of Chinese Geography* 6 (2), 97–102.

Harrison, R., Kinnaird, V., McBoyle, G., Quinlan C. and Wall, G. (1986) Climate change and downhill skiing in Ontario. *Ontario Geographer* 28, 51–68.

IPCC (Intergovernmental Panel on Climate Change) (1990) *Impacts Assessment of Climate Change – Report of Working Group II*. Geneva: United Nations Intergovernmental Panel on Climate Change.

IPCC (1996) *Climate Change 1995: Impacts, Adaptations and Mitigation of Climate Change: Scientific-Technical Analyses*. Geneva: United Nations Intergovernmental Panel on Climate Change.

IPCC (2001) *Climate Change 2001: Impacts, Adaptation and Vulnerability*. Geneva: United Nations Intergovernmental Panel on Climate Change.

Lamb, P. (2002) The climate revolution: A perspective. *Climatic Change* 54, 1–9.
Loomis, J. and Crespi, J. (1999) Estimated effects of climate change on selected outdoor recreation activities in the United States. In R. Mendelsohn and J. Newmann (eds) *The Impact of Climate Change on the United States Economy* (pp. 289–314). Cambridge: Cambridge University Press.
Lorenzoni, I., Jordan, A., Hulme, M., Turner, K. and O'Riordan, T. (2000) A co-evolutionary approach to climate change impact assessment: Part I. Integrating socio-economic and climate change scenarios. *Global Environmental Change* 10, 57–68.
Masterton, J. (1980) Applications of climatology to the ski industry. In W. Wyllie and L. Magiure (eds) *Proceedings of a Workshop on the Application of Meteorology to Recreation and Tourism*. Toronto: Atmospheric Environment Service.
Mendelsohn, R. and Markowski, M. (1999) Estimated effects of climate change on selected outdoor recreation activities in the United States. In R. Mendelsohn and J. Newmann (eds) *The Impact of Climate Change on the United States Economy* (pp. 267–88). Cambridge: Cambridge University Press.
Perry, A. (1997) Recreation and tourism. In R. Thompson and A. Perry (eds) *Applied Climatology* (pp. 240–8). London: Routledge.
Perry, A. (2000) Tourism and recreation. In M. Parry (ed.) *Assessment of Potential Effects and Adaptations for Climate Change in Europe* (pp. 217–26). Norwich: Jackson Environment Institute, University of East Anglia.
Scott, D. (2003) Climate change and sustainable tourism in the 21st century. In J. Cukier (ed.) *Tourism Research: Policy, Planning, and Prospects*. Waterloo: Department of Geography Publication Series, University of Waterloo.
Smith, K. (1990) Tourism and climate change. *Land Use Policy* April, 176–80.
Smith, K. (1993) The influence of weather and climate on recreation and tourism. *Weather* 48, 398–404.
US National Assessment Team (2000) *Climate Change Impacts on the United States: The Potential Consequences of Climate Variability and Change*. US Global Change Research Program. New York: Cambridge University Press.
Viner, D. and Amelung, B. (2003) *Climate Change, the Environment and Tourism: The Interactions*. Proceedings of the European Science Foundation Workshop, Milan 4–6 June. Norwich: Climate Research Unit.
Wall, G. (1992) Tourism alternatives in an era of global climate change. In V. Smith. and W. Eadington (eds) *Tourism Alternatives* (pp. 194–236). Philadelphia: University of Pennsylvania Press.
Wall, G. (1993) *Impacts of Climate Change for Recreation and Tourism in North America*. Washington, DC: Office of Technology Assessment, US Congress.
Wall, G. (1998) Climate change, tourism and the IPCC. *Tourism Recreation Review* 23 (2), 65–8.
Wall, G., Harrison, R., Kinnaird, V., McBoyle, G. and Quinlan, C. (1986) The implications of climatic change for camping in Ontario. *Recreation Research Review* 13 (1), 50–60.
Wall, G. and Badke, C. (1994) Tourism and climate change: An international perspective. *Journal of Sustainable Tourism* 2 (4), 193–203.
WTO (World Tourism Organization) (2003) *Climate Change and Tourism*. Proceedings of the First International Conference on Climate Change and Tourism, Djerba 9–11 April. Madrid: World Tourism Organization.

Part 2: The Effects of Climate Change on Tourist Flows and Recreation Patterns

Chapter 4

Climate and Policy Changes: Their Implications for International Tourism Flows

SUE MATHER, DAVID VINER AND GRAHAM TODD

Introduction

The global climate is already changing. A key element in leisure travel demand is the degree of comfort (or discomfort) to be experienced at the traveller's destination. Subject to issues such as humidity and precipitation, human comfort becomes harder to maintain once air temperatures exceed around 31°C. This 'comfort factor' is also affected by other elements such as disease risk, extended rainfall and changes in extremes. These factors all affect the choice of destination by leisure travellers. Climatic factors, especially those of reliable summer weather, are the prime motivation for mass leisure travel. Their impact on other forms of leisure travel varies in importance.

This chapter focuses at a 'macro' level on the likely influences on the tourism industry of climate change. A broad approach is adopted that can hide important local or regional variations. The limitations of this approach are discussed as well as some broad lines of action to be taken by tourism authorities in assessing the likely impact on their own tourism sector.

Climate Change

It is accepted that climate change induced by human activity is underway. Of all the effects on the world's climate, greenhouse gas (GHG) emissions are having the greatest impact. While human-induced climate change is manifesting itself in a warming of the planet, more rapid changes may arise as a result of unquantifiable surprises and shocks.

Even if it were possible to stop all human GHG emissions and stabilise GHG concentrations in the atmosphere at today's level, the effects on climate change of earlier emissions would continue to be felt for at least the next 1000 years due to the large thermal inertia in the oceans. The Kyoto Protocol (the first step on the ladder towards stabilisation of the climate system), aims to reduce CO_2 emissions from developed countries by 5% of their 1990 emissions. The current atmospheric concentration of CO_2 is approximately 380ppmv and is rising at 1% a year.

Complex three-dimensional General Circulation Models (GCMs) are used to predict how the climate system will respond to changes in GHG concentrations. These models are used to make predictions as far ahead as the year 2100, and use the 2020s (2010–2039), the 2050s (2040–2069) and the 2080s (2070–2099) as dates to illustrate the likely impacts of climate change.

Climate change will vary according to the rate of global economic growth, population change and other socio-economic factors. The Intergovernmental Panel on Climate Change (IPCC) (1999, 2001) has defined four scenarios of future economic and social development, each having equal probability, to take account of these factors. These scenarios, known as SRES (Special Report on Emissions Scenarios) in turn influence the quantitative predictions of climate change, especially the variations around mean predictions. Clearly, all sectors of the world's economy, including tourism, will be influenced by economic and social factors, and thus all long-term predictions inevitably contain a degree of uncertainty.

The central climate change assumptions for the period 1990–2100 can be simply stated:

- average global temperatures will rise by at least 1.4°C and 5.8°C, a rate unprecedented in past 10,000 years;
- sea levels will rise by between 9 and 88cm by 2100, with a central forecast of 48cm, implying a rate of increase between two and four times greater than during the 20th century;
- there will, however, be regional and seasonal variations around these global average annual changes.

The implications for the world's major tourism regions are summarised in Table 4.1. The impacts of climate change will be diverse and wide ranging and will depend upon the location, geography and society of the region, key impacts on the tourism industry are summarised in Table 4.2.

Development of Tourism Demand Hypotheses

Summary of main global tourism flows

The concentration of tourism activity is heavily focused on a relatively few markets and destinations. In terms of pure volume, three regions – Europe, North and South East Asia, and North America – account for a large proportion of both demand and supply. These flows are

Table 4.1 Summary of climate change predictions

	Temperature	*Rainfall (precipitation)*	*Other key variables*
Global	0.2–0.6°C rise in temperature per decade Decrease in diurnal temperature range	Increase in rainfall by 3–10% by 2050 Increase in droughts over continental areas	4–10cm increase in sea level per decade
Northern Europe	0.4–0.8°C rise in temperature per decade Increase in winter and summer temperatures Decrease in frost days	Increase in amount and intensity of winter rainfall Decrease in summer rainfall	Summers become ‘better’ and appear ‘more reliable’
European Alps	Increase in winter temperatures Snow line increases in altitude by up to 100m per decade	Increase in winter snowfall	Increased risk of avalanches due to combination of higher temperatures and increase in snow Shorter ski season
Mediterranean basin	0.3–0.7°C rise in temperature per decade Increase in heat index Increase in number of days over 40°C	Decrease in summer rainfall (–15%) Increase in desertification Increase in winter rainfall Increase in erosion and runoff	Increased risk of forest fires Increased risk of flash floods Water resource pressures increase Coastal areas and infrastructure vulnerable to sea level rise
Middle East/ North Africa	0.3–0.7°C rise in temperature per decade Increase in number of very hot days	Slight decrease in annual rainfall Rainfall and convective activity become more intense	Increased pressure on water resources Increase in flash floods and erosion Sea level rise threatens beaches Increased sea surface temperatures

Table 4.1 continued

	Temperature	*Rainfall (precipitation)*	*Other key variables*
Sub-Saharan Africa	0.3–0.7°C rise in temperature per decade	10–15% increase in winter rainfall	Wetter warmer winters Drier more intensely hot summers
North America	0.3–0.7°C rise in temperature per decade	Slight increase in annual rainfall Rainfall and convective activity become more intense	Increased rainfall and hurricane activity over Pacific states associated with El Niño events
Caribbean	0.2–0.6°C rise in temperature per decade	Slight increase in annual rainfall Rainfall and convective activity become more intense	Increased rainfall and hurricane activity associated with La Niña
Central/South America	0.2–0.6°C rise in temperature per decade	Decrease of 3% in annual rainfall	Decrease in hurricane activity associated with El Niño events
North East Asia	0.4–0.8°C rise in temperature Increase in winter and summer temperatures Decrease in frost days	Increase in amount and intensity of winter rainfall Decrease in summer rainfall	Summers become 'better' and appear more 'reliable'
South Asia	0.1–0.5°C rise in temperature per decade	Little change in rainfall	Coastal areas vulnerable to erosion
Far East/Pacific	0.1–0.5°C rise in temperature per decade	Little change in rainfall	Small island states and coastal areas vulnerable to sea level rise

Source: Climatic Research Unit, University of East Anglia

Table 4.2 Summary of climate changes and their probable impact on other world regions

Destination region	***Climate change predictions***	***Implications/consequences for tourism industry***
Alpine and mountain regions	– Warmer winters – receding snow line – Wetter winters ie more snow at high altitudes, more rain at lower levels – Shorter snow season – Warmer, drier summers	*(Travel to mountain regions for winter sports is predominantly intra-regional)* – Shorter skiing season – Greater demand for high altitude resorts for skiing – Greater risk of avalanches – Extended season for non-ski mountain activities – Possibly less overall demand for skiing?
Tropical island states	– Relatively small temperature rises – Relatively small rainfall changes – Sea level rises critical – Storm frequency and intensity increases – Coral bleaching	– Beach erosion and coastal flooding – Possible submergence – Salinisation of aquifers – Reef damage and erosion – Increasingly untenable beach resorts – Ingress of new tropical diseases – Increased energy costs for air-conditioning – Reduced demand for holidays to worst affected islands – Reduced demand for dive holidays
Middle East and North Africa	– Warmer winters – Much warmer summers – Drier summers in North Africa – Wetter summers in Arabian peninsula	– Probable decline in peak summer visits – Stronger winter tourism market – Stronger tourism demand in shoulder months

Table 4.2 continued

Destination region	*Climate change predictions*	*Implications/consequences for tourism industry*
South America	– Below average winter warming for Amazon region – Warmer summers in Amazon – Little change in annual rainfall in Amazon region – Warmer summers in southern cone, inconsistent predictions for winter – Inconsistent predictions for southern cone rainfall	– South America likely to be among the least-affected regions globally for tourism demand – Possible rise in skiing's importance in Andean region (Chile, Argentina)?
Sub-Saharan Africa	– Inconsistent temperature predictions except (1) warmer winters in West Africa and (2) warmer 'summers' (June–Aug.) in southern Africa – Increased rainfall in West and East Africa in winter – Small decrease in rainfall in 'summer' (June–Aug.) in southern Africa – Wetter summers in Sahara	– Little clear indication of climate changes affecting tourism – Hotter and drier 'summer' months in southern Africa could diminish demand slightly? – Wetter 'winters' in East Africa could diminish demand for 'safari' and beach holidays?
Australia, New Zealand, Pacific Islands	– Inconsistent temperature predictions, except warmer 'summers' (June–Aug.) in northern Australia – Small decrease in 'summer' (June–Aug.) rainfall	– Winter demand (Oct.-Mar.) from northern hemisphere to remain strong? – Pacific islands very vulnerable to sea level rises and greater storm activity (see 'Tourism to tropical island states' above)

Source: Travel Research International (2003)

summarised in Figure 4.1. Around 58% of all international arrivals take place in Europe, 16% in North and South East Asia and around 12% in North America. This represents almost nine out of ten of the world's international tourist arrivals. While travel to other regions of the world may be highly significant to the destinations concerned – such as to the Pacific islands – it is negligible in terms of the global flows. Tourism is also highly concentrated: the top four markets – the USA, Germany, the UK and Japan – account for over one-third of all international demand, and the top ten (the previous four plus France, Italy, China, the Netherlands, Canada and Belgium/Luxembourg) for well over half.

Most international travel takes place intra-regionally, and most of this is within sub-regions: approximately 87% of all international arrivals in Europe are from Europe itself (some 350 million arrivals). Figure 4. 2 highlights the major flows within Europe, with a corresponding figure of 71% in the Americas (92 million arrivals) and 77% in the Asia Pacific region (88 million). In addition to this intra-regional activity, there are six major tourism flows that dominate international travel and account for around a quarter of total arrivals:

- Northern Europe to the Mediterranean 116 million
- North America to Europe 23 million
- Europe to North America 15 million
- North East Asia to South East Asia 10 million

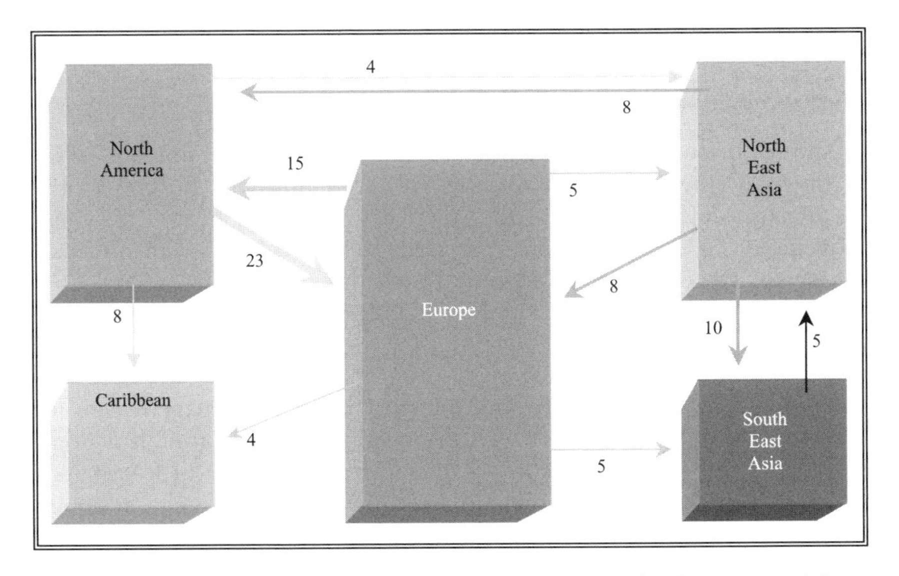

Figure 4.1 Major international tourism flows, 2000 (million arrivals)
Source: Travel Research International (2003)

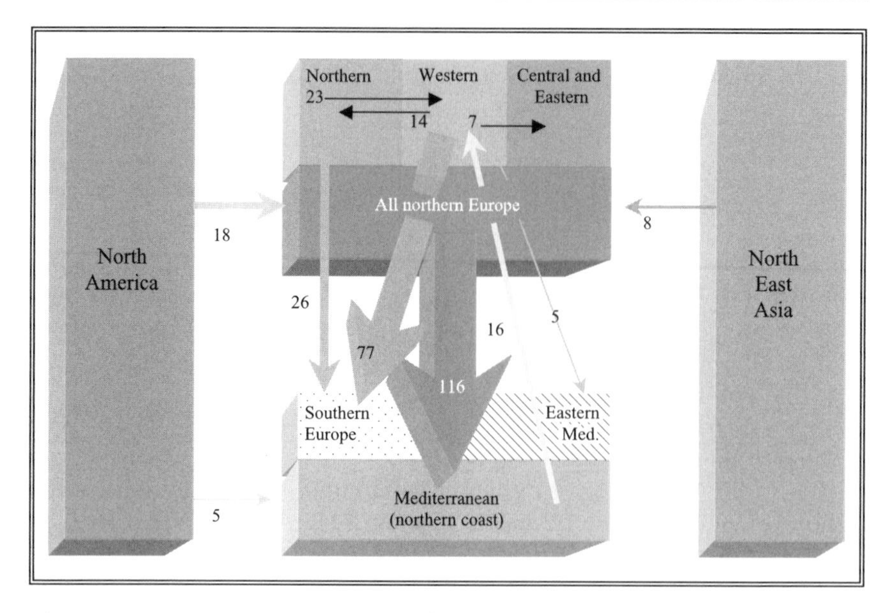

Figure 4.2 Major tourist flows within and into Europe, 2000 (million arrivals)

Source: Travel Research International (2003)

- North East Asia to North America 8 million
- North America to the Caribbean 8 million

Climatic influences on tourism flows

Among these major bilateral flows, the climate is a dominant factor in much of the travel that takes place from northern Europe to the Mediterranean and from North America to the Caribbean. This mass movement of people is not only primarily leisure-based, but the intrinsic reason for travel is to visit a sunny beach destination.

The climate is a far less significant influence on travel from North America to Europe. In part there is a far higher proportion of business travel within this flow. However, even for the leisure traveller, the weather is not a major determining factor since the primary reason for travel is more likely to be to visit the destinations' cultural attractions, rather than for the appeal of its weather. The flow from North East Asia to South East Asia has a large sun, sand and sea component, although there is also a significant element of business and VFR (visit friends and relatives) travel. Equally, the composition of travel between North East Asia and North America and between Europe and North America encompasses a variety of types of traveller.

The principal regions

The largest single flow is the mass transfer of tourists from the colder northern regions of Europe southwards to countries bordering the northern coast of the Mediterranean – primarily the search for summer sun. This amounted to around 116 million arrivals in 2000 – about one-sixth of all tourist trips worldwide. Europe is also a major origin market for other regions of the world, with a large number of European tourists crossing the Atlantic to North America – a flow of some 15 million in 2000. There are smaller but nonetheless significant flows of tourists to North East Asia and South East Asia (5 million in each case) and to the Caribbean (4 million).

North America is the next major centre of activity. Europe attracts a major flow of tourists from the USA and Canada – some 23 million in 2000. The Caribbean is also a major attraction for North Americans, primarily a market seeking the warm Caribbean climate during the winter months. There were around 8 million North Americans visiting the Caribbean in 2000.

The third centre of activity is Asia. Again, intra-regional travel predominates, with 10 million North East Asians visiting South East Asia in 2000, and 5 million South East Asians travelling in the reverse direction. There are also two relatively large flows from outside the region. North East Asia received 5 million European visits and 4 million visits from North America in 2000. Smaller numbers are recorded in South East Asia, which received 5 million Europeans and around 2 million North Americans.

Assumptions Regarding Potential Effects on Tourism Demand

Table 4.3 examines the likely impact of climate change on tourism market behaviour for the six largest international tourism flows identified above. The analysis includes the effects of climate change in both origin and destination regions and suggests how tourism markets might behave as a result. The composition of each travel market (i.e. the reason for travel), the level of climate change in a tourist's origin market and the changing weather conditions in the destination will each have a bearing on the way in which tourism behaviour alters. These are discussed below:

Northern Europe to the Mediterranean

Broadly, this is a market with a single purpose, the search for sunshine. Altered weather patterns induced by climate change could mean that northern Europe becomes more attractive and reliable during the summer

Table 4.3 Summary of climate changes and their probable impact on major international travel flows

Major tourism flow	*Origin market climate change*	*Destination region climate change*	*Implications for destination region*	*Possible market reactions*
Northern Europe to Mediterranean	– Much warmer, wetter winters – Warmer, drier summers – More 'reliable' summers	– Warmer, wetter winters – Much warmer, drier summers – Changes more marked in Eastern Mediterranean – Increased heat index – More days above 40°C – More arid landscape – Small tidal range means greater sea level rise impact	– Greater drought and fire risk – Much warmer, drier shortages – Greater personal heat stress – Beach degradation and habitat loss due to sea level rises – Vulnerability to more tropical diseases (e.g. malaria) – More flash floods – Poor urban air quality in cities	*Overwhelmingly a leisure travel market* – Improvement of northern European summers triggers more domestic holidays – Decreased incentive for Mediterranean summer holidays – Increased incentive for shoulder month Mediterranean holidays – Increased incentive for southerners to go north
North America to Europe	– Warmer winters – Warmer summers – Slight rainfall increases – S.E. USA (Florida) at risk from beach erosion, greater storm risk	**Northern Europe** – Much warmer, wetter winters – Warmer, drier summers – More "reliable" summers	**Northern Europe** *(80% of flow is to this sub- region)* – More attractive climate for summer holidays – Possibly greater congestion at key sites and cities	*Approx. 70% leisure, 30% business* – Too hot for peak summer cultural visits to southern Europe – Shoulder months travel may increase

Table 4.3 continued

Major tourism flow	*Origin market climate change*	*Destination region climate change*	*Implications for destination region*	*Possible market reactions*
Southern Europe	– Pacific coast greater storm risk and higher rainfall	**Southern Europe** – Warmer, wetter winters – Much warmer, drier summers – E. Med. esp. sharp changes – Increased heat index – More days above 40°C – Sea level rises	*(20% of flow is to this sub-region)* – Greater drought risk – Increased water shortages – Greater fire risk – More beach degradation due to sea level rises	– Little change foreseen for travel to northern Europe
Europe to North America	**Northern Europe** – Much warmer, wetter winters – Warmer, drier summers – More 'reliable' summers **Southern Europe** – Warmer, wetter winters – Much warmer, drier summers – E. Med. esp. sharp changes – Increased heat index – More days above 40°C	– Warmer winters – Warmer summers – Slight rainfall increases – S.E. USA (Florida) at risk from beach erosion, greater storm risk – Pacific coast greater storm risk and higher rainfall	– Sea level rise damages Florida coast and Everglades – Risk of Pacific coastal damage – Geomorphic damage to south-eastern coast – Increased heat index – Coastal erosion and storm damage risk on east coast – Rising health costs as tropical disease risk rises	*Biggest destinations are Florida, California and New York* – Florida may become less attractive at peak times – Possibly greater attraction of Carolina coast? – E. coast US and Canadian cities too hot in summer – Stronger winter ski market due to reduced capacity in Europe

Table 4.3 continued

Major tourism flow	*Origin market climate change*	*Destination region climate change*	*Implications for destination region*	*Possible market reactions*
North East Asia to South East Asia	– Warmer all year round – Small year round rainfall increase	– Little change in rainfall – Relatively little change in temperatures – Coastal areas vulnerable to sea levels rises	– No dramatic climatic changes foreseen – Islands and tourist coasts vulnerable – Coral bleaching	– Climatic factors unlikely to influence travel patterns greatly – Possible decline in dive and beach markets
North East Asia to North America	– Warmer all year round – Small year round rainfall increase	– Warmer winters – Warmer summers – Slight rainfall increases – Pacific coast greater storm risk and higher rainfall	– Risk of Pacific coastal damage – Geomorphic damage to south-eastern coast – Increased heat index – Coastal erosion and storm damage risk on east coast	– E. coast US and Canadian cities too hot in summer? – Sightseeing travel not likely to be greatly affected by climate change

Table 4.3 continued

Major tourism flow	*Origin market climate change*	*Destination region climate change*	*Implications for destination region*	*Possible market reactions*
North America to Caribbean	– Warmer winters – Warmer summers – Slight rainfall increases – S.E. USA (Florida) at risk from beach erosion, greater storm risk – Pacific coast greater storm risk and higher rainfall	– Warmer winters – Warmer summers – Small decrease in rainfall – Sea level rises	– Particularly vulnerable to sea level rises – Increased beach erosion – Coral bleaching and reef damage – Salinisation of aquifers – Higher energy costs for air conditioning – Greater need for sea defences and flood control – More tropical diseases (e.g. malaria) – Increased pressure on natural resources and eco-systems	– Beach product offering becomes less attractive (heat index, beach erosion, sea and coral quality) – Less need to escape northern climate – Loss of confidence in destination health risks

Source: Travel Research International (2003)

months, while the Mediterranean generally deteriorates in its appeal for the holidaymaker: the temperatures may become too hot. The limited tidal range of the Mediterranean makes it more vulnerable to sea level rise. As a result, this mass movement of tourists is likely to gradually slow, with northern Europeans holidaying either domestically or at least increasingly within northern Europe. Equally, southern Europeans may travel north to escape uncomfortable summer conditions at home.

North America to Europe

The North American market to Europe is more cosmopolitan with 70% travelling for leisure purposes and 30% on business. The business travel component generally is unlikely to be much affected by the climatic changes. Leisure travel also is likely to be relatively unaffected since culture and sightseeing are the prime motivations for an American trip to Europe. However, cities such as Rome and Florence could become too hot during the summer peak and therefore for southern Europe there could be a greater shift to shoulder season travel. Equally, more favourable summer weather conditions in northern Europe could mean greater congestion in cities such as London, Paris and Stratford-upon-Avon, during the summer peak and again travel from North America could see a greater spread into the shoulder months.

Europe to North America

The significant business component within this market is unlikely to be greatly affected by climatic changes. The leisure sector contains a number of sub-sections, the largest of which are travel to Florida, California and New York in the USA and to the West Coast in Canada. As climate change begins to have an impact, it is likely that European travel patterns to Florida – which is largely an organised tour beach market from the UK – may shift its seasonality or location as the region suffers from coastal degradation, an increasing likelihood of tropical diseases (already malaria prevention costs are rising) and it becomes uncomfortably hot during the peak summer season. Equally, travel to the Pacific coast, though not dependent on the 'sun, sea and sand' market to the same extent as Florida, may be impacted by the increasing risk of unpredictable weather. Summer travel to East Coast cities such as New York, and to California's Los Angeles and San Francisco, are likely to shift to spring and autumn as a result of increasing temperatures. Warmer winters and summers in the Rockies will decrease the effective length of the skiing season, increase avalanche risk and also affect lower lying ski resorts, although summer activity holidays are likely to benefit.

North East Asia to South East Asia

Climate changes in both origin and destination region are predicted to be small relative to other regions and as a result this large flow is likely to be relatively unaffected within the time-frame covered. Sea level rises and warming sea temperatures will, however, have an impact on the region's islands and coasts, which attract considerable numbers of visitors from countries such as Japan and Taiwan.

North East Asia to North America

Arrivals in North America from North East Asia are predominantly from Japan, South Korea and Taiwan. The large market travelling for business purposes will remain relatively unaffected, as will the sizeable VFR component. The leisure market is predominantly visiting the USA and Canada for sightseeing purposes; again, climate change is unlikely to have a major impact. Greater storm frequency on the Pacific coast and excessive heat in cities such as New York, Los Angeles and San Francisco may well instigate seasonal alteration to travel.

North America to Caribbean

As one of the world's main sun, sea and sand playgrounds, the Caribbean's tourism offering is totally dependent on its climate and beach product. Its main market, North America, is on the one hand escaping from its own cold and grey winter climate, while on the other moving to warmth, sunshine and coastal pursuits. As a result, the Caribbean is especially vulnerable to climate change since both factors may change – parts of the USA may become warmer, thus obviating the need to escape, while rising sea levels make the islands especially vulnerable, damaging beaches and causing infrastructural damage to the predominantly low lying coastal regions. Rainfall decreases and an increased need for air conditioning will put additional pressure on the islands' water and energy resources.

Other destination regions

In addition to the major international tourism flows discussed above, it is also useful to summarise the climate changes predicted for other world regions, which, while not featuring among the major flows, nevertheless have important tourism sectors. Also, two types of destination in which tourism is especially important – mountain regions and small island states – need consideration (see Chapters 9 to 11, this volume, on alpine areas; and Chapter 12, this volume, on Oceania). It is worth noting that, by and large, the rate of climate change in the southern hemisphere will be less than in the northern hemisphere.

Estimates of Tourism Growth

The World Tourism Organisation's (WTO) (1998) forecasts contained in their 2020 vision provide significant long-term tourism predictions that can be used as a basis for forecasting. Over that period, and in the longer term, the performance of the tourism sector will clearly be influenced by social change, political developments, economic growth, environmental change and demographic trends. Because there is no tourism forecast beyond 2020, no analysis has been done of the effect of these various factors on tourism growth. The 2020 vision predicts that global international arrivals will rise on average by 4.1% a year to a total of 1.56 billion arrivals in 2020, with the Middle East growing fastest (7.1% per annum) and Europe slowest (3.0% per annum). In the longer run beyond 2020, these rates seem certain to decline. (If the global growth rate of 4.1% a year were to continue to 2100, total international arrivals would reach 40,000 million, probably four times the global population at that date.)

It is nevertheless instructive to look at just two examples – the main flows from northern Europe to the Mediterranean (the world's largest tourism flow), and from North America to the Caribbean (because economically, tourism is more vital in the Caribbean than in any other region on earth).

The 'base case' in Table 4.4 shows the effects of extrapolating tourism growth in these two regions at 3% a year on average, based broadly on WTO forecasts to 2020, for a further 30 years. However, it is possible that climate changes may cause these rates of growth to be lower over the long term than they would otherwise have been. Purely to illustrate the possible effects, Table 4.4 also shows two lower rates of growth – 2.5% and 2% a year on average.

WTO data show also that in 2000, on average, each international arrival in Europe generated expenditure of US$580 excluding transport costs, and each North American arrival in the Caribbean generated US$1,000. Thus the northern Europe–Mediterranean flow was worth around US$70 billion in 2000, and that from North America to the Caribbean US$8 billion. By 2050 these would be US$300 billion and US$35 billion respectively. While such analysis can be no more than hypothetical, if climate change caused the decreases in growth rates shown in the table, the Mediterranean in 2050 might have foregone between 111 and 198 million arrivals, worth (in constant 2000 prices) between US$64–110 billion in receipts. The Caribbean (from North America only) in 2050 might have forgone between 8 and 13 million arrivals worth US$8–11 billion.

Table 4.4 Hypothetical results of alternative growth rates in international tourism flows to two key regions, 2000–2050 (million arrivals)

Flow	*Growth rate (% pa)*	*2000 (the base year for these flows)*	*2020*	*2030*	*2040*	*2050*
'Base case' extrapolation						
N. Europe to Mediterranean	3.0	117	211	283	382	513
N. America to Caribbean	3.0	8	14	19	26	35
Half point growth rate reduction						
N. Europe to Mediterranean	2.5	117	192	245	314	402
N. America to Caribbean	2.5	8	13	17	21	27
Full point growth rate reduction						
N. Europe to Mediterranean	2.0	117	174	212	258	315
N. America to Caribbean	2.0	8	12	14	18	22

Source: Travel Research International

While these tourists would not necessarily be 'lost' to global travel since they might simply go to other regions or convert to domestic tourism, the destination regions concerned would nevertheless have to deal with radically different economic circumstances, especially in employment, than would have been the case without these effects.

Development of Policy Options

The consequences of climate change – such as rising temperatures and sea levels, increased storm frequencies, and increasing or diminishing precipitation will impact on tourism destinations. It is therefore important that the tourism industry recognises potential impacts and lays plans to mitigate the adverse effects as far as possible. The following summarises some of the actions that may be open to some destinations in this regard.

The physical environment

Coasts and beaches

The effects of climate change on islands, coasts and beaches – the natural playground of a significant percentage of the world's leisure tourists – probably produce the most severe repercussions for the tourism industry. There are a number of different impacts:

- sea level rises cause coast and beach erosion, inundation of flood plains, rising water tables, destruction of coastal eco-systems, salinisation of aquifers and, at worst, the total submersion of islands or coastal plains;
- warmer sea temperatures of 1–2°C cause coral bleaching. It is predicted that coral bleaching will increase in intensity and frequency to the extent that in the Caribbean and South East Asia it will occur annually by 2020 and in the Pacific by 2040;
- change in storm frequency, especially in conjunction with rising sea levels, leads to damage to sea defences, protective mangrove swamps and shoreline buildings, to beach erosion and causes storm-surge damage to coral reefs.

Adaptation measures A rise in sea level means that not only does the coast itself needs protection but also that measures may be required to protect the hinterland from flooding. It is commonly accepted that strategies for adaptation to sea level rises fall into three main categories – delineated as 'protect', 'accommodate' or 'retreat' – and that all need to be considered in the broader context of coastal management. Measures might include:

- the building of sea wall defences and breakwaters: this has been the traditional response, and may be the only practical option, but it has been found sometimes to create as many problems as it solves and can ultimately destroy a location's natural beauty;
- enhancement and preservation of natural defences (such as the replanting of mangrove swamps or raising the land level of low lying islands);
- adapting to the changed conditions by building tourism infrastructure and resorts further back from the coast;
- importing sand to beaches in order to maintain their amenity value: this is costly and temporary and may damage the area from which the sand is drawn;
- dealing with degradation of coral reefs, which are not only a source of vital tourist income for island destinations but also essential

in protecting low-lying islands, such as those in the Pacific and Indian Oceans: when coral bleaching occurs and the reefs die, alternative man-made protection is unlikely to succeed; ultimately, abandonment may be the only option.

Mountain regions

Climate change is affecting the world's mountain regions, including the popular ski regions of the Alps and across the USA, in a broadly similar way: the snow line is receding due to warmer winters (for every 1°C increase, the snow line recedes by 150m), the ski season is becoming shorter, and there is greater winter precipitation – which means more snow at higher altitudes. The threat to winter sports is already manifesting itself in the skiing regions of Switzerland and Austria for example (see Chapter 10, this volume), while Scotland is having to address increasingly frequent snow-deficient winters (see Chapter 9, this volume). At the same time, mountain summers are becoming warmer and drier, presenting opportunities for extending the non-ski market.

Adaptation measures A number of adaptation strategies for ski resorts present themselves, ranging from mitigation of the detrimental effects to a complete change in the tourism product. These include:

- at the margin, increased use of artificial snow can help to extend and supplement natural snow cover as temperatures rise;
- high altitude resorts are likely to become more popular and may have to adapt to greater demand;
- lower altitude resorts, with reduced snow cover, may need to introduce an increased range of alternative attractions to skiing during the winter season;
- as a result of less stable (wetter) snow, greater avalanche protection will be required;
- as mountain summers become drier and warmer, the summer tourist season may be extended into the shoulder months; changing demographic patterns, particularly an ageing population with more leisure time, may prove beneficial to attracting this market.

The built environment

Climate change is already causing planning authorities to revise many aspects of policy, such as redefining flood plain risks and discouraging development on them. In the tourism sector there will be an increasing need to take climatic factors into account specifically in tourist areas, of which coastal areas are likely to be the most important.

Adaptation measures

Policy options will have to be considered for tourism infrastructure in a variety of areas. These may include:

- traditional designs may have to be encouraged to deal with alternative methods of cooling buildings in increasingly hot climates to counteract rising energy costs;
- physical planning issues will require building lines to be moved back from eroding coasts;
- coastal infrastructure, such as drainage, waste disposal, electricity, water supply, railways and roads may also have to be moved back from eroding coastal areas;
- water supply itself will have to be re-examined for many, increasingly arid areas;
- in the financial sphere, newly-built tourism infrastructure (e.g. hotels) in vulnerable areas may have to be written off over shorter-than-usual periods, which in turn could have an effect on the prices tourists have to pay; and
- increased insurance costs will have to be factored into resort profitability.

The tourism sector

While the impacts of climate change vary and are likely to manifest themselves in a variety of ways, what is certain is that the tourism sector has to show itself adaptable to changing conditions. Some changes may indeed favour an increase in tourism – such as warmer, drier summers in northern Europe leading to greater domestic travel and a reverse flow of tourists from the south – so that one destination's loss is another's gain. However, the majority of changes – rising temperatures and sea levels and increasing storm frequency, in particular – present the industry with real challenges.

Adaptation measures

Each situation calls for its own individual solution, but mitigation tactics that the industry might employ include:

- the introduction of built attractions to replace natural attractions if the appeal of the latter diminishes – e.g. the installation of an ice rink and spa facilities, if skiing becomes less reliable at lower altitudes;
- the development of alternative marketing strategies to cope with an expanding or a diminishing market (including stronger promotion of domestic tourism);

- adaptation to changes in the seasonality of tourist arrivals – e.g. the increased heat of the Mediterranean during the summer months may lead to reduced visitor numbers during the peak season but an increase during the shoulder months;
- co-operation with governments in order to deal with problems such as those associated with health, availability of water and vulnerability of infrastructure;
- recognition of the vulnerability of some eco-systems – e.g. wetland areas such as the Everglades in Florida – and the adoption of measures to protect them as far as possible; and
- the recognition that the tourism industry will be required to meet more stringent insurance conditions.

Issues of government policy

Government policies on climate change cover all sectors of the economy. The following is a suggested list of adaptation measures aimed explicitly at the tourism sector. These might include:

- introduce fiscal incentives (e.g. accelerated depreciation) or financial assistance for changes to the built tourism infrastructure (e.g. expanding flood drainage provisions in hotel properties, or redevelopment further back from beaches), to deal with the consequences of climate change;
- consider changing the fiscal regime where necessary (e.g. new hotels in vulnerable coastal zones may have to be written off in a shorter-than-normal period due to sea level rises, and thus may also require accelerated depreciation provisions);
- greater public investment in infrastructure for new tourism developments (e.g. land preparation, coastal defences or supporting infrastructure investment) to meet climate change impacts;
- passing legislation to change planning policies, zoning, land use priorities, as necessary;
- introducing changes to the school year in order to change peak holiday times (e.g. if traditional mid-summer periods become too hot or ski resorts have shorter snow seasons);
- providing direct training to the tourism sector in dealing with the consequences of climate change, including assistance with practical issues e.g. the 'hazard' mapping of sites and zones;
- revising policies on the financing of national tourism offices in order to ensure that promotional and marketing activities are tailored to new climatic realities (e.g. promoting 'shoulder' seasons); and
- recasting national transport policies where, say, aviation should be given a lower priority and internal transport higher priorities.

Conclusions

The 'macro' approach adopted in this chapter can do no more than scratch the surface of a complex subject. In reality, the effects of changing climatic conditions on the global tourism industry will be influenced at local and sub-regional levels by factors such as:

- the impact of individual climatic characteristics on local destinations;
- the physical environment;
- topographical characteristics;
- local geological factors;
- changing local health risks as a result of climate change;
- the nature of the tourism markets being served;
- the types of tourism facility and attractions offered.

Against this background, individual tourism destinations need first of all to interpret these global trends in the context of their own localities. Although the timescale during which climate change will have an impact may seem long, the impact on the tourism industry (and, of course, many other sectors) has already begun.

While tourism's adaptations to climate change need to be made at a local level, there are a number of conclusions that the industry as a whole must recognise. Climate change is happening now and its impact on tourism has to be taken seriously. Because of the importance of weather and environment to leisure demand, tourism is one of the sectors most likely to be affected by climate change. The greater the economic importance of tourism, the greater is the importance of understanding potential impacts and planning for appropriate action. The potential impact of climate change has very important implications for employment, investment policies, government policies and for the livelihoods of local residents. It therefore seems essential that the tourism industry should get involved in joint initiatives – with governments, the research community, local authorities or the international agencies – in assessing the implications of climate change.

Central governments and tourism ministries will always have a key role to play in defining and promoting action in the tourism sector in response to climate change. However, it is likely that such initiatives will need to be implemented at a sub-regional or local level in order to take into account local conditions and needs. The precise content of local initiatives will depend on local circumstances of course. However, the industry needs to:

- assess whether/how climate change has already begun to have an impact on tourism;
- prepare physical plans and 'hazard maps' of vulnerabilities due to climate change (e.g. impending rises in sea level or growing water

shortages in peak summer periods) that will affect the industry directly;
- co-operate with the research comunity, physical planners, public authorities and other appropriate specialists to prepare outline plans of the potential impact;
- define and cost whatever mitigation measures and actions may be appropriate in local circumstances;
- assess the issue of whether additional/new products need to be introduced to cope with changing circumstances;
- define and cost the benefits to the tourism industry that might be gained from appropriate remedial measures and from new opportunities that may arise for the industry;
- act as a focal point for lobbying actions to bring to the attention of policy makers the issues that will affect the tourism sector arising from the potential impact of climate change;
- adopt an integrated approach to tourism management in order to accommodate medium and long term concerns;
- initiate discussions with tourism planners, national tourism offices and tourism ministries on the changes to tourism policy and promotional efforts, which may be required as the effects of climate change become more pronounced;
- undertake an ongoing monitoring of changes that may be emerging as a result of climate change – e.g. physical changes to destination, visitor health, changing markets, changing product.

Acknowledgements

We would like to thank the World Tourism Organisation for support in preparing this chapter. This work is based upon a Report prepared for the WTO, *Climate Change and Tourism*, ISBN: 92–844–0632–3 (copyright WTO, 2003).

References

Climatic Research Unit, University of East Anglia, Norwich. Available at: www.cru.uea.ac.uk.

Intergovernmental Panel on Climate Change (IPCC) (1999) *Aviation and the Global Atmosphere.* A Special Report of IPCC Working Groups I and III. Cambridge: Cambridge University Press.

Intergovernmental Panel on Climate Change (IPCC) (2001) *Climate Change 2001: Impacts, Adaptation and Vulnerability.* Geneva: United Nations Intergovernmental Panel on Climate Change.

Travel Research International (2003) *The Impact of Climate Change on the Tourism Sector.* Study undertaken for the World Tourism Organisation in association with Climatic Research Unit of the University of East Anglia. Beaconsfield: Travel Research International.

World Tourism Organisation (WTO) (1998) *Tourism Vision 2020 Executive Summary.* Madrid: WTO.

Chapter 5

The Mediterranean: How Can the World's Most Popular and Successful Tourist Destination Adapt to a Changing Climate?

ALLEN PERRY

The climate of the Mediterranean is perceived, quite erroneously, by many tourists as idyllic, benign and delightful. It is the renowned radiance and clarity of light, rather than the heatwaves, droughts, storms and floods that can plague the region at times, that have made the area seductive to north Europeans for many centuries. The Mediterranean is currently the world's most popular and successful tourist destination with over 120 million visitors every year. Climate constitutes an important part of the environmental context in which all recreation and tourism takes place and because tourism is a voluntary and discretionary activity, participation will depend on perceived favourable conditions. For many sporting and leisure activities there are critical threshold levels beyond which participation and enjoyment levels fall and safety or health may be endangered.

Whilst in the 18th and 19th centuries it was the winter that was 'the season' with the aristocracy of northern Europe fleeing the cold and dark conditions, today it is mass 'sun-lust' package-tourism that leads to a seasonal peak in high summer. A UK survey suggested that for over 80% of overseas holidaymakers, better weather than can normally be found in the UK in summer was the primary reason for choosing an overseas holiday. Concern about skin cancer and worries about UV-B radiation has so far tended merely to modify behaviour (e.g. the use of more effective sunscreen treatments), rather than cause a change in destination preference. It is still the case that for many the acquisition of a suntan and the purchase of a holiday is as important as buying consumer durables for the home. The beach has become a fun place. Smith (1990) showed that the level of tourism from the UK to the Mediterranean was influenced by precipitation in the UK during the previous summer.

A new study from the Netherlands (Lise & Tol, 2003) has extended this analysis and shown how models can be built linking tourist demand to climate. If the current mass travel movement is viewed as a kind of import substitution, then such trends as growth of the domestic short-holiday market in northern European countries could have an impact on the balance of payments of several countries. An interesting study of the amenity value of climate has been carried out by Maddison (2001) who finds that British tourists are attracted by climates that deviate little from an averaged daytime maximum of 29 degrees C.

The Future Climate

The Intergovernmental Panel on Climate Change has shown that a rise in maximum temperatures and an increase in the frequency of heat-waves and hot days is likely during the 21st century. The Mediterranean will probably become less attractive for health reasons in the summer. Apart from the dangers increasingly associated with skin cancer, many Mediterranean beach resorts may simply be too hot for comfort in the peak season, with a much higher frequency of severe heatwaves (Perry, 1987). Carter (1991) has used an approximate index of climatic favourability to investigate changes of seasonal climate in Europe under possible future climate change. Results suggested that a climate warming of 4 degrees C would lead to a shift in the optimum summertime climate from the traditional southern coastal resorts northwards to currently less fashionable regions. This result holds true regardless of whether the warming is associated with moderate decreases or increases of precipitation. Mieczkowski (1985) proposed a tourism climate index (TCI) as a means of evaluating world climates for tourism. While he used five climate variables in the TCI formulae, thermal comfort was considered the most important and given a 50% weighting in the formulae. Using the ACACIA (2000) A2 High scenarios the index was calculated for the recent good summer of 1995 and an average summer 1999, together with the expected index value in 2020, 2050 and 2080 for the UK resort of Bournemouth. By the middle of the 21st century most summers are likely to have a preponderance of very good, excellent or ideal days for the holidaymaker in the UK. However, the attractiveness of the Mediterranean coastal zone in spring and autumn would be enhanced relative to the present. It is in the months of October-November that the lingering warmth and sunshine of the Mediterranean provides the biggest contrast with the weather in northern Europe. At this season maximum temperatures at present are 8–10 degrees C higher than in London whilst in April this difference is only 5–7 degrees C. Rotmans *et al.* (1994) suggest that the area suitable for sun-related tourism will decline in much of Italy and Greece as summer temperatures make beach tourism too uncomfortable. Indeed, it is single-product destinations that are most vulnerable to climate change.

Key Sensitivities to Weather and Climate

Major holiday decisions within many of the 'tourist exporting' countries of northern Europe are subject to a push and pull effect. The higher temperatures and settled weather of the Mediterranean summer exerts a big attraction, but better summers at home will reduce overseas holiday bookings. Giles and Perry (1998) have shown that the exceptional summer of 1995 in the UK led to a drop in outbound tourism and a big reduction in demand in the peak summer season for Mediterranean package holidays. In hot years there is a suggestion that Dutch tourists too prefer domestic to foreign beach holidays (WISE, 1999). Large numbers of people indulge in short-term opportunistic decision-making and switch their normal holiday preferences to take account of the unusually favourable conditions at home. Such limited evidence does suggest that climate warming might alter the competitive balance of holiday destinations with adverse effects on high season tourism in the Mediterranean. A limited survey of UK travel agents revealed that their customers most of all wanted guaranteed fine warm weather. Press reports about adverse health conditions, terrorism threats and devastating forest fires was more likely to concern customers than reports of very high temperatures.

Drought

The Spanish drought of the early 1990s showed how island resorts like Majorca could become dependent on water being transported from the mainland with attendant political tensions (Wheeler, 1995). In the last three decades there is evidence of the wet season ending earlier and the dry season onset also occurring earlier at many Mediterranean locations. High natural variability of rainfall helps to mask overall trends but endemic water scarcity is a very likely future scenario, especially south of 40 degrees north.

Small islands, for example in the Aegean, could be particular affected if tourism is allowed to continue to grow and it might be necessary to severely restrict tourist growth in such environments. Nicholls and Hoozemans (1996) have shown that in the Mediterranean there are 162 islands exceeding 10 square kilometres in size. Most have a low resource base but significant tourist development. Decline in rainfall and water supply availability, together with beach erosion could undermine their tourist industries and hence their local economies. It has been suggested (Karas, 1997) that Crete could experience serious water shortages in five years out of six by 2010. There is likely to be an increase in friction, with a conflict of interest between local people and tourist authorities on the use of scarce water. It has been calculated that a luxury hotel consumes around 600 litres of fresh water per guest per night. Water-hungry land uses like golf courses and water parks will be seen as water-stealers by local people. Projected decreases in runoff will exacerbate the problem of salinisation of water resources. Increased degradation of the environment and spreading desertification is likely to make some areas less scenically attractive to tourists. Comprehensive adaptation strategies, including reuse of wastewater and high technology solutions such as desalination plants or water grids are likely to be needed and would leads to greater intervention by government in tourism planning.

Heatwaves

Two major factors have interacted to impede the development of a definition of what a heatwave is, namely, the absence of a simple meteorological measure representing the complex interaction between the human body and the thermal environment, and the lack of suitable homogeneous time series of the meteorological variables likely to be involved (Robinson, 2001). Should we use exceedance of fixed absolute values, or deviation from the normal local climate as the basis for a definition? There are clearly several dimensions to very hot weather that need to be considered.

Extended heatwaves, defined as ten days or more, appear to be becoming more frequent in the Mediterranean. In the 15 years to 1994 Italy endured eight such heatwaves. In addition short-duration heatwaves of three to five days with temperatures 7 degrees C or more above normal have occurred on 33 occasions in the central Mediterranean between 1950–95. Individual heatwave days have increased from 52 days in the decade 1950–9 to 230 in the decade 1980–9 (Conte *et al.*, 1999). Heatwaves cause rises in the death rate, especially in urban areas, for example in one episode from 13 July to 2 August 1983 in Rome 450 deaths above the normal average occurred. In 1987 more than 1100 residents died in Greece between 20–31 July (Katsouyanni *et al.*, 1988) with a combination of temperatures above 40 degrees C and poor air quality. In 1998 in Cyprus 45 deaths attributable to heat were noted when the maximum temperature exceeded 40 degrees C on eight successive days. In Athens the National Weather Service of Greece forecasts heatwave emergencies and warnings are disseminated to the public. Extreme heatwaves and the deaths involved frequently get reported in the media of foreign countries and give a negative image to potential holidaymakers. Emotive phrases like 'killer heatwave' have been used. Even reports by reputable organisations can use hyperbole to get their message across. The World Wide Fund for Nature reported that some tourist destinations could be turned into 'holiday horror stories'. It has to be remembered that holidaymakers from northern Europe will be unused to temperatures as high as 40 degrees C and may be more at risk than local people, who are used to long hot summers. Gawith *et al.* (1999) have shown that at Thessaloniki in northern Greece the temperature–humidity index (THI) which assesses the impact of high temperatures and humidity will rise above a value of 84 (when nearly everyone feels uncomfortable) for more than twice as long as at present by 2050. In addition there will be significant increases in the shoulder warm periods suggesting a lengthening of the summer season. Forest fires, such as were very widespread in August 1994 in Tuscany, Corsica, Sardinia and France, and in southern France and Portugal in 2003 can lead to evacuation from tourist facilities such as campsites. Pinol *et al.* (1998) found that in coastal eastern Spain there has been increased fire activity and the number of days of very high fire risk is likely to increase further since there is a correlation between summer heat and fire occurrence. In Italy a strong association has been found between the number of forest fires and both higher summer temperatures and lower summer precipitation. Measures such as the closure of forest and parkland in summer may become increasingly necessary.

The tourist industry is very vulnerable to natural disasters. The publicity given to heatwave deaths in Greece in summer 1998, if repeated regularly, could act as a deterrent to tourism. In that year there were stories in the UK press of holidaymakers staying in their hotel rooms to try to escape the intense heat on the beaches. Queues of Britons were reported at hospitals and pharmacies suffering from heatstroke and burns while others cut short their holidays and returned home early. Rising mean summer temperatures will inevitably be accompanied by more occasions of extreme maximum temperatures. Extreme weather episodes are likely to have a stronger impact than average on weather changes. Heatwave conditions are also implicated in the development and proliferation of algal blooms, which can lead to closure of beaches, disfiguration of the coastal environment, and kills of fish, as has happened in the Adriatic.

Drought and Heat Combined: The Example of Summer 2003

The three summer months, June–August broke both heat and drought records over a large area of western Europe from Germany in the north to Spain in the south, and from Portugal in the west to Switzerland in the east. National temperature records were broken in the UK, Switzerland, Germany and Belgium but not in France, Spain, Portugal or Italy. The first heatwave in Italy and Spain began in mid-June with Majorca recording 39.5 degrees C, a new record for the month. There was a slight lull in the hot weather in early July but by the middle of the month temperatures of between 38–40 degrees C were recorded in several Mediterranean countries. The most intense phase of the heat-wave occurred from 4–13 August and during this period 70 out of 180 stations in France broke all-time records and in 15% of towns temperatures exceeded 40 degrees C. The intense heat was protracted, severe and in many places in central, southern and western Europe unprecedented. Even by the final weeks of September the summer drought continued, despite some thunderstorms, and with temperatures above 30 degrees C there was little seasonal relief. This type of summer is entirely consistent with what the computer models of the climate are saying will become more frequent so a study of its major impacts on tourism can provide valuable lessons for the future.

The main impacts of the extreme weather on tourism appear to have been the following:

- The most vulnerable tourists appear to have been campers and caravaners. Forest fires threatened sites and actually destroyed some and there were a number of injuries and fatalities. At several sites emergency evacuations were required. The worst fires were in southern France, Portugal, southwestern Spain and southern Italy. These low-cost holidays are also especially vulnerable to heat-waves since there is no obvious access to air-conditioning for campers and many mobile home users and there were many reports of holidaymakers abandoning their holidays and returning home early to escape the great heat.
- The excess death rate in many southern European countries was very considerable, reaching 15,000 in France alone, 6000 in Spain and more than 4000 in Italy. Overall summer 2003 was a major public health incident in Europe where the total excess death rate probably exceeded 40,000 and might have been as high as 70,000. The numbers of deaths to tourists is not recorded separately, so the impact is difficult to assess.
- Local people, especially those living in cities like Rome and Milan tended to abandon their cities whenever possible and retreat to the coasts, rivers and lakes joining the normal tourist influx and increasing congestion on roads and beaches.
- Infrastructure problems including power cuts in Spain and Italy as a result of excessive demand for air-conditioning and train cancellations because of buckled rails also affected tourists.
- British tourists travelling to the Mediterranean received very little advice and warnings before their departures. It was often left to tour representatives, themselves with very little medical knowledge, to warn of the dangers, especially from dehydration from excessive alcohol consumption.

The extreme conditions of the summer provided a further opportunity to monitor the demand for Mediterranean holidays in the light of the evolving conditions. In the UK there was a double heatwave with peaks of temperature in mid-July and again in early August. Between these peaks the weather in late July, at the beginning of the traditional holiday period coinciding with the school holiday period, was more unsettled although still warm. The return of settled and hot weather in the first few days of August was well forecast by the UK Meteorological Office but it was clear that holiday demand was more related to the actual than the forecast weather. The prices of late-availability holidays was monitored on a day-by-day basis using the websites of both an upmarket specialist tour operator specialising in Greece and Turkey and a holiday consolidator, for the period from 16 July to 20 August. At the beginning of the survey period holiday demand was lower than normal with prices quite weak, but after a delay of just a few days as the more unsettled weather began, prices began to rise by between 25% to 40% on both websites. Peak prices were achieved in the first few days of August, following nearly two weeks of unsettled weather and continued until the hot, settled weather was re-established, after which they fell dramatically and by mid-August, normally an extremely busy time, demand had fallen and prices had eased considerably, in some cases to lower levels than had been prevailing at the start of the survey period. This limited survey suggests that for a wide range of differently priced Mediterranean holidays catering to different price brackets, late-bookers tend to be highly influenced by the prevailing, as opposed to the forecast, conditions in the home country. Such tactical booking is probably a result of several factors acting in tandem, for example the desire to achieve a bargain holiday by very late booking, the inability to plan holidays ahead because of work commitments, or possibly a dislike of foreign travel unless there are perceived weather advantages to be derived from it. By mid-August there were numerous press reports that campsites in many holiday areas of the UK were full, and camping and holiday centre operators reported exceptionally high booking rates. The continued fine weather in northern Europe in September continued to depress prices and demand in the latter part of the season, which is normally a favourite holiday period for older people.

Disease

Higher temperatures could lead to some Mediterranean holiday areas becoming a suitable habitat for malaria-bearing mosquitoes. Spain, for example, is currently seen as a friendly, easily accessable, no risk destination not requiring immunisation, or courses of treatment against exotic diseases for intending visitors. It is anticipated that by the 2020s suitable habitats for malaria will have spread northward from North Africa into Spain. Increases in the incidence of food poisoning and food-related diseases linked to enhanced microbiological activity e.g. salmonella and e coli are likely to increase as temperatures rise. There will be a higher risk of epidemics of cholera and typhoid as well as other infectious diseases. Adverse publicity would follow such public health scares and frighten tourists away, as happened at Salou, Spain a few years ago. Extra costs will be involved in maintaining and strengthening public health defences and in health and hygiene education programmes.

Tourists Reactions to the Changing Climate and Adaptive Responses

Considerably more research has been done on the likely changes that Mediterranean climates may experience than on the possible impact of those changes on tourists in the future. It is not always easy to tease out the impact of climate from the many other factors influencing holiday choice (Perry, 2000). Tourism is a continuously adapting industry, responding to changing demographic and economic conditions as well as to new demands and technologies. Climate change will present new challenges but also lead to opportunities for tourist investment to capitalise on the new environmental conditions. Work has only just begun on 'translating' the suggested future climate scenarios into their impacts on tourism but already some interesting adaptations are emerging:

- Higher air and sea temperatures are likely to encourage a longer tourist season. If the summer becomes widely perceived as too hot the season could become 'doughnut shaped', with peaks in spring and autumn months and a hole in high summer. Such a pattern might resemble the current profile of visitor demand for a resort like Dubai. Maddison (2001) indicated that a lengthening and flattening of the tourist season is likely in Greece although with overall tourist numbers almost unchanged. With this in mind, resorts need to discourage a 'closing down' attitude at the end of summer. Higher temperatures will allow a prolongation of the season and if possible added cultural and sporting attractions such as arts festivals, regattas, food or drink events and local fiestas can help this process. Breaking the traditional seasonal patters has as much to do with changing consumer attitudes as with developing new attractions and more targeted advertising could help in this respect. A longer tourist season would allow quicker returns on investment with more intensive utilisation of facilities over a longer period. What in the UK is called the short-haul beach package has almost certainly peaked, but beach holidays will still be popular. They will be price-sensitive and probably booked later and we are likely to see greater segregation between resorts who continue to cater for this market and those who choose to chase other markets and become more diversified. Some parts particularly of the Spanish coasts, have an inheritance of many 30-year-old hotels, devoid of modern amenities and catering for a declining number of holidaymakers, many of whom will be low-spending, low-yielding Eastern European tourists. The demand will be for more individual 'bespoke packages' offering a little more excitement than the 'identikit' traditional inclusive tour (Middleton, 1991).

- The larger numbers of older people in the population will still wish to escape the dark, dreary winters of northern Europe. More are likely to consider moving permanently to, or buying second homes in Mediterranean areas. King *et al.* (1998) have shown that in several retirement destinations, including the Costa del Sol and Malta, the most important reason given for moving to the chosen destination was climate. Thus the climate of the receiving region for these migrants has been considered to be the most important pull factor. There are considerable planning implications if the growth of new apartments, villas and bungalows is not to cause environmental blight in some of these coastal areas. Along with this development will come increased demand for leisure pursuits e.g. golf courses, marinas.
- Tourists will increasingly expect holiday accommodation to be air-conditioned. Such accommodation will attract a premium price, whilst poorer quality self-catering apartments and rooms without air-conditioning will be much less attractive in the summer. At present only a fifth of rooms in hotels in Mediterranean countries are in the four- and five-star categories. Increased demands will be made on electricity supplies from the demand for additional cooling systems.

Conclusions

Tourism in the Mediterranean may become less sustainable as a result of climate change. Issues that are coming to the foreground include what are the safe and tolerable limits of the Mediterranean climate to sustain tourism? What is the likelihood that these limits will be breeched and at what time? Predicting climate change is complex but even more complex is predicting how people will respond to that change. Socio-economic scenario analysis is now being used to study possible future pathways of tourism and their regional implications (Amelung *et al.*, in press). Significant climate change could occur within the lifetime of many current tourist investment projects, such as large-scale marinas (also see

Scott *et al.*, Chapter 7, this volume). Advice on how to construct tourist facilities in harmony with the local climate and to provide the least stress to users is needed. At present we can see a hierarchy of flexibility to climate change. Namely, tourists are most flexible, tour operators have a degree of short-term flexibility, e.g. altering flight destinations, and local tourist managers are the least flexible with committed capital installed and not always transportable.

References

ACACIA (2000) The changing climate of Europe. In *Assessment of Potential Effects and Adaptations for Climate Change in Europe* (pp. 47–84). Norwich: University of East Anglia.

Amelung, B., Martens, P. and Rotmans, J. (in press) Tourism in transition: Destinations and itineraries. *Journal of Sustainable Tourism*.

Carter, T.R. (1991) *The Hatch Index of Climatic Favourability*. Helsinki: Finnish Meteorological Institute.

Conte M., Soarani, R. and Piervitali, E. (1999) Extreme climatic events over the Mediterranean. In J. Brandt, N. Geeson and J. Thornes (eds) *Mediterranean Desertification: A Mosaic of Processes and Responses* (Vol. 1 *Thematic Issues*). London: Wiley.

Gawith, M.K., Downing, T. and Karacostas, T.S. (1999) Heatwaves in a changing climate. In T. Downing, A. Oissthoorn and R. Toll (eds) *Climate, Change and Risk* (pp. 279–307). London: Routledge.

Giles, A. and Perry, A.H. (1998) The use of a temporal analogue to investigate the possible impact of projected global warming on the UK tourist industry. *Tourism Management* 19, 75–80.

Karas, J. (1997) *Climate Change and the Mediterranean*. Available at: www.greenpeace.org.

Katsouyanni, K., Trichopoulos, D., Zavitsanos, X. and Touloumi, G. (1988) The 1987 Athens heat wave. *Lancet* 3, 573.

King, R., Warnes, A.M. and Williams, A.M. (1998) International retirement migration in Europe. *International Journal of Population Geography* 4, 91–111.

Lise, W. and Tol, R.S. (2003) Impacts of climate on tourism demand. In C. Giupponi and M. Shechter (eds) *Climate Change and the Mediterranean: Socio-economic Perspectives of Impacts, Vulnerability and Adaptation*. Aldershot: Edward Elgar Publishing Co. (in press).

Maddison, D. (2001) In search of warmer climates? The impact of climate change on flows of British tourists. *Climatic Change* 49, 193–208.

Middleton, V. (1991) Whither the package tour? *Tourism Management* 12 (3), 185–92.

Mieczkowski, Z. (1985) The tourism climatic index: A method of evaluating world climates for tourism. *Canadian Geographer* 29, 220–33.

Nicholls, R.J. and Hoozemans, F.M. (1996) The Mediterranean vulnerability to coastal implications of climate change. *Ocean and Coastal Management* 31, 105–32.

Perry, A.H. (1987) Why Greece melted. *Geographical Magazine* 59, 199–203.

Perry, A.H. (2000) Impacts of climate change on tourism in the Mediterranean: Adaptive responses. Fonazione Eni Enrico Mattei (FEEM) Research Paper Series No. 35. Milan: FEEM.

Pinol J., Terradas, J. and Lloret, F. (1998) Climate warming, wildfire hazard and wildfire occurrence in coastal eastern Spain. *Climate Change* 38, 345–57.
Robinson, P.J. (2001) On the definition of a heat wave. *Journal of Applied Meteorology* 40, 762–75.
Rotmans, J., Hulme, M. and Downing, T.E. (1994) Climate change implications for Europe. *Global Environmental Change* 4, 97–124.
Smith, K. (1990) Tourism and climate change. *Land Use Policy* 7, 176–80.
Wheeler, D. (1995) Majorca's water shortages arouse Spanish passions. *Geography* 80, 283–6.
WISE (1999) *Workshop on Economic and Social Impacts of Climate Extremes Risks and Benefits*. Briefing document. Amsterdam: WISE Workshop.

Chapter 6

Greenhouse Gas Emissions from Tourism under the Light of Equity Issues

GHISLAIN DUBOIS AND JEAN-PAUL CERON

Tourism is affected by global warming, but it is also a major contributor to greenhouse gas emissions, by emitting CO_2, and other gases (mainly CH_4, N_{20}), and through specific phenomena (e.g. contribution to the formation of cirrus clouds by airplanes). This second aspect is less studied. Moreover research up to now has mainly focused on the impacts of energy consumption, and of transport, without assessing the specific contribution of tourism. This chapter discusses the methodological and practical difficulties of such an assessment. This is followed by results for France, which show that tourism is increasingly dependent on transport. Given the overwhelming importance of the stakes linked to global warming and the far from negligible contribution by tourism, attention is increasingly focused on mitigation policies. The second part of the chapter points out the importance of the problems that will have to be dealt with in any mitigation strategy and the equity issues that stem from them.

An Assessment of the Contribution of French Tourism to Global Change

Within an environmental assessment of the tourism sector at a national scale, the authors conducted an evaluation of the contribution of tourism transport to global warming for the French Institute for the Environment (IFEN). This work forms the basis of much of the following chapter.

Hypothesis and methodological choices

In order to simplify the evaluation, and after taking into account the reliability of data, a number of decisions were made with respect to methodology. First, to take as a starting point the tourist rather than the economic activities related to tourism (i.e. to privilege a consumption approach rather than a production approach). Therefore, the impacts of sub-sectors, such as travel agencies, or the operating of air companies (independently of the air trip itself), are not included in this assessment. Second, because of practical difficulties, not to utilise a life-cycle approach. In such an approach the impacts of the construction and dismantling of airplanes, hotels, equipment or energy plants should be added, as well as the impacts of implementing 'clean' energy production processes. Third, to evaluate tourism impacts at a national level rather than at the destination level. Research on environmental indicators for tourism is essentially destination oriented (Ceron & Dubois, 2001), whereas most international discussions and the Kyoto Protocol on the reduction of greenhouse gas emissions refer to a national basis. Therefore, a national evaluation was considered as necessary to define priorities. Fourth, to calculate a *total* contribution rather than a *net* contribution to greenhouse gas emissions. If a tourist stays at home instead of spending vacations, s/he consumes water, and energy, produces waste and greenhouse gas, which should be subtracted from the total tourism impacts in order to calculate the net emissions due to tourism. On the one hand, *total* estimates are most appropriate to establish benchmarks, or to facilitate regional or place-based analysis (EPA, 2000). They do not depend on the variability of alternative activities, which may change over time and places. On the other hand, n*et* estimates are useful to point out what is environmentally friendly in a tourist's way of life, compared to everyday life (staying in a camp for instance). At home, households also contribute to greenhouse gas emissions, through personal or professional movements. However, in the case of transport, net emissions seem very close to total emissions, since the distances travelled are much more important during vacations owing to the travel between home and the destination. Finally, to focus only on the contribution of tourism transportation to global warming. The main components of the tourism product/experience are transport, lodging and catering, the use of equipments (e.g. ski lifts, swimming pools), and activities (e.g. walking, swimming). Generally, the impacts of tourism on the environment can be separated between on-site impacts (including on-site transport) and transportation (to the destination) impacts. Each step of this consumption pattern contributes to global warming and should be evaluated, although it is not the case here.

Calculating the total contribution of the tourism sector remains uncertain because of the lack of required data. The methodology for accommodation and equipment would require a knowledge of the number of overnight stays/visitors for each type of accommodation/ equipment, allowing the multiplication by ratios, such as the average use of energy per overnight stay/visitor. The breakdown between the different energy sources used (e.g. electricity, fuel, gas) should be known in order to calculate greenhouse gas emissions. As far as electricity is concerned, it is possible to convert energy consumption into greenhouse gas emissions thanks to national data on electricity production and its split between sources, e.g. coal, nuclear and renewable. For accommodation, these ratios depend on the standard of comfort, the age of accommodation, and the climate of the location. This is why data based on local surveys are not very helpful. Only a few surveys provide such ratios for hotels, campsites and secondary homes on a national basis. Current research on eco-labelling and its forthcoming monitoring will provide more ratios in the near future.

There is also a great lack of data concerning equipment. Few data are available for theme parks and water parks. IFEN (2000) estimated the energy consumption of 4000 lifts during a casual winter season as between 571 to 734GWh, e.g. from one-quarter to one-third of the annual energy production of a nuclear plant. Furthermore, when lodging, catering and equipment are included in an overall evaluation, it really seems necessary to consider both *total* and *net* impacts on climate change. It would not be fair to the tourism sector to just consider the *total* estimate. One should rather focus on the incremental emissions caused by tourism, and on ways to reducing them. This is why the idea of an evaluation of the contribution to climate change of the overall tourism activity was abandoned.

In spite of a considerable amount of research devoted to the environmental impacts of day-to-day household travel, until recently only little work has specifically focused on the environmental impacts of household *tourism* travel. Indeed, according to the OECD (2001), one source of tourism-related environmental impacts – travel – remains consistently and conspicuously absent from the general discourse on sustainable tourism. With regards to global warming, the evaluation of transportation impacts should be considered as a priority. The Environment Protection Agency (EPA, 2000), estimated that for the United States, 76.5% of greenhouse gas emissions of the tourism and recreation sector are caused by transportation (against 15% for lodging, 2.7% for restaurants, 1% for retail, and 4.8% that are activity-specific). On-site travels usually have a lower impact than the travels from home to destination. In Calvia (Balearic Islands) on-site tourist movements represented 73,000 tons of CO_2 in 1995, when air transport to the destination contributed eight times more to greenhouse gas emissions (534,000 tons).

Detailed methodology

Estimation of the total contribution of tourism transport

The first step was to calculate emissions of greenhouse gas related to domestic tourism by road transport, for which very precise data were available. Then, the results were extended to the whole of tourism transport, taking into account the modal distribution of domestic and adding international tourism to France.

Evaluating the contribution of domestic tourism road transport to global warming required very precise data concerning tourist movements (in kilometres), and emission factors for various pollutants. Some very consistent data are available for road transport, thanks to the National transport survey for 1994. Households were asked to part their annual road trips (in kilometres) into five categories: home to work; professional trips; weekends; holidays; others private trips. The selected categories do not exactly match with a definition of tourism, since some weekend trips can be undertaken without an overnight stay. They do not match either with recreation, since the category 'other private trip' (including daily leisure trips) has not been taken into account. The survey enabled us to determine which type of vehicles were used for the different types of trip (petrol or diesel, age and capacity of vehicles), which is important since holidays and weekend trips appear to use diesel engines more frequently, and more recent and higher capacity vehicles. In a policy-making approach, these very detailed results will help point out the main factors (e.g. technical, socio-economic, load factors of cars) that influence the current emission profile of tourism and leisure.

The emission factors per kilometre were provided by the Copert III program (computer program to calculate emissions from road transports), for an average speed of 100km/h, which correspond to the highway or national road trips linked to tourism transportation. Copert III provided results for energy consumption and for nine pollutants: CO_2, CH_4, N_{20} (all greenhouse gases), CO, NOX, COV (light organic compounds), PM (particles), NH_3, as well as the global warming potential for the 100 coming years (GWP100) which is an aggregated index of the three main gas contributing to greenhouse effect (GWP100 (in kg carbon-equivalent) = $12/44.(CO_2$ emissions (kg) + $21.CH_4$ emissions (kg) + $310.N_20$ emissions (kg)).

Estimate of greenhouse gas emissions for specific travel

The indicators provided by the previous evaluation are quite technical and hard to communicate to the general public. They do not insist enough on individual responsibilities towards the greenhouse effect, and do not provide satisfactory information about the implication of the modal choice on greenhouse gas emissions. To present these results in a more attractive way, it was decided to calculate a specific indicator: a comparison of price, energy consumption and greenhouse gas emissions of a Paris to Nice trip, according to the mode of transport (see Figure 6.3). The indicator was calculated both for a family trip (four persons), and for an individual trip, during a peak period (influencing prices and load factor). The indicator provides a range (minimum and maximum) between the most and the least polluting vehicle for each mode of transport, or between the most and least polluting kind of travel (in the case of airplanes, a charter flight pollutes less per person, because of better load factor, for instance).

As noted above, the Copert III methodology was used for road transport, with a distinction between diesel and petrol vehicles, and precise data about the characteristics of travel (different types of roads, and their respective speed). For air transport, data were provided by Airbus industries (Airbus Aircraft performance program) and completed by the Emep/Corinair (European Environment Agency, 1996) simplified methodology for other airplanes. Emissions were calculated for a load of factor of 100%. For rail transport, the Jørgensen and Sorenson (1997) methodology was used (emissions depend on the weight of the train, slope and average speed). French railways provided data for TGV (high speed trains). Average emission factors for diesel train were collected in Zinger and Hecker (1979, cited in Jørgensen and Sorenson, 1997). Emissions were calculated for a load factor of 100%.

Results

Estimation of the total contribution of tourism transport

The first result is the important contribution of tourism to the emission of air pollutants: the contribution of tourism to the single road transport emissions (Figure 6.1) varies from 6% for COV, to 26% for NOX. Compared to the French overall emissions in 1994, the contribution of domestic tourism transports is negligible for methane (CH_4: 0.1%), which mainly comes from agricultural sources, or ammoniac (NH_3), but is still important for air pollutants such as carbon monoxide (CO: 4.1%), nitrogen oxides (NOX: 11.8%), light organic compounds (COV: 2.4%) and carbon dioxide (CO_2: 5.5%). Domestic tourism road transport represents (Figure 6.2) 24% of personal vehicles emissions of CO_2, 14.7% of road transport, 12% of the overall transport sector, and 5.5% of French emissions. It reaches annually 105 billions of kilometres and emits 17 million tons of CO_2. The potential of global warming (GWP100) of domestic tourism road transportation accounts for 3.8% of French total emissions (4.8 million tonnes carbon-equivalent).

The total contribution of tourism transport (domestic and international tourism, all modes of transports included) implies:

- Air transport for domestic (within France and abroad) tourism reaches 15% of the distance travelled by road for tourism purposes,

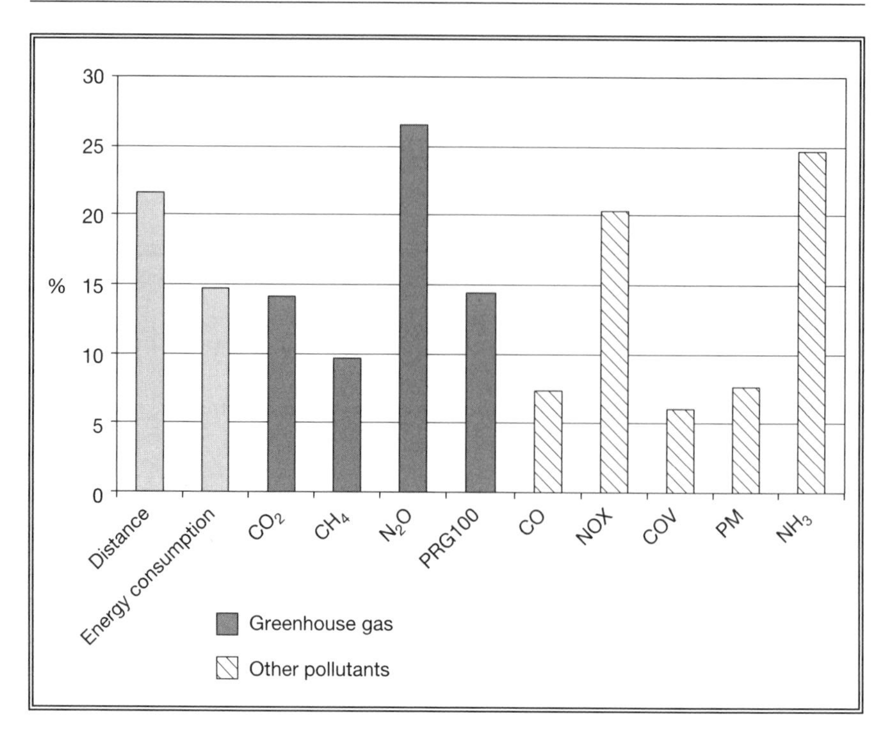

Figure 6.1 Contribution of domestic tourism travel to emissions of road transport

Source: IFEN (2000) based on SNCF, European Environment Agency (Copert III and MEET programs), IPCC, Airbus Industries, EDF

with emission per passenger/km from two to four times more important. This leads to a rough estimate of 45% of domestic road tourism transport.

- Rail transport for domestic tourism represents 20% of the distance travelled by road, with emissions per passenger/km three times less important (National transport survey), this comprises about 7% of domestic tourism road transport emissions.
- International tourism travel to France represents 30% of domestic tourism travel, with more air travel than domestic tourism. This comprises at least 45% of domestic tourism road transport emissions. However, owing to long-distance travel this figure is underestimated.

The total contribution of tourism transport to French greenhouse gas emissions is thus roughly about 7% to 8%.

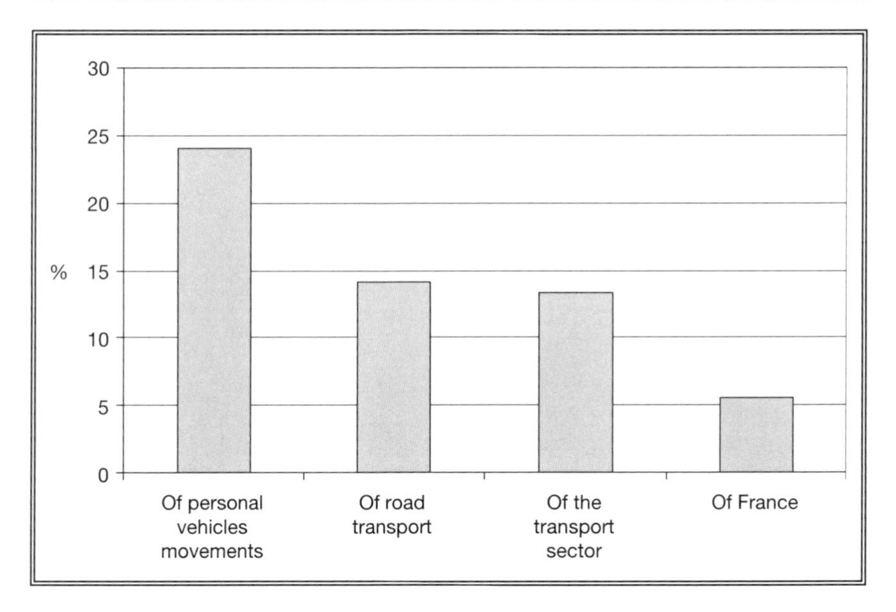

Figure 6.2 Contribution of domestic tourism road transport to CO_2 emissions of personal vehicle movements, road transport, transport sector and France

Source: IFEN (2000) based on SNCF, European Environment Agency (Copert III and MEET programs), IPCC, Airbus Industries, EDF

Estimation of greenhouse gas emissions for specific travel

In the example of a Paris to Nice trip, a family will have contributed three times more to global warming with an airplane than with a car, and five times more than with a train. In the case of an individual traveller – responsible for the total of the car emissions, but only responsible for one fourth of the airplane and train emissions calculated for a family – airplane and car emissions are almost equal. In all cases, the train is the winner of this competition, with very low emissions when the electricity comes from nuclear or renewable sources. For that range of travel distance, air transport represents 2% of personal trips taken in France, the train 18%, and cars 80% (INSEE, National transport survey).

Transport has a growing responsibility in greenhouse effect. The contribution of transport in French CO_2 emissions climbed from 8% to 39% between 1960 and 1990 (see *Citepa*, http://www.citepa.org/emissions/index_en.htm). The modal choices, and consequently the infrastructure choices, have a strong impact on this contribution. This issue is all the more important for tourism as it is highly dependent on transport.

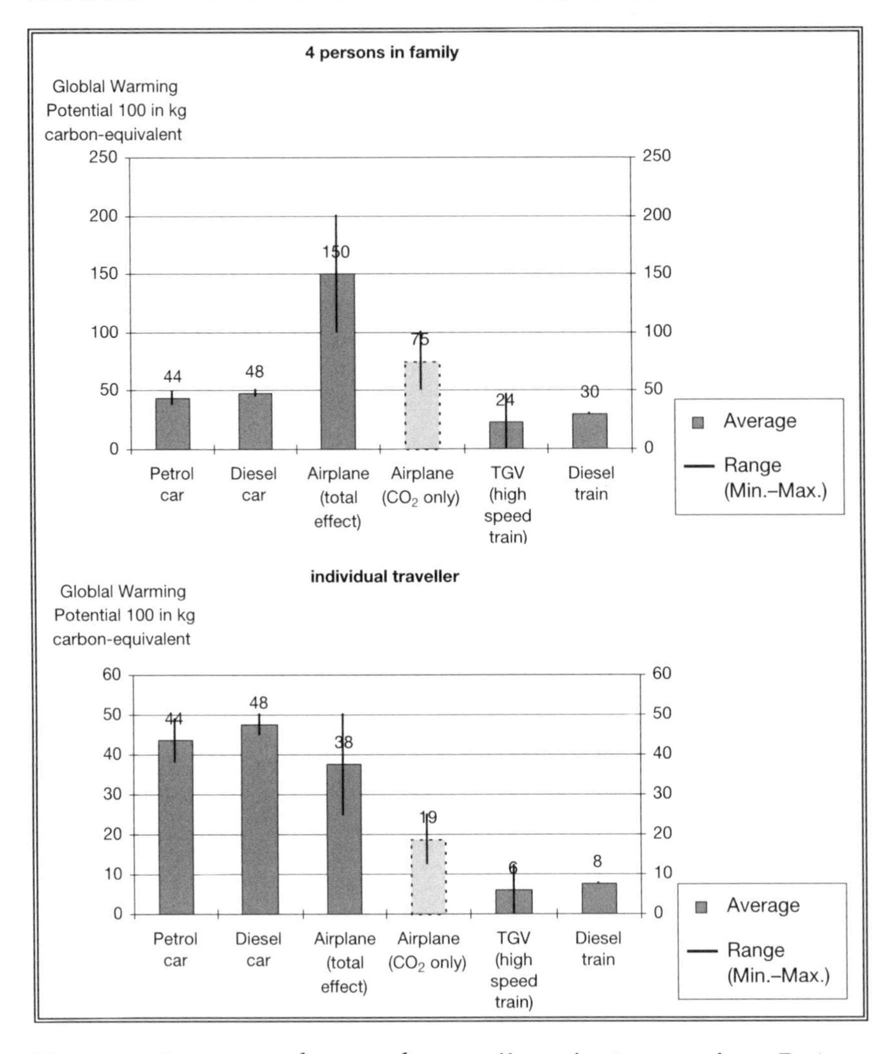

Figure 6.3 Impact on the greenhouse effect of a journey from Paris to Nice, depending on the mode of transport (family and individual traveller)

Source: IFEN (2000) based on SNCF, European Environment Agency (Copert III and MEET programs), IPCC, Airbus Industries, EDF

Notes:
Range: from the most to the least polluting vehicle in each category.
Cars – variables: age, horsepower, type of journey (motorway or main road).
Planes – variables: type of airplane. Two estimations are given: one for the effects of carbon dioxide (CO_2), which are well known, and the other for the impact on the greenhouse effect of all pollutants emitted during the flight. In this case, the effects of nitrogen oxides, water vapor, sulphur oxides and jet trails are all taken into account and sum up by applying a factor 2 to the CO_2 emissions; this is relatively low and some will consider multiplying by 2,7 (Hoyer, 2001: 457).
Trains – variables: type of energy used to produce electricity for a TGV, from hydraulic power (0 or near 0) to coal (47).

The Dependence of French Tourism on Transport

Growing transport intensity of French tourism

Changes in households travel behaviours, rather than the growth of overnight stays, are the main reason for the growth of tourism mobility in France. Current trends show that greenhouse gas emissions follow the increase in mobility. Moreover, it can be expected that they will grow even faster because of the evolution of the modal partition of tourism departures. Tourism and transport surveys show more frequent departures, for shorter stays and longer distances. The recent French law on the reduction of working time (35 hours), adopted in 1999, certainly reinforces this trend, since it enables more departures and shorter stays:

- From 1979 to 1999, the number of overnight stays in France by French tourists staggered from 733 million to 709 million (–3%), whereas the number of tourist trips increased from 43.8 million to 62.1 million (+ 41%) (INSEE: Vacances survey).
- The average length of stay for a French tourist dropped from 18 to 12 days between 1975 and 1999 (INSEE: Vacances survey).
- The number of personal annual departures rose from 3.1 to 4.8 between 1982 and 1994 (National transport survey), and was around 4.5 in 2000 (SDT survey).
- The average annual distance travelled for personal long-distance trips (> 100km), is 5230km per person, whereas the average length of one personal trip is 1430km (SES, 2002).

In the future, the contribution of French tourism to the greenhouse effect should increase even faster than the number of departures, since, as Figure 6.4 shows, the most polluting modes of transport (airplane and cars) take a greater share in departures. French tourism seems structurally dependent on road and air transport. The attitude of tourists to transport, the spatial distribution of resorts within France, past infrastructure choices (highways rather than train service), and current trends of the tourism demand (e.g. the attraction of French tourists for remote areas) lead to this high-impact situation.

Dependence on air transport

French tourism is a healthy economic sector. However, its growth relies on inbound and outbound (international) tourism. Outbound overnight stays increased by 32% while international departures increased by 76% between 1979 and 1999. Meanwhile, inbound overnight stays increased by 104% (Enquête aux frontières survey, 1996). France benefits from the proximity of other European countries as the main international

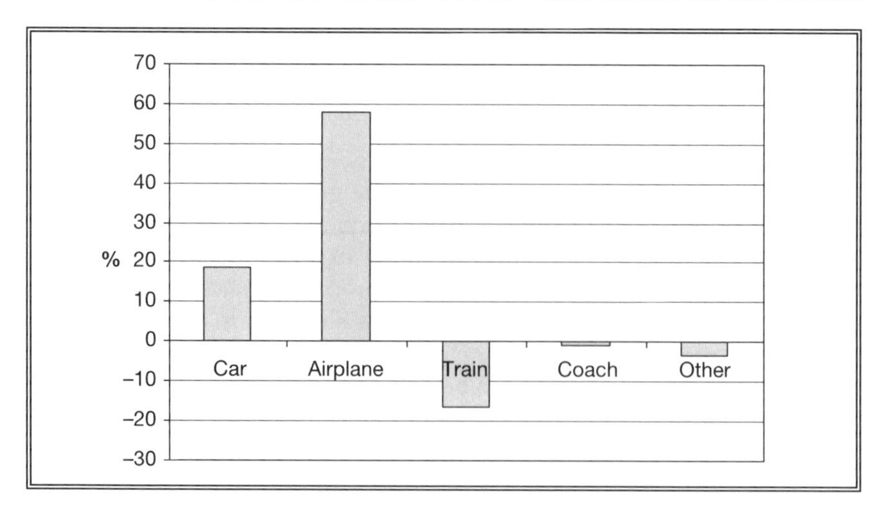

Figure 6.4 Evolution of the modal repartition of holiday departures (> 4 nights), 1986–1999

Source: INSEE: Vacances survey

markets. This is why 57% of foreign tourists used their car to reach France in 1996. However, international tourism remains strongly dependent on air transport: 44% of French departures to international destinations (including French territories) use planes (SDT survey, 1997), as do 15% of international arrivals to France (Enquête aux frontières survey, 1996).

Dependence on road transport

For various reasons, French domestic tourism also relies on road transport. First, the French tend to organise their holidays themselves, and tour operators (which are more likely to use collective means of transportation, be it coach, train or plane) only hold a low market share. Second, rural tourism accounts for one-third of domestic tourism. It is obviously more difficult to provide a coach or rail service in rural areas than in waterside or urban resorts (87% of stays in rural areas use car, against 73% in urban destinations). Third, tourist lodgings are dispersed rather than concentrated, especially in coastal areas.

Future emission prospects

What are the prospects if demand is allowed to develop freely? Is the demand for holiday travel going to stagnate in a predictable future (just as the demand for cars or for domestic appliances in some countries) or expand significantly with the associated environmental impacts? To assess this the authors defined four types of tourism/leisure patterns corresponding to the present trends of tourism and leisure in France (Ceron & Dubois, 2003a, 2003b):

- a conventional pattern, maintaining the present situation;
- a 'great traveller pattern' with frequent and varied departures (e.g. partly for long distances, frequently for short stays);
- a 'domestic leisure' pattern associating more home leisure with occasional departures for exotic destinations; and
- a 'bi-residential' pattern in which leisure mobility is essentially due to the travel between two homes.

These patterns link the characteristics of mobility (frequency, distance, means of transport) with the emission impacts. The methodology is described in Ceron and Dubois (2003a, 2003b) and the results summarised in Figure 6.5. These results show that, compared to present mobility reflected in the conventional pattern, none of the alternative patterns permit a reduction of the emissions of greenhouse gases from tourism transport: these are multiplied by a factor from two to five. Air travel is the most powerful variable leading to these results but tourism depending less on a merchant organisation and more on self-organisation does not necessarily imply less mobility: weekly migrations to a second home can lead to high-impact situations.

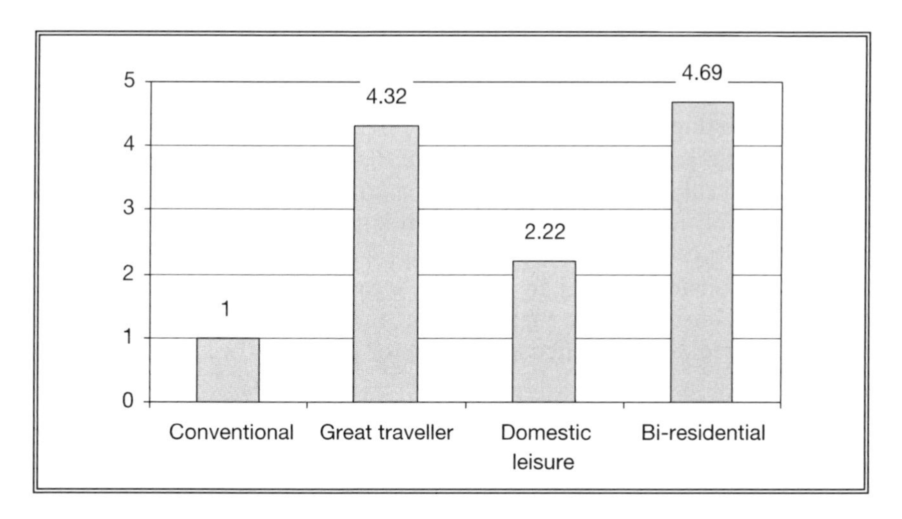

Figure 6.5 Impacts on climatic change for different mobility patterns
Source: Ceron and Dubois (2003a, 2003b)

Future Transport Policies

Tourism contributes significantly to the greenhouse effect, and relies more and more on transport: these two statements show the sensitivity of tourism to future transport policies in the context of a reduction of greenhouse gas emissions under the Kyoto Protocol, and more restrictive transport policies that might emerge in the future. For tourism, the level of impact will strongly depend on the mode of distribution of the constraints between miscellaneous stakeholders (economic sectors and countries). Referring to sustainable development implies that the distribution should be fair. The difficulty is that there is no unique criterion by which fairness or equity can be judged. As Boltanski and Thévenot (1991) demonstrate, within a society, different systems of values coexist regarding which choices can be justified: civic, industrial, merchant and domestic justification contexts, *cités* as they call them.

Referring to a 'civic city' context leads to attributing an equal share of emission rights for each inhabitant of the planet. Referring to an industrial context would privilege sharing the emission rights proportionately to GDP so as to minimise the negative impacts on activities. Referring to a domestic context would promote an approach taking into account the rights acquired by previous generations ('grandfathering'), e.g. the inhabitants of the developed countries can claim rights on the atmosphere they have been using these last centuries, not knowing that they were compromising mankind's future environment. Using the emissions in 1990 for a benchmark as the Kyoto Protocol suggests, in fact utilises this last frame.

The above examples show that calling upon equity to share the burden of the greenhouse gases linked to tourism or to determine emission rights for tourism can lead to dramatically different results. It nevertheless seems interesting to dwell on the first reference to an interpretation of equity attributing an equal share of emission rights for each inhabitant of the planet. First, because this equity criterion is often put forward as self-evident, or at least as a criterion superior to all others. Though nobody really suggests it could be applied immediately, it is more frequently considered that in the long term a worldwide convergence of emissions rates could be a legitimate goal. Secondly, because it appears that the consequences of such a criterion for tourism would be absolutely devastating.

An equalitarian perspective

At present, the use of energy generates roughly six billion tonnes carbon-equivalent for some six billion people: one tonne per head. Such a rate of emissions leading to a warming of the earth might induce severe consequences for humankind. Some argue that reducing the pace of warming to an acceptable level, in the context of a moderate demographic growth would imply diminishing emissions per head by a factor of two, that is to 0.5Tc (i.e., see Jancovici, 2003). Since the global warming potential of domestic tourism is somewhat over 5% of the global French contribution, the question is what could each of us do with the quota of 25kg carbon (5% of 500kg) equivalent if it could be devoted to tourism-related travel? Figure 6.5 indicates that this quantity would not even allow a family to travel by car from Paris to Nice; using a train fed by nuclear or renewable energy would be the only alternative.

Such an approach might be considered as extremist. The theme of an equalitarian access to resources is nevertheless supported by groups of intellectuals both in the north and in the south (Agarwal, 1991). One can also recall that radical discourses might be listened to (i.e. the maximalist interpretations of the precautionary principle), especially when they meet the frustrations that large groups of the population are confronted with.

Such drastic implications for tourism are difficult to accept, let alone to cope with. Some might say that this leads to questioning the criterion on the grounds of realism though it remains a valid equity principle. In fact this opinion itself can also be questioned. The main reason for this is that the equalitarian principle does not really provide for the determination of the level of emissions the present generation could afford. The burden should be shared not only by the all individuals of the present generation, but also fairly between present and forthcoming generations. If, as the IPPC (2001) scenarios suggest, the world population could range from 6 to 17 billion when the concentration in the atmosphere reaches a level of 450 to 650ppm (corresponding to 630 to 1080 billion tonnes of carbon equivalent), abiding by the equalitarian principle would allow for an individual belonging to the present generation to emit from 0.5 to 1.6Tc. This first means that under such premises, the quantity of emissions we can afford now is largely undetermined (Godard, 2003). Second, it shows that in the low population combined to high emissions scenario, the rate of emission is not very far from the French reality in 1990 (1.8Tc), but somewhat more for other developed countries: 7Tc per head in the US (0.9Tc in China). The nuclear origin of electricity in France (which poses other environmental and ethical problems) explains much of the gap with other European countries. From this, one could conclude that the changes in our present way of life should not be as dramatic as the first figures above suggested. A reasonable level of tourism activity could survive, though at a level lower than the present one and unquestionably this would not avoid breaking with the present trend of a growing transport intensity of the activity. Finally, an equalitarian view of the share of emission rights essentially contributes to show strikingly that the present situation and the present trends are not sustainable, it does not really lead to solutions that must be sought in other directions.

The prospects of international negotiations

An alternative approach to tackling the problem is to look at 'what can be done' at the expense of taking some distance with 'what should be done'. What can we expect from the negotiation processes within the international community and, taking into account the way they suggest to share the efforts, and what could be the effects on tourism?

On the one hand, the Brundtland Report (World Commission on Environment and Development, 1987) refers to common but differentiated responsibilities, which clearly shows a will to introduce an principle of equity into the negotiation process. On the other hand the characteristics of the international negotiations on climate change do not fundamentally differ from those on other topics, such as trade. However, there is no world government that could enforce measures reflecting superior interests of mankind, and equity principles, which can stand within a constitutional democratic state, cannot be applied in this context. Instead, we deal with an assembly of independent states, each one being lead by its own interests. In such a context any proactive stakeholder (a state for instance) that would consider itself as a victim of climatic change or wish to counteract the negative effects on climatic change of an activity, such as tourism, would have to convince its partners to implement measures they are not interested in and thus to compensate them for doing so (Godard, 2003). This means that in a context where no superior institution can enforce measures and make the polluter pay, as is the case in a country, the only alternative left is for the victim to pay. This is contradictory to the common sense of equity and anyway not very appealing for any stakeholder conscious of climatic issues. Thus ethics and equity remain are only paid lip service throughout the negotiation process (Godard, 2003).

Arguably, it was legitimate to consider the Kyoto Protocol as the successful outcome of such a process, as long as the USA did not reject it. Notwithstanding this last event, nor the fact that, on the contrary, the Protocol is considered as insufficient by some important stakeholders (the UK has already adopted a 20% reduction target by 2010 beyond the Kyoto Protocol and is working towards a 60% cut of carbon dioxide by 2050 (British Secretary of State for Trade and Industry, 2003)), it remains interesting to examine its potential implications for tourism.

By signing the Kyoto Protocol, France has committed itself to bring back its emissions to the 1990 level. This can be seen as the first step towards reducing emissions in the long run, at a moderate rate. In this case also, it is worthwhile looking at how an expanding tourism activity can cope with such prospects. Naturally, this issue concerns France and Europe where growth prospects are moderate, but also other regions of the world where tourism is expanding at a much higher rate. According to World Trade Organization (WTO) forecasts, international tourist arrivals should almost treble within the next two decades and by 2020, 1.6 billion tourists should travel to foreign destinations. The growth should be more moderate in Europe with a rate of 3% and 717 million visitors (WTO, 2002). France is not in the worst position to maintain its greenhouse gas emissions at a constant level, nevertheless, tourist mobility is following an upward trend. How can this problem be dealt with?

The latest assessment of greenhouse gas emissions shows that, for the moment, France is complying with the Kyoto objective, releasing 2.1% less (in terms of global warming potential) than in 1990. As long as this situation goes on, why should tourism be questioned, as it seems hardly relevant to discuss how to mitigate its impacts? This situation could go on for some years. However, the context appears to change completely if one looks at the long term. If the most ambitious national strategies (UK) are to be extended on an international basis, then it is obvious that tourism will have to seriously diminish its impact.

Calling for technology to rollback ultimate environmental constraints is by no means new (e.g. Sachs *et al.*, 1973). To what extent can productivity gains, which diminish the quantity of energy used per kilometre, help? As far as cars are concerned, for a given category of vehicle, gains around 25% in energy consumption are expected in the next decade. This gain will be partially offset by safety requirements, which increase the weight of cars, and by the consumption of more frequent air-cooling equipment. This figure can also be compared to the 31% and 38% increase in holiday and weekend distances between 1982 and 1994 (National transport survey). With respect to planes, the decrease of energy consumption per passenger/km is roughly 25% from one generation of planes to the next (every 25 years). Before the next technological leap occurs (use of liquid hydrogen), the gains are likely to be less important than previously. The OECD believes that during the next 20 years, the contribution of air transport to total greenhouse gas emissions due to energy use will increase from 3% to more than 7%. At world level it could, somewhere between 2010 and 2030, catch up with that of road transport. One is reminded that the number of holidays taken by the French in a foreign country increased from 3.8 to 11 million between 1964 and 1994 (INSEE: Vacances survey) and that the figure of personal trips of the French, using air transport increased by 16% from 1996 to 2000 (Direction du tourisme, SDT survey). For air and road transport, technical progress does not permit offsetting the effect of the increase of tourist movements and it seems it will be even less likely to be the case in future decades.

Aviation is one area of transport that is under pressure for change given the consequences of greenhouse gas emissions. Both the white paper of the European Commission on transport and the Royal Commission on pollution in the UK recommend taxing kerosene, which since the Washington Convention in 1937 has escaped any form of tax. This preferential treatment should not hold indefinitely and, if one considers the suggestion of the Royal Commission i.e. a tax of a hundred euros on each European departure, one can imagine the consequences for low cost companies and on short stays, though long-distance flights should be less affected.

Such measures will not suffice in the long run to curb the emissions of tourism and it will be inevitably concerned with the debates on tradeable permits. This type of mechanism, allowing the more and the less environmentally efficient industries to exchange emission rights, is to be set up for the European industry. One can imagine tour operators buying extra emission quotas from other sectors so as to maintain their activity. Another possibility is to try to have just as pleasant a life with less long-distance tourism (Peeters, 2003). We have great difficulties imagining ways of life radically different from present ones, which is after all surprising if we consider how they have changed and what we have experienced over the last half a century. Take, for instance, time spent watching TV. Who, in 1950, would have expected that 50 years later the French would spend on average two hours a day doing so (Dumontier and Pan Ké Shon, 1999)? The key point is the part tourism takes within leisure time; forward thinking on that point implies that we should admit that leisure activities and the uses of leisure time will probably change considerably over the next decades. The important point is not so much to predict what will change, but to know that the change will be considerable and might both influence the demand for tourism mobility and allow a rethink of the place of tourism mobility within leisure time.

French public policies during the last 20 years (since the short-lived Ministère du temps libre in 1981) have focused essentially on tourism and left aside leisure. They appear to have been led mainly by the search of the economic benefits of tourism (notably the inflow of foreign currencies) and by the effects on employment of shorter working hours (35-hour working week). Do they not somehow miss the point of more ambitious leisure/quality of life policies? The need for tourism is often linked to a bad quality of life, to a desire to escape, especially from urban areas (the Parisian syndrome). Would a better quality of life (e.g. the possibility of outdoor recreation, green belts, leisure activities) undermine the need for such tremendous mobility?

We are living in times where technological, economic and social changes are opening new opportunities but also where global constraints must be now dealt with. In a globalised world, sustainable mobility is one of the major challenges that has to be faced and tourism mobility is, although not alone, part of the problem.

A Note on Surveys Mentioned in the Text

The chapter utilises a number of French national enquiries or surveys, which are undertaken periodically by the major French statistical institution (INSEE), sometimes in conjunction with ministries (which can also separately conduct surveys). These enquiries/surveys include the following:

- Enquête Vacances (INSEE) noted as 'INSEE: Vacances survey'
- Enquête Suivi des déplacements touristiques des Français (SDT) (Sofres-ministère du tourisme) noted as 'SDT survey'
- Enquête aux frontières
- Enquête Transports et communications (INSEE-Ministère des transports) noted as 'National transport survey'
- Enquête Emplois du temps (INSEE)
- Enquête Logement (INSEE) noted as 'Housing survey'.

The results of these surveys lead to various publications which provide synthesised conclusions or analysis of specific issues arising from the survey documents.

References

Agarwal, A. (1991) *Global Warming in an Unequal World: A Case of Environmental Colonialism*. New Delhi: Centre for Science and Environment.

British Secretary of State for Trade and Industry (2003) *Energy White Paper. Our Energy Future. Creating a Low Carbon Economy*. Report presented to Parliament. Norwich: TSO.

Boltanski, L. and Thévenot, L. (1991) *De la justification. Les économies de la grandeu*. Paris: Gallimard.

Ceron, J.P. and Dubois, G. (2001) Tourism and sustainable development indicators: Two French experiments facing theoretical demands and expectations. Presentation at the International Sustainable Development Research Conference, Manchester, April.

Ceron, J.P. and Dubois, G. (2003a) Changes in tourism/leisure mobility patterns facing the stake of global warming. Paper presented at IGU Commission on Mobility conference, Human Mobility in a Globalizing World, Palma de Mallorca, 2–5 April 2003.

Ceron, J.P. and Dubois, G. (2003b) Draft proposal for a research agenda. Final report of the European Science Foundation workshop, Tourism, Climate Change and the Environment, Milan, 4–6 June.

Dumontier, F. and Pan Ké Shon, J.L. (1999) En 13 ans, moins de temps contraints et plus de loisirs. *INSEE-Première*, No. 675.

EPA (Environmental Protection Authority) (2000) *A Method for Quantifying Environmental Indicators of Selected Leisure Activities in the United States*. EPA-231-R-00–001. Washington, DC: Environmental Protection Authority.
European Environmental Agency (EEA) (with Emep/Corinair) (1996) *Atmospheric Emission Inventory Guidebook* (CD-ROM). Copenhagen: EEA.
Godard, O. (2003) L'équité dans les négociations post-Kyoto: Critères d'équité et approches procédurales in D4E-MEDD Les engagements futurs dans les négociations sur le changement climatique. Paris: La Documentation Française.
Hoyer, K.G. (2001) Conference tourism: A problem for the environment, as well as for research? *Journal of Sustainable Tourism* 9 (6), 451–70.
IFEN (2000) *Tourisme, Environnement, Territoires: Les Indicateurs*. Paris: Dunod.
IPCC (Intergovernmental Panel on Climate Change) (2001) *Climate Change 2001: Impacts, Adaptation and Vulnerability*. Geneva: United Nations Intergovernmental Panel on Climate Change.
Jancovici, J. (2003) *A quel niveau faut-il stabiliser le CO_2 dans l'atmosphère?* Available at: http:/www.manicore.com/documentation/serre/reduction.html.
Jørgensen, M.W. and Sorenson, S.C. (1997) *Estimating Emissions from Railway Traffic*. Report for the Project MEET (Methodologies for Estimating Air Pollutant Emissions from Transport). Lyngby: Technical University of Denmark.
OECD (2001) *Household Tourism Travel: Tends, Environmental Impacts and Policy Responses*. Report ENV/EPOC/WPNEP (2001)14. Paris: OECD.
Peeters, P. (2003) The tourist, the trip and the earth. In NHTV Marketing and Communication Departments (ed.) *Creating a Fascinating World* (pp. 1–8). Breda: NHTV.
Sachs, I., Ceron, J.P., Godard, O., Hourcade, J.-C., Théry, D. *et al.* (1973) *Suggestions pour un programme Environnement/Développement, étude effectuée pour le Programme des Nations Unies pour l'Environnement*. Paris: Ecole Pratique des Hautes Etudes Vième section.
SES (2002) Les voyages à longue distance des Français en 2000. *SES Infos rapides*, No. 143.
World Commission on Environment and Development (1987) *Our Common Future (the Brundtland Report)*. London: Oxford University Press.
World Tourism Organization (WTO) (2002) *Développement durable du tourisme. Contribution au sommet de Johannesburg*. Report by the Secretary General of WTO. Madrid: WTO.

Chapter 7

Climate Change and Tourism and Recreation in North America: Exploring Regional Risks and Opportunities

DANIEL SCOTT, GEOFF WALL AND GEOFF MCBOYLE

Introduction

The countries of North America are an important component of the global tourism industry. The United States and Canada were among the top ten nations for international tourist arrivals in 2001, ranking third (46 million) and eighth (20 million) respectively (World Tourism Organization, 2002). In terms of international tourism receipts, the United States led all countries generating US$112 billion while Canada ranked eighth (US$16.2 billion) (World Tourism Organization, 2002). Both countries also possess strong domestic tourism markets, several times greater than their international tourism markets. Combined domestic and international tourism spending in Canada reached C$54.6 billion in 2001 (Canadian Tourism Commission, 2002) and until the cross-border flows were interrupted by terrorism, the war in Iraq and SARS, was one of the fastest-growing industries in the nation. In 2001, combined tourism expenditures in the United States were US$559.6 billion (Travel Industry Association of America, 2003). Such statistics confirm the importance of the tourism sector to the North American economy.

Tourism in North America is as diverse as the communities and landscapes it occurs in, from the urban centres of New York and Las Vegas, to the beaches of the Florida Keys and ecotourism in the Arctic. As varied as is the tourism sector of North America, so too are the potential impacts of projected climate change (US National Assessment Team, 2000; IPCC, 2001). Consequently, it is beyond the scope of this chapter to provide a comprehensive assessment of the implications of climate change for the tourism industry in North America. Instead, the chapter will illustrate potential impacts of climate change on tourism in three regions of North America (Great Lakes, Rocky Mountains, Gulf of Mexico

Coast), chosen for the diverse characteristics of their tourism sectors and the varied nature of projected climate change impacts. Nature-based tourism is a very important component of North American tourism and because tourism regions that rely on their natural resource base to attract visitors are likely to be more at risk than those that depend on cultural or historical attractions (Wall, 1992), particular emphasis will be placed on this dimension of North American tourism.

Great Lakes Region

Some of the earliest research to examine the impact of climate change on tourism was on the skiing industry in the Great Lakes region. McBoyle *et al.* (1986), using the climate change scenarios available at the time, found that the ski season to the north of Lake Superior would be reduced by 30% to 40%. Skiing conditions would also be curtailed in south-central Ontario, resulting in the contraction or possible elimination of the ski season (40% to 100% reduction). Skiing in the Lower Laurentian Mountains of Quebec was projected to experience a 40% to 89% reduction in season length (McBoyle & Wall, 1992). Lamothe and Periard Consultants (1988) similarly projected that the number of skiable days would decline by 50% to 70% in southern Quebec. Comparable results were also projected for ski areas in the Great Lakes region of the United States. For example, Lipski and McBoyle (1991) estimated that Michigan's ski season would be reduced by 30 to 100%.

An important limitation of these early studies on climate change and skiing in North America (and indeed the international literature) has been the omission of snowmaking as a climate adaptation strategy. In order to reduce their vulnerability to current climate variability, ski areas in eastern Canada and the Midwest, Northeast and Southeast regions of the US have made multimillion dollar investments in snowmaking technology and many now have 100% snowmaking coverage of skiable terrain. Scott *et al.* (2002) were the first study to examine snowmaking as an adaptation strategy. Using a range of climate change scenarios based on the Intergovernmental Panel on Climate Change's (IPCC) Special Report on Emission Scenarios (SRES), Scott *et al.* (2003) found that with current snowmaking capabilities, doubled-atmospheric CO_2 equivalent scenarios (relating to the 2050s) projected a 7% to 32% reduction in average ski season in the central Ontario study area. With improved snowmaking capabilities, modelled season losses were further moderated to between 1% and 21%. The findings clearly demonstrate the importance of snowmaking, as the vulnerability of the ski industry was reduced relative to previous studies that projected a 40% to 100% loss of the ski season in the same study area under doubled-CO_2 conditions (McBoyle & Wall, 1992). Similar reassessments of widely cited

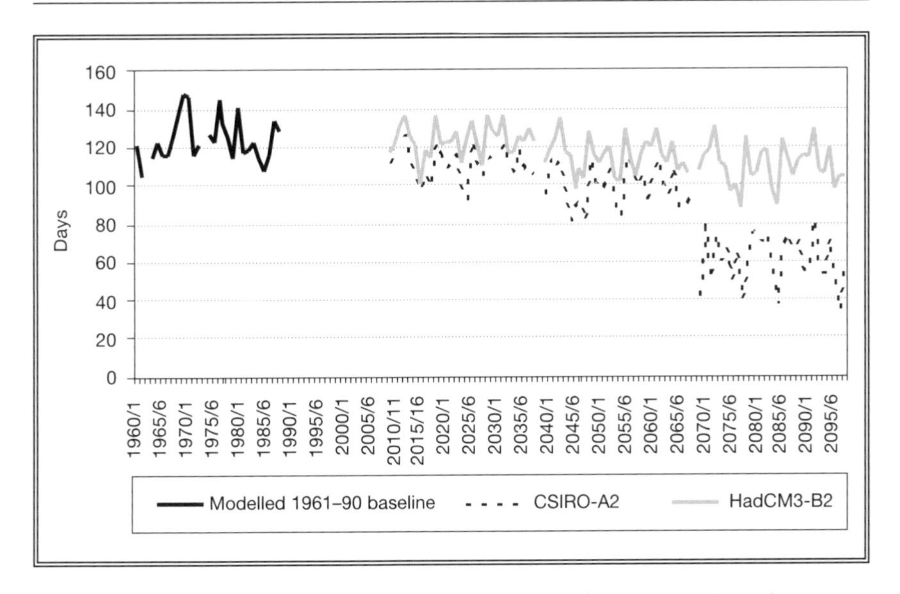

Figure 7.1 Modelled ski season length at Horseshoe Resort (south-central Ontario, Canada)

earlier climate change studies on the Quebec ski industry and other areas of North America are required.

Importantly, snowmaking requirements to minimize ski season losses in the study area were projected to increase 191% to 380% by the 2080s (Scott *et al.*, 2003). However, it should be recognized that while snowmaking is an effective adaptation strategy, it is not without associated challenges, for both capital and operating costs are substantial and there are large water requirements. The additional snowmaking requirements and greater energy required to make snow in warmer than average temperatures would be an important cost increase that could affect the profitability of some ski areas. Thus, it may not be the inability to provide snow on ski hills, but the cost of making additional snow and the negative perceptions related to no snow conditions in ski market areas that could cause adverse economic impacts within the Ontario ski tourism industry. Large corporate ski entities in the region such as Intrawest and American Skiing Company may be less vulnerable to the impacts of climate change than single ski operations because they generally have more diversified business operations (real estate, warm-weather tourism resorts and four-season activities), are better capitalized (so that they can make substantial investments in snowmaking systems) and, perhaps most importantly, are regionally diversified (which reduces their business risk to poor snow conditions in one location).

Other important components of winter tourism in North America have been overlooked by climate change impact assessments. The International Snowmobile Association (2003) has estimated the annual economic value of the North American snowmobile industry at over US$10 billion and, in some regions, it exceeds the economic importance of skiing. Notably, climate change was not considered in Canada's recent National Snowmobiling Tourism Plan (Pannell Kerr Forster, 2001), despite the potentially important implications for the sustainability of snowmobile-based tourism in the plan's long-term time-frame (starting in 2008).

In the only study that has examined the implications of climate change for snowmobiling, Scott *et al.* (2002) found that snowmobiling was more vulnerable to climate change than alpine skiing because of its greater reliance on natural snowfall. Snowmaking has very limited application in snowmobiling and Nordic skiing because of the technical and economic barriers associated with producing snow over tens or hundreds of kilometres of trail. When the climate change impact scenarios for seven snowmobiling areas in south-central Ontario were compared, the average projected reduction in season length was substantial (29% to 49%) as early as the 2020s. The average snowmobiling season was projected to decline by approximately 50% at most locations by the 2050s, with further average season reductions of between 70% and 79% by the 2080s. Similar assessments are required in other regions of Canada and the United States where snowmobiling is a key component of winter tourism.

Winter festivals in the Great Lakes region have experienced problems over the past few winters due to lack of snow and warmer than normal temperatures. The Canadian Ice Fishing Championship was cancelled in 2002 because of the lack of lake ice on Lake Simcoe (Ontario), while some winter festivals in the states of Minnesota and Wisconsin had to cancel events due to dangerous ice conditions and a lack of snow in early 2003. Snow and ice cover scenarios in the Midwest and New England regions of the United States (US National Assessment Team, 2000) indicate that both snow and ice cover will decline and become more variable under climate change, putting community winter festivals at even greater risk in the future.

Implications of climate change are not confined to the winter season. The Great Lakes have long been a Mecca for recreational boating and fishing, and with the longest freshwater beaches in the world, are the location of recreational facilities such as cottages and parks. The Great Lakes constitute a dramatic example of the implications of fluctuating water levels and, hence, climate variability, for recreational activities. Marinas and recreational boating are harmed by extremes of both high and low water, particularly the latter, which according to Mortsch *et al.* (2000) is projected under most climate change scenarios. Surveys undertaken in 1992 of marina operators and recreational boaters on the

Canadian side of the Great Lakes indicated that almost all had incurred costs at some time or other associated with fluctuating water levels (Bergmann-Baker *et al.*, 1995). Since they had been operating their marinas (approximately one-third had been a marina operator for less than five years although most marinas were considerably older), in times of low water, 67% of respondents had experienced problems of access to docks or berths, with inadequate channel depths or had ramp access difficulties, and smaller proportions were forced to use fewer slips, experienced short boating seasons or had dry rot in wooden structures. In response to these problems, 55% had dredged, 45% had adjusted their docks, 44% restricted the sizes of boats, 44% had to relocate boats, 27% closed slips, 19% constructed floating docks, and 7% replaced rotted structures. Unfortunately, it is not possible to put a precise dollar value on these adjustments but clearly it has been substantial. In addition, other adjustments were made in periods of high water. In fact, there are examples of marina operators experiencing low water problems at times when they are still paying off loans acquired to build breakwaters to protect themselves from high water. A similar survey of cottage owners along the shores of the Great Lakes (Scott, 1993) found that water levels projected by two climate change scenarios were substantially lower than those preferred by cottagers, particularly on Lakes Huron, Erie and Ontario. Boaters also accrue a variety of costs (most commonly hull and propeller damage) but they are more mobile than marina operators and, thus, can adapt more easily by using alternative boat launch facilities and travelling to other bodies of water.

Below average water levels on the Great Lakes during the summers of 1999–2002 once again revealed the sensitivity of marinas and the recreational boating industry to climate variability. Water levels on Lakes Huron and Michigan that were approximately one metre below the long-term average brought about the creation of a C$15 million Great Lakes Water-Level Emergency Response Programme by the government of Canada to assist marinas with emergency dredging costs.

Lower than average water levels on the Great Lakes also have implications for the character of shoreline ecosystems and their tourism potential. Studies of altered Great Lake shorelines under climate change scenarios (Lee *et al.*, 1996; Schwartz *et al.*, 2004) reveal the vulnerability of some key wetland complexes that are important waterfowl habitat and a source of recreation for many people. For example, in naturally confined marshes such as Point Pelee National Park, lowered lake levels will cause the marsh to revert to marsh meadow and, eventually, to dry land. Because of the protective sand spits, the marsh will be prevented from moving lakewards and vegetation will shift from hydric to mesic conditions (Wall, 1998). Some plant species may change growth form to accommodate to drier conditions but vegetation will change dramatically as species intolerant of drying die and are replaced by species emerging from buried seeds. The trees, which mark the landward edge of the marsh, may advance due to a lowering of the flood line. Wetland species diversity will decline and the suitability of the marshes as a habitat for recreationally and commercially valued species, such as migrating waterfowl and muskrats, will be reduced. Sport fishing may also be affected by the reduced quality of shoreline marshes where fish feed and spawn. In time, the marsh may lose its wetland character and, under extreme conditions, key bird migration routes may change, diminishing the quality of North America's sixth ranked birdwatching site (by annual visitation and economic impact) (American Birding Association, 2003).

The multi-billion dollar freshwater sport fishery of North America and the associated tourism market (estimated at over US$11 billion) (American Sportfishing Association, 2001) would also be impacted by climate change. A number of cold-water fish species are particularly sought by anglers. Studies of the potential impact of changes in water temperatures for selected cold-water species have projected negative impacts throughout the United States, including the lower Great Lakes. A study by the US Environmental Protection Agency (US EPA) (1995) projected substantial losses (50%–100%) in cold-water fish habitat in the Great Lakes states. Tourists that are attracted to this region for cold-water species may have to travel to other regions, such as northern Ontario, Quebec or New Brunswick, where these preferred species may still be available or more prolific. This situation is further complications by the proliferation of non-native warm-water species in the region. Notably, the US EPA (1995) study estimated annual economic damages to the sport fishing industry in the United States as a whole at US$320 million in the 2050s. This study also found that when alternative modelling assumptions were used, the estimated damages increased substantially, suggesting the need for further research to narrow the range of uncertainty.

Rocky Mountain Region

The natural environment is very important in determining the attractiveness of a region for tourism and the scenic landscapes and parks of the Rocky Mountain region are internationally renowned tourism destinations. A study of the nature-based tourism market (HLA Consultants & ARA Consulting Group Inc., 1995) found that the natural setting was the most critical factor in the determination of a quality tourism product. Consequently, if climate change adversely affects the natural setting of mountain destinations (the loss of glaciers, special flora or fauna, increased fire and disease impacted forest landscapes) the quality of the tourism product could be diminished with implications for visitation and local economies.

A number of studies has examined the potential biophysical impacts of climate change on mountain environments in North America and provide insight into the implications for tourism. Vegetation modelling suggests that the Rocky Mountain region experience both latitudinal and elevational ecotone changes, with the potential for species reorganizations and implications for biodiversity. The upslope migration of the tree line has already been documented in Jasper National Park (Alberta, Canada). Similar impacts are expected in Yellowstone National Park (Wyoming, USA), where vegetation modelling results project that the range of high-elevation species will decrease, some tree species will be regionally extirpated, and new vegetation communities with no current analogue will emerge through the combination of existing species and non-native species (Bartlein *et al.*, 1997). Vegetation modelling in Glacier National Park (Montana, USA) projected a 20m per decade upslope advance of forest through 2050, with considerable spatial variation determined by soil conditions and aspect (Hall & Farge, 2003). A study of mammal populations in the isolated mountain tops of the Great Basin in the western United States, projected that regional average warming of 3°C would cause a loss of 9% to 62% of species inhabiting each mountain range and the extinction of 3 to 14 mammal species in the region (McDonald & Brown, 1992).

Like glaciers around the world, those in western North America have been retreating over the past century. Glacier National Park (Montana, USA), which early visitors referred to as the 'little Switzerland of America', has lost 115 of its 150 glaciers over the past century and scientists estimate the remaining 35 glaciers will disappear over the next 30 years (Hall & Farge, 2003). Similar projections have been made for glaciers in Canada's Rocky Mountain parks (Brugman *et al.*, 1997). The loss of glaciers has a direct impact on tourism operations such as Snowcoach Tours in Jasper National Park (Alberta, Canada), which currently provides glacier tours to over 600,000 visitors annually. The loss of Glacier National Park's namesake would be a significant heritage loss, but could serve an important educational role to inform visitors how the landscape the park was established to protect has changed in only 100 years. Scott and Suffling (2000) suggested that the indirect impact of the loss of natural beauty associated with glacial landscapes for tourism also remains an important uncertainty.

The drought and wildfires in the state of Colorado and province of British Columbia during the summers of 2002 and 2003 may provide an important analogue of the potential impacts of climate change in the mountainous regions of western North America, as several studies have projected increases in wildfire severity and frequency in large areas of western Canada and the United States (Stocks *et al.*, 1998). The statewide drought in Colorado created dangerous wildfire conditions and the park closures and media coverage of major fires in some parts of the state had a significant impact on summer tourism. Visitor numbers declined by 40% in some areas of the state and reservations at state campgrounds dropped 30%. The Colorado drought also affected fishing and river-rafting tourism in the state. Anglers were restricted from fishing in many state rivers because fish populations were highly stressed by low water levels and warmer water temperatures. Low water levels also shortened the river-rafting season substantially. Some outfitter companies lost 40% of their normal business and statewide economic losses exceeded US$50 million.

Drought and wildfires also affected the tourism industry in the province of British Columbia in the summer of 2003. Several parks had to be closed at various times during the peak tourism season due to extreme fire hazard conditions. The Kettle Valley Mountain Railway trail, a national historic site that attracts over 50,000 tourists annually, lost five of its wooden trestle bridges to the fires. The estimated cost to restore the bridges is C$30 million. The net impact of the fires for tourism in British Columbia has yet to be determined.

Many of the most popular parks in Canada and the United States are located in western mountain ranges and projected climate change has the potential to extend the park visitation season. An assessment of the implications of climate change for seasonal park visitation in Rocky Mountain National Park (Colorado) projected increased annual visitation of 6.8% to 13.6% (Richardson & Loomis, 2003). Subsequent economic analysis indicated that this increased visitation would generate a 6% to 10% increase in local economic output and 7% to 13% increase in local jobs. Similar or even greater opportunities for increased visitation are anticipated in most of the parks in western Canada and the Northwest US, as the visitor seasons are currently equally climate limited throughout this region. Changes in visitor numbers and seasonal visitation patterns are important for park revenues and the economies of nearby communities, but also have ecological implications. Visitor numbers and related tourism infrastructure were identified by Parks Canada as a significant ecological stressor in 24 of Canada's 38 National Parks and increased visitation has the potential to heighten visitor pressures in certain parks.

The Rocky Mountains are home to some of North America's best-known ski resorts and are an international winter tourism destination. Although snow cover modelling in the mountains of Northwestern United States projected a 75cm to 125cm reduction in average winter snow depth under two climate change scenarios and an estimated upward shift in the snow line from 900masl (metres above sea level) to 1250masl (US National Assessment Team, 2000), the implications for major ski areas in the region have not yet been examined.

Gulf of Mexico Coastal Region

Coastal zones are among the most highly valued recreational areas and are primary tourist destinations. The economy of many coastal communities in this region is dominated by the exploitation of sea, sun and sand for recreation. Climate change has important implications for this region both through the redistribution of climate resources for tourism and sea level rise.

One of the most direct impacts of projected climate change on tourism will be the redistribution of climatic assets among tourism regions, with subsequent implications for tourism seasonality, tourism demand and travel patterns. Changes in the length and quality of tourism seasons would have considerable implications for the long-term profitability of tourism enterprises and competitive relationships between destinations, particularly those where climate is the principal tourism resource like the 'winter getaway' holiday destinations of the United States Sunbelt and the Caribbean.

Using a tourism climate index (TCI) as a standardized metric to assess climate conditions for tourism, Scott *et al.* (2004) investigated change in the spatial and temporal distribution of the climate resource for tourism in North America under two climate change scenarios. Analysis of the number of cities with 'excellent' or 'ideal' TCI ratings (TCI > 80) in the month of January as illustrative of the winter getaway holiday season (Figure 7.2), revealed that the number of cities in the United States increased from two in the baseline period (1961–90) to four or seven in the 2050s and seven or nine in the 2080s. The implication is that southern Florida and Arizona would face increasing competition for 'winter getaway' travellers and the 'snowbird' market (retirees from Canada and the northern US states who spend two to six months in winter peak and optimal climate destinations). In contrast, the number of Mexican cities with TCI ratings of 80 or greater decreased from six to four and one in the 2080s, suggesting that Mexico could become less competitive as a winter getaway destination.

While travellers would have greater choice of 'winter getaway' destinations in the Gulf of Mexico coastal region, with shorter and less severe winters, they may be less compelled to travel to warm-weather destinations as a winter escape. In addition, if, as Last (1993) suggests, cultural attitudes toward sunbathing shift in response to increased risk from UV radiation associated with ozone depletion and ongoing health education campaigns, 'winter getaway' destinations may be competing for a diminished travel market.

An important consequence of global climate change is sea level rise. In the Gulf of Mexico region, the mid-range estimate for sea level rise is 50cm by 2100 (IPCC, 2001). This projected rise in sea level is a threat

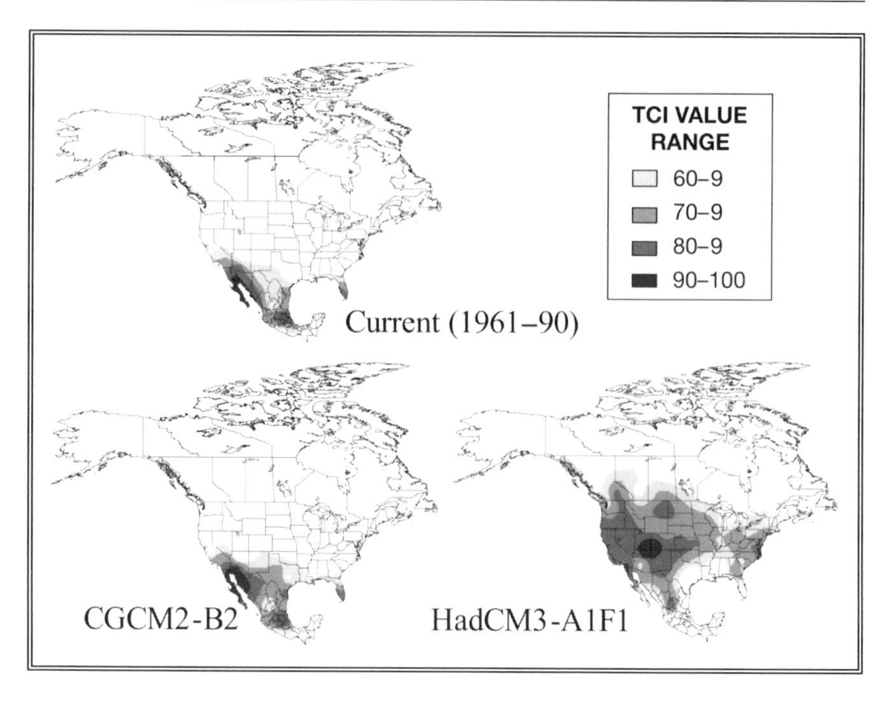

Figure 7.2 Projected changes in tourism climate index ratings in January

to coastal tourism infrastructure from south Florida to Texas, but is a particularly serious problem in the low-lying beaches and wetlands of the Florida Everglades and the compacting (and thus subsiding) sediments of the Mississippi River Delta in Louisiana.

The health of south Florida's tourism industry is strongly dependent on the region's beaches. Extensive investments have been made to extend and renourish highly valued recreational beaches in the past. Sea level rise in beach areas backed by sea walls or other development that precludes landward migration, would lead to the loss of beach area through inundation or erosion and pose an increased threat to the recreation infrastructure concentrated along the coast (sea-front resorts, marinas, piers). The US EPA (1999) reported that a 60cm sea level rise would erode beaches in parts of south Florida 30m to 60m unless beach nourishment efforts were expanded. The cumulative cost of sand replenishment to protect Florida's coast from a 50cm rise in sea level by 2100 is estimated at US$1.7 to $8.8 billion (US EPA, 2003). A study of beach nourishment as an adaptation strategy to preserve major recreational beaches throughout the United States estimated the cost at US$14.5 billion for a 50cm sea level rise and US$26.7 billion for a one metre sea level rise (Leatherman quoted in Smith & Tirpak, 1990).

Sea level rise would also exacerbate the impact of major storms on the Florida tourism industry. The US EPA (1999) indicated that the ten–day closure and clean-up period from Hurricane Georges (September 1998) resulted in tourism revenue losses of approximately US$32 million in the Florida Keys region. A similar ten–day closure in the Miami Beach area would cause losses of approximately US$44 million. The Federal Emergency Management Agency in the United States estimated that a storm of similar strength, imposed on a 30cm sea level rise, would cause 35% to 60% more damage, extending the clean-up period and related tourism losses.

New Orleans is a major tourism destination in the region that is severely threatened by sea level rise. The city is, on average, already 2.4m below sea level and, with the current rate of subsidence and a 50cm sea level rise, most of New Orleans and vicinity would be 3.2m to 3.8m below sea level by 2100 (Burkett, 2001). The estimated cost of protecting the city over the next 50 years alone is US$14 billion.

The insurance implications of increased risk to tourism infrastructure from tropical storms and associated flooding in the region are also an important consideration. Munich Re and other large insurance companies have determined that worldwide economic losses due to natural disasters have been doubling every ten years over the past three decades (IPCC, 2001). Although the trend in insured losses has not increased as dramatically as economic losses, it has increased almost tenfold between 1975 and 2000. The implication of the significant increase in insured losses for insurance premiums is obvious. As Scott (2003) has discussed, the insurance costs for tourism infrastructure or business interruption in high-risk areas like hurricane prone coastlines could increase substantially if the above trends continue. In some regions, insurance coverage may no longer even be available, exacerbating the impacts of climatic extreme events and restricting new tourism investment in high-risk regions. Climate change risk is beginning to be considered in credit assessments and investment valuation. Innovest Strategic Value Advisors (2003) noted that Hypovereinsbank includes climate change in the general environmental risk audit of its credit risk assessment and Credit Suisse now considers climate change in the financial assessments of projects.

The ecological impacts of sea level rise on wetlands and coral reefs in the region also have significant implications for sport fishing and diving-related tourism. Sea level rise could become a major cause of wetland loss throughout the coastal zone of the region. Coastal wetlands in Louisiana are currently converting to open water at a rate of 80 square kilometres a year and a 30cm to 90cm increase in sea level is projected to submerge at least 70% of Louisiana's remaining salt marshes (US EPA, 2003). Even freshwater marshes located far inland may convert to brackish or salt marsh. Similarly, large areas of the Florida Everglades are projected to be inundated as a result of a 50cm sea level rise (US National Assessment Team, 2000).

The reefs of the Florida Keys support a large diving and fishing industry. These activities generated an estimated US$4.4 billion in tourism revenues in a four-county area of south Florida in 2000–1 (Johns *et al.*, 2001). Like reef systems around the world, the reefs across this region have been under considerable human-induced stress (overfishing, pollution). Coral reefs across the Caribbean have suffered an 80% decline in cover over the past 30 years (Gardner *et al.*, 2003). Recent coral bleaching events caused by high water temperatures and scenarios for future water temperatures in the region portend an imperiled future for coral reefs and related tourism.

Conclusion

This chapter has examined how the impacts of climate change could affect the sustainability of the tourism industry in three regions of North America during the 21st century. Although the examples of potential climate change impacts discuss are no means exhaustive (see also Wall, 1992, 1998; Scott, 2003), it is clear from the above discussion that climate change has far-reaching consequences for North American tourism. Future climates will influence the viability of alternative types of tourism, providing challenges and threats to some destination areas and enhanced opportunities for others. The magnitude of the impact of climate change will depend upon the importance of the tourism industry in the regional economy, the characteristics of climate change and its affect on the natural environment, the adaptive response of tourists, the capacity of the tourism industry itself to adapt to climate change, and how the impacts of climate change interact with other long-term influencing variables in the tourism sector (globalization and economic fluctuations, fuel prices, ageing populations in industrialized countries, increasing travel safety and health concerns, increased environmental and cultural awareness, advances in information and transportation technology, environmental limitations, water supply and pollution, and so on).

The critical uncertainties regarding the magnitude of projected climate change, the subsequent environmental impacts, and how these environmental changes will affect different segments of the tourism marketplace and alter the competitive relationship between tourism destinations, precludes any definitive statement regarding the net impact of climate change on the North American tourism sector at this time.

Research to understand the implications of global climate change for the tourism sector is still in its infancy. The development of a strategic research agenda is required to assess the implications of climate change for North American tourism and to target the key informational needs of tourism decision makers (both in government and business). In particular, adaptation is not a well-developed theme in the research on climate change and tourism and, because it is critical to understanding the vulnerability of tourism to climate change, requires greater attention in the future. As Wall and Badke (1994) indicated, almost none of the research on climate change and tourism has been done by tourism experts. Consequently, increased collaboration between climate change scientists, government tourism officials and the tourism industry should be paramount in the development of such a research programme.

References

American Birding Association (2003) *The Growth of Birding and the Economic Value of Birders*. Colorado Springs, Colorado: American Birding Association. Available at: http://americanbirding.org/programs (accessed 1 September 2003).

American Sportfishing Association (2001) *Sportfishing in America: Values of Our Traditional Pastime*. Alexandria, Virginia: American Sportfishing Association.

Bartlein, P., Whitlock, C. and Shafer, S. (1997) Future climate in the Yellowstone National Park Region and its potential impact on vegetation. *Conservation Biology* 11 (3), 782–92.

Bergmann-Baker, U., Brotton, J. and Wall, G. (1995) Socio-economic impacts of fluctuating water levels for recreational boating in the Great Lakes basin. *Canadian Water Resources Journal* 20 (3), 185–94.

Brugman, M., Raistrick, P. and Pietroniro, A. (1997) Glacier related impacts of doubling atmospheric carbon dioxide concentrations on British Columbia and Yukon. In E. Taylor and B. Taylor (eds) *Canada Country Study: Climate Impacts and Adaptation – British Columbia and Yukon*. Ottawa: Environment Canada.

Burkett, V. (2001) Trends in sea-level rise, subsidence and precipitation: Implications for New Orleans and vicinity. Conference paper given at the US Geological Survey's Subsidence and Aquifer Mechanics Meeting, Galveston Island, Texas, 27–9 November.

Canadian Tourism Commission (2002) *Canadian Tourism Facts and Figures 2001*. Ottawa: Canadian Tourism Commission.

Gardner, T.A., Côté, I.M., Gill, J.A., Grant, A. and Watkinson, A.R. (2003) Long-term region-wide declines in Caribbean corals. *Science* 301, 958–60.

Hall, M. and Farge, D. (2003) Modeled climate-induced glacier change in Glacier National Park, 1850–2100. *BioScience* 53 (2), 131–40.

HLA Consultants and ARA Consulting Group Inc. (1995) *Ecotourism/Nature/Adventure/Culture: Alberta and BC Market Demand Assessment*. Vancouver: Department of Canadian Heritage.

Innovest Strategic Value Advisors (2003) *Carbon Finance and the Global Equity Markets*. London: Carbon Disclosure Project.

IPCC (Intergovernmental Panel on Climate Change) (2001) Chapter 15 – North America. In *Climate Change 2001: Impacts, Adaptation and Vulnerability*. Third Assessment Report. Geneva: United Nations Intergovernmental Panel on Climate Change.

International Snowmobile Association (2003) *International Snowmobile Industry Facts and Figures*. Available at: http://www.ccso-ccom.ca/contente.htm (accessed 28 February 2003).

Johns, G., Leeworthy, V., Bell, F. and Bonn, M. (2001) *Socioeconomic Study of Reefs in Southeast Florida. Final Report for Broward, Palm Beach and Miami-Dada and Monroe Counties*. Florida Fish and Wildlife Conservation Commission and National Oceanic and Atmospheric Administration. Available at: http://marineeconomics.noaa.gov/Reefs/02–01.pdf.

Lamothe and Periard Consultants (1988) *Implications of Climate Change for Downhill Skiing in Quebec*. Climate Change Digest 88–03. Ottawa: Environment Canada.

Last, J. (1993) Global change: Ozone depletion, greenhouse warming, and public health. *Annual Review of Public Health* 14, 115–36.

Lee, D., Moulton, R. and Hibner, B. (1996) *Climate Change Impacts on Western Lake Erie, Detroit River, and Lake St. Clair Water Levels*. Toronto: Environment Canada and the Great Lakes Environmental Research Laboratory.

Lipski, S. and McBoyle, G. (1991) The impact of global warming on downhill skiing in Michigan. *East Lakes Geographer* 26, 37–51.

McBoyle, G. and Wall, G. (1992) Great Lakes skiing and climate change. In A. Gill and R. Hartmann (eds) *Mountain Resort Development* (pp. 70–81). Burnaby, British Columbia: Centre for Tourism Policy and Research, Simon Fraser University.

McBoyle, G., Wall, G., Harrison, K. and Quinlan, C. (1986) Recreation and climate change: A Canadian case study. *Ontario Geography* 23, 51–68.

McDonald, K. and Brown, J. (1992) Using Montane mammals to model extinctions due to global change. *Conservation Biology* 6 (3), 409–15.

Mortsch, L., Hengeveld, H., Lister, M., Lofgren, B., Quinn, F. *et al.* (2000) Climate change impacts on the hydrology of the Great Lakes-St. Lawrence system. *Canadian Water Resources Journal* 25 (2), 153–79.

Pannell Kerr Forster (2001) *National Snowmobile Tourism Study*. Prepared for Canadian Tourism Commission, Canadian Council of Snowmobile Organizations and Provincial/Territorial Partners. Ottawa: Canadian Tourism Commission.

Richardson, R. and Loomis, J. (2003) The effects of climate change on mountain tourism: A contingent behavior methodology. *Proceedings of the First International Conference on Climate Change and Tourism*. 9–11 April, Djerba, Tunisia. Madrid: World Tourism Organization.

Schwartz, R., Deadman, P., Scott, D. and Mortsch, L. (2004) GIS modelling of climate change impacts on a Great Lakes shoreline community. *Journal of the American Water Resources Association* 40 (3), 647–62.

Scott, D. (1993) Ontario cottages and the Great Lakes shoreline hazard: Past experiences and strategies for the future. Unpublished Masters thesis, Department of Geography. Waterloo: University of Waterloo.

Scott, D. (2003) Climate change and sustainable tourism in the 21st century. In J. Cukier (ed.) *Tourism Research: Policy, Planning, and Prospects*. Department of Geography Publication Series. Waterloo: University of Waterloo.

Scott, D. and Suffling, R. (2000) *Climate Change and Canada's National Parks*. Toronto: Environment Canada.

Scott D., McBoyle, G. and Mills, B. (2003) Climate change and the skiing industry in Southern Ontario (Canada): Exploring the importance of snowmaking as a technical adaptation. *Climate Research* 23, 171–81.

Scott D., McBoyle, G. and Schwartzentruber, M. (2004) Climate change and the distribution of climatic resources for tourism in North America. *Climate Research* 27, 105–17.

Scott, D., Jones, B., Lemieux, C., McBoyle, G., Mills, B. *et al.* (2002) *The Vulnerability of Winter Recreation to Climate Change in Ontario's Lakelands Tourism Region*.

Occasional Paper 18, Department of Geography Publication Series. Waterloo: University of Waterloo.
Smith, J. and Tirpak, D. (1990) *The Potential Effects of Global Climate Change on the United States*. New York: Hemisphere Publishing Corporation.
Stocks, B., Fosber, M., Lynham, T., Merans, L., Wotton, B. *et al.* (1998) Climate change a forest fire potential in Russian and Canadian boreal forests. *Climatic Change* 38, 1–13.
Travel Industry Association of America (2003) *Travel Statistics and Trends*. Washington, DC: Travel Industry Association of America, Bureau of Economic Analysis, US Department of Commerce. Available at: www.tia.org/travel/econimpact.asp (accessed 12 September 2003).
US EPA (United States Environmental Protection Agency) (1995) *Ecological Impacts from Climate Change: An Economic Analysis of Freshwater Recreational Fishing*. US EPA #220-R-95–004. Washington, DC: Environmental Protection Agency.
US EPA (1999) *Global Climate Change: What Does it Mean for South Florida and the Florida Keys?* Report on the Environmental Protection Agency Public Consultations in coastal cities, 24–8 May. Washington, DC: Environmental Protection Agency.
US EPA (2003) *Climate Change and Louisiana*. EPA Climate Change State Impacts Information Reports. Available at: http://yosemite.epa.gov/oar/globalwarming.nsf/content/ImpactsStateImpactsLA.html (accessed 1 September 2003).
US National Assessment Team (2000) *Climate Change Impacts on the United States: The Potential Consequences of Climate Variability and Change*. US Global Change Research Program. New York: Cambridge University Press.
Wall, G. (1992) Tourism alternatives in an era of global climate change. In V. Smith and W. Eadington (eds) *Tourism Alternatives* (pp. 194–236). Philadelphia: University of Pennsylvania Press.
Wall, G. (1998) Impacts of climate change on recreation and tourism. In *Responding to Global Climate Change – National Sectoral Issues* (pp. 591–620). Volume XII of the Canada Country Study: Climate Impacts and Adaptation. Toronto: Environment Canada.
Wall, G. and Badke, C. (1994) Tourism and climate change: An international perspective. *Journal of Sustainable Tourism* 2 (4), 193–203.
World Tourism Organization (2002) *Tourism Highlights 2001*. Madrid: World Tourism Organization.

Chapter 8
Nature Tourism and Climatic Change in Southern Africa

R.A. PRESTON-WHYTE AND H.K. WATSON

Introduction

Tourism in southern Africa features as one of the economic activities engaged in the complex manoeuvring towards a sustainable future. Among the various kinds of tourism available in the region there is continuing demand for the array of spectacular and varied landforms and landscapes as well as for the immensely diverse and largely endemic flora and fauna. It is not surprising, therefore, that governments and the tourist supply sector perceive nature tourism to be an important contributor to regional prospects for continuous, durable, self-generating and ecologically tolerable wealth creation. The purpose of this chapter is to consider these perceptions in the light of possible modifications to ecosystems caused by changes in regional climate. The impacts of such changes on nature tourism are discussed with respect to selected well-known and much frequented national parks and conservation areas.

There are two reasons for choosing large parks to assess the effects of climatic change. First, wildlife in most parks will have to adapt or die as fences and a whole range of other land transformations and activities associated with human habitation restrict migration from them. From this it could be argued that the larger the spatial extent and/or altitudinal range of a park, the greater is the probability of species adaptation to change. The implication then is that if parks are large and well managed they may remain sufficiently ecologically viable to merit their continued conservation status despite changing climates. Second, there is consensus that the trend towards warmer conditions evident over the past century will continue (Joubert & Kohler, 1996; Hulme *et al.*, 2001; Midgley *et al.*, 2001). The climate change predictions are less confident about rainfall changes although there is increasing agreement that large parts of southern Africa may become drier (Joubert, 1997; Hulme *et al.*, 2001; Midgley *et al.*, 2001). However, even if this trend did not materialize, rising temperatures would increase evapotranspiration rates leading to a decrease in potential water availability. In line with this scenario much of the assessment in this chapter is based on Hulme's (1996) assessment of effects on biodiversity of warmer and effectively drier conditions in national parks. Given that the effects of increased aridity and temperature are likely to differ from one broad ecological region to another, the discussion is directed towards possible impacts in selected national parks and conservation areas (Figure 8.1) located in a range of ecozones (Figure 8.2).

The chapter is divided into two sections. First, climatic change is placed in the context of previous variations in temperature and rainfall over

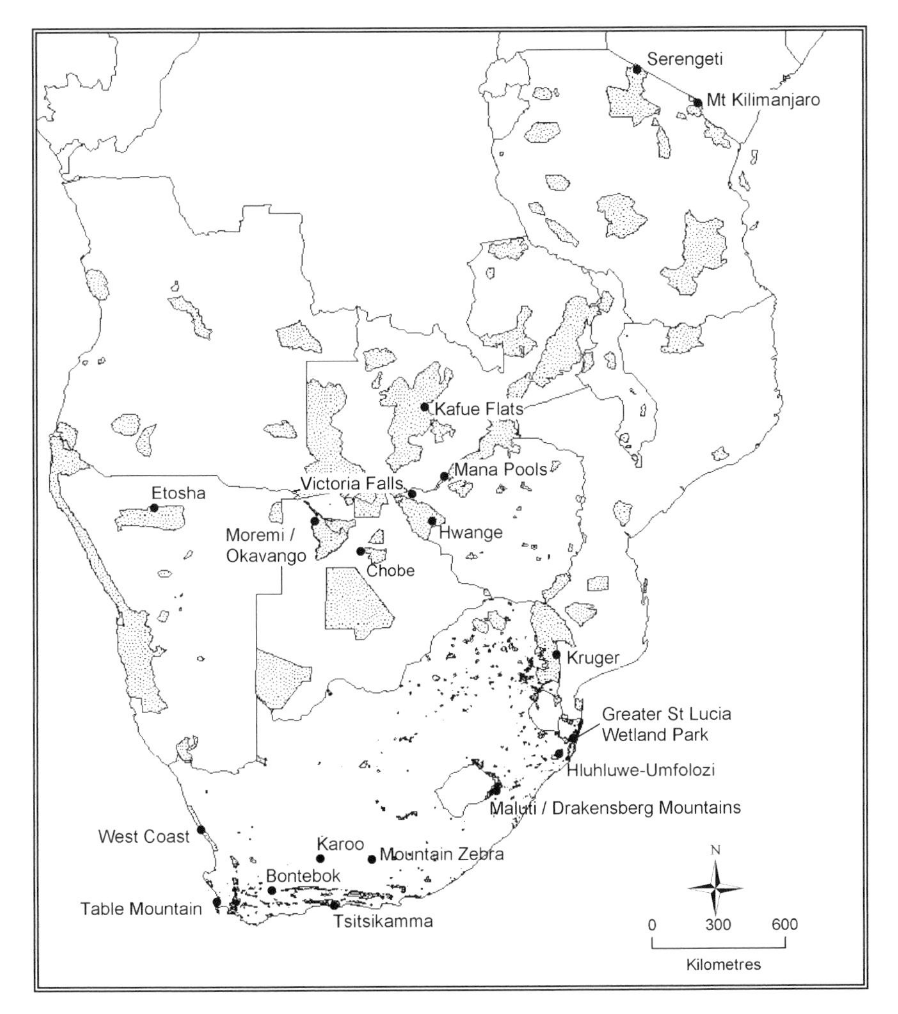

Figure 8.1 National parks and state conservation areas in southern Africa

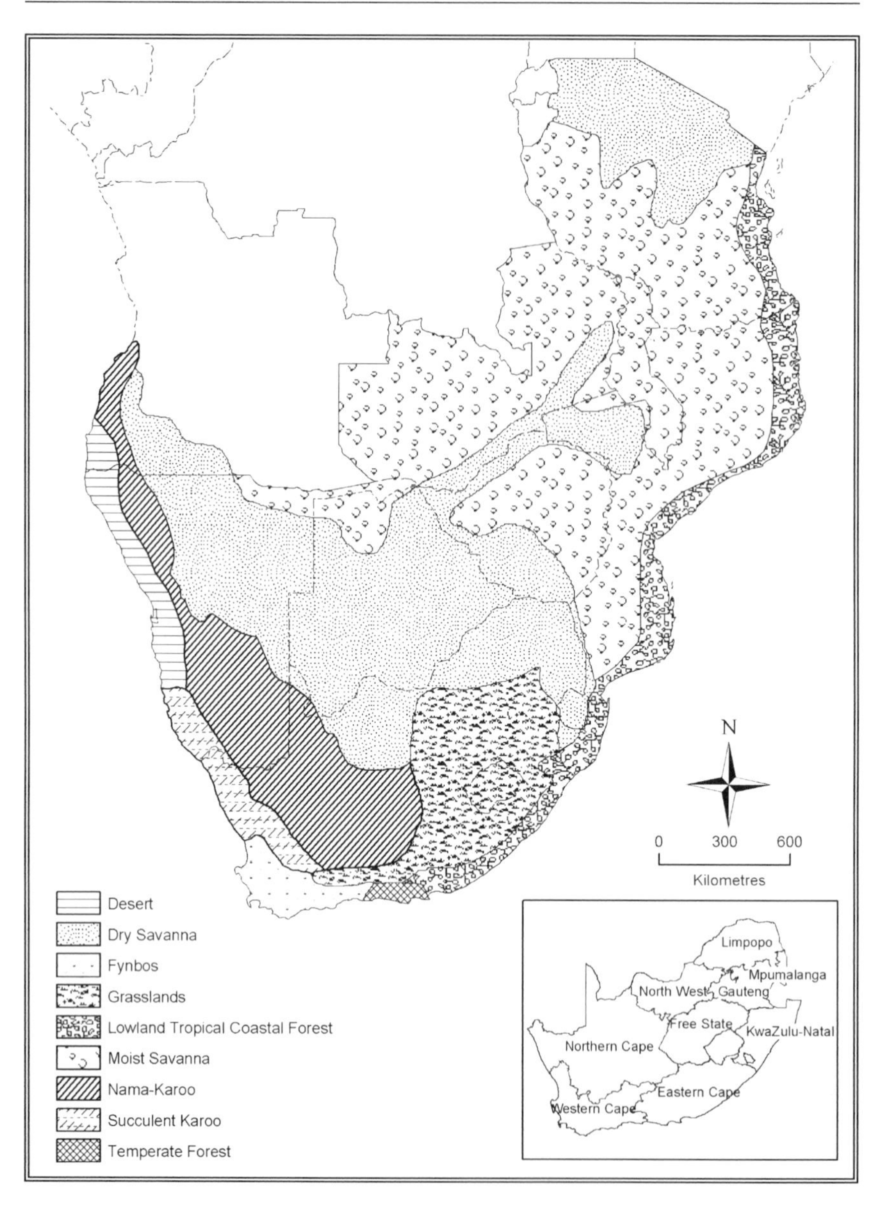

Figure 8.2 Ecozones of southern Africa

southern Africa with concomitant impacts on the inhabitants. Our rapidly improving understanding of atmospheric processes is also acknowledged along with the development of predictive climate models (e.g. Tyson, 1991; Hernes *et al.*, 1995; Hulme, 1996; Joubert & Kohler, 1996; Ringius *et al.*, 1996; Hulme *et al.*, 2001). However, it is recognized that much work still needs to be done on future climate change scenarios for Africa and most carry caveats that specify their limitations and recommend caution. Second, with this in mind the findings of Hulme (1996), Hulme *et al.* (2001) and Midgley *et al.* (2001) are used to inform the nature of predicted climate change on national parks in the region's ecozones and possible tourist reaction to these changes.

Regional Climate Trends

There is nothing new in changing climates. Climate and its variability are placed in the context of an environmental determinant that has impacted on the inhabitants of southern Africa since prehistoric time (Tyson, 1986). This means that they would also have had to cope with changing ecosystems. For example, Johnson *et al.* (1996) records that prior to 13,000 BP Lake Victoria was a dry, grass-covered depression. A lake then began to form and overflowed around 7500 BP (Beuning *et al.*, 1998). Over the last millennium, oxygen isotope records from stalagmites in the Makapansgat valley in northern South Africa provide evidence of alternating warming and drying periods. The period of medieval warming that occurred between 1100 to 700 BP can be identified along with the cooling associated with the Little Ice Age from 700 to 200 BP (Holmgren *et al.*, 1999). Also interesting is evidence of an inverse relationship in temperature and rainfall between Lake Naivasha in Kenya and Makapansgat in South Africa (Holmgren *et al.*, 1999; Verschuren *et al.*, 2000). For example, during the period of medieval warming South Africa was wet while Kenya was dry.

There is evidence that these changes in climate influenced human activities. Prior to 19th-century colonization, low levels of Lake Naivasha were associated with famine, population migrations and political unrest (Verschuren *et al.*, 2000), while prosperity, agricultural expansion and population growth occurred when levels were high (Webster, 1980). Similar conclusions have been drawn regarding the changes in Iron Age settlement patterns over South Africa during medieval times (Hall, 1984; Huffman, 1996), and the collapse at the onset of the Little Ice Age of the Mapungubwe state in the Shashi-Limpopo Basin (Vogel, 1995; Huffman, 1996).

While ongoing climatic variations and the concomitant impacts on human activities is a recognized phenomenon, what *has* changed is our improving ability to predict how climates may alter in the future. This is partly because we have a better understanding of how atmospheric systems work and what is likely to trigger variations. We know for instance that small changes in the subtropical high-pressure system that dominates much of southern Africa can have profound effects on circulation types and influence variations in precipitation. Wet and dry spells respond to changes in the general circulation of the atmosphere as shown by pressure anomaly fields and the concomitant frequency of perturbations associated with them (Tyson, 1986; Tyson & Preston-Whyte, 2000). These changes are particularly evident at the 500hPa level where pressure anomaly fields show distinctive regional gradients and opposite patterns for wet and dry months. Given that pressure change modulations are associated with changes in the wind field, it is perhaps not surprising to find areal correlation between annual atmospheric circulation controls of wet and dry spells.

We also know that wet and dry spells over southern Africa are substantially influenced by events in the South Pacific Ocean and the teleconnections that translate changes in pressure, temperature and wind anomalies through the Walker Circulation and the El Nino-Southern Oscillation (ENSO) events. During wet spells, an ascending limb of the linked series of zonal cells that forms the African Walker Circulation is situated over tropical Africa (Lindesay, 1987; Tyson & Preston-Whyte, 2000). This tends to occur during non-ENSO or La Nina events and is associated with enhanced easterlies, negative pressure anomalies over the subcontinent and moisture advection from the Indian Ocean. During dry spells, often associated with El Nino conditions, the African Walker Circulation withdraws eastwards reducing the moisture flux via the tropical easterly. The primary moisture conveyor is now from the southwest, and pressure anomalies over the subcontinent become positive with the establishment of anticyclonic circulations and associated dry subsiding air.

Our ability to predict future variations in regional climate has also been advanced by general circulation model (GCM) climatic change experiments. Although the results of these models need to be approached with caution, they do provide an indication of changes that are likely to occur as a consequence of carbon dioxide and sulphate loading of the atmosphere. The report edited by Hulme (1996) describes the potential impacts of climatic change on southern Africa's hydrology and biota based on such models. A key feature of their main or 'core' scenario is a predicted average warming by 2050 of 1.7°C. When applied regionally, greatest warming in mean surface air temperatures is in excess of 2°C over the interior upland plateau incorporating eastern Namibia, Botswana, western Zimbabwe and Zambia, and about 1.5°C along the coastal margins. For most, the region warming is greatest during the May to September dry winter season. Under this scenario rainfall increases in some areas during the wet season but is more than offset by rainfall decreases in the remaining months of the year. The net result is increased aridity over most of the region.

More recently, Hulme *et al.* (2001) produced a range of climatic change scenarios for the continent as a whole, all of which predict warming.

However, they admit to less confidence in predicting future rainfall scenarios although 'significant' decreases in rainfall over South Africa and Namibia are mentioned. Similar results are predicted by Midgley *et al.* (2001) who conclude that by 2050 the most probable climatic change scenario for South Africa would be an increase in January temperatures along the coast of 0.5 to 1.0°C and in the central interior and northern Cape of 2.5 to 4.5°C. This would be accompanied by a decrease in summer rainfall of between 5% in the northern part of the country to 25% in the eastern and southern Cape. There would also be a substantial decrease in winter rainfall in the western Cape.

While these model scenarios indicate the direction of temperature change and point with some ambiguity to rainfall variations, the current developments lack a number of important inputs that would improve their predictive power. These include adequate representation of the influence of El Nino on climatic variability, the effect of dust and biomass loading in the atmosphere, and feedbacks between changes in vegetation and land use. There is, therefore, still uncertainty regarding the trends and magnitudes of temperature and rainfall changes over the region. The general consensus, however, is that conditions will be warmer and somewhat drier.

The Impact of Warming and Aridification on Tourism in National Parks in Specific Ecozones

Temperate forest

The Tsitsikamma Forest in the mountains surrounding Knysna in the Cape Province is the only large park in this ecozone (Figures 8.1 and 8.2). Aridification could see as much as a fifth of this forest disappear by 2050 (Hulme, 1996). While this loss will negatively effect popular hiking trails through the forest, the impact of climate change on the popular Otter Trail could be more severe. This trail extends over 67 kilometres of extremely rugged coastline and is the park's key attraction. Hiking it takes five days and, in addition to magnificent vistas, it offers good birding and possible encounters with otters when crossing river mouths and sightings of dolphins and whales off the coast.

Several studies reviewed by Hulme *et al.* (2001) predict a 25cm rise in mean sea level by 2050. This could have the effect of increasing the salinity of water in the river mouths. This would reduce their suitability as otter habitats. In addition, substantial portions of the coast's wave-cut terraces would be submerged. This would impact on the popular hiking activity of exploring pools and crevices at low tide.

In common with the coastlines of the Eastern Cape and KwaZulu-Natal (Figure 8.1), many of the sightings of dolphins and birds along this stretch of coast are associated with the annual migration of sardines. During the austral winter sardines migrate eastwards for hundreds of kilometres as cold South Atlantic waters push the Agulhus current back. The density and size of sardine shoals and the predator activity associated with them is now recognized as a tourist attraction equivalent to the vast ungulate herds that migrate between Tanzania's Serengeti and Kenya's Masai-Mara parks. The predicted warming of the oceans is likely to reduce the distance these fish migrate and the size of their shoals (Turnbull, 2002).

Lowland tropical coastal forest

The Greater St Lucia Wetland Park on the east coast of South Africa is the largest park in this ecozone (Figures 8.1 and 8.2). Although dominated by Lake St Lucia, the park contains a wide range of habitats that include smaller lakes, islands, several estuaries, mangrove swamps, marshes, grasslands, forests, the Ubombo Mountains, coastal dunes, sandy beaches, rocky shores and coral reefs. These habitats support a diverse biota and offer a wide range of tourist attractions including angling, boating, hiking, birding, snorkelling and scuba-diving.

Sadly, there has been considerable human interference with the hydrological system that sustained Lake St Lucia. The Mfolozi River was diverted, the Hluhluwe and Mkuze Rivers were dammed and exotic plantations were established along its shores. As a result of siltation and water deprivation, the mean depth of Lake St Lucia over the past few decades has been reduced to about two metres. During dry spells, usually associated with El Nino periods (Walker, 1990; Mason, 1995; Landman & Mason, 1999), the flow of water out of the lake into the sea has repeatedly been inadequate to keep the mouth open. Once closed, the evaporative loss of water has resulted in a corresponding increase in salinity and an associated change in avian fauna from predominantly ducks to pelicans and flamingos (Begg, 1978).

Aridification could result in the expansion of this ecozone's forests by as much as 150% by 2050 (Hulme, 1996). This will have both positive and negative consequences for tourism. The increase in the parks forests would most likely be at the expense of the marshes and grasslands. Lake St Lucia, as well as all the other aquatic systems in the park would be saline to hypersaline. This would offer an enhanced attraction for birders. However, under these saline conditions launch cruises would be affected detrimentally by the demise of hippos, crocodiles and mangrove swamps. There would also be restrictions on the size of boats that could be used in extremely shallow waters.

Although sightings of marsh and grassland ungulates such as reedbuck, waterbuck and zebra will decrease, game viewing in the woodland and forest habitats could improve provided they are correctly managed.

Conditions in these habitats will be more favourable for buffalo, bushbuck, bushpig, duiker, elephant, impala, and both the square-lipped and hook-lipped rhinoceros (Hulme, 1996). While the decreased flow of water, and hence sediment yield, of the large number of rivers opening into the sea along KwaZulu-Natal and Mozambique are likely to enhance snorkelling and scuba-diving conditions, the rise in sea level will detrimentally affect coral reef biota (Levy, 2001; Turnbull, 2002).

Savanna

Several important geomorphic attractions and most of the region's large parks occur in the Savannas ecozone (Figures 8.1 and 8.2). The intrigue of a snow-capped mountain on the equator, the challenge of climbing the continent's highest mountain and the opportunity of seeing vast herds of wildebeest and zebra, as well as the 'Big Five' on the plains surrounding it, have long attracted tourists to Mount Kilimanjaro (Figure 8.1). According to Desanker (2001), since 1912 when its glaciers were first thoroughly surveyed, warming has reduced their spatial extent by 82%. Continued warming is likely to melt most of the remaining ice and reduce its attraction to visitors.

Further south, the Victoria Falls (Figure 8.1) was described by the first European to set eyes on it as, 'a scene so lovely (that it) must have been gazed upon by angles in their flight' (Livingstone, 1858). Besides the magnificent spectacle of the Zambezi River forming a curtain of water up to 1700m wide as it thunders into a narrow fractured gorge over a hundred metres deep, this locality offers sedate game viewing while cruising the river above the falls and rainforest hikes, bungee jumping and white-water rafting below the falls.

All of the attractions on the Zambezi River will be detrimentally affected by a warmer and drier future climate. According to Desanker and Magadza (2001), by 2050 the river's flow will be substantially diminished by the combined effects of up to 15% less rainfall, 25% more evaporation and 40% less runoff in its catchment. These effects could also significantly influence tourism in the Etosha Pan in northern Namibia; Moremi/Okavango Delta and Chobe in northern Botswana; Hwange and Mana Pools in northern Zimbabwe and parks in central Zambia's Kafue Flats (Figure 8.1). In addition, during the dry season large herds of ungulates and their associated predators quench their thirst in a daily sequence at water sources in these parks. Under an aridification scenario wildlife will concentrate around these resources for longer periods and will have to be carefully managed to ensure that they do not deplete both the diminishing water and vegetation resources.

While opportunities to view wetland species such as hippo, lechwe, otto and sitatunga from tranquil cruises along rivers and waterways through swamps will decrease, these parks will become more favourable habitats for blue wilderbeest, eland, giraffe, kudu and warthog (Hulme, 1996). Analysis of historic records of discharge from the Quito and Zambezi Rivers, led McCarthy *et al.* (2000) to suggest that, in response to instabilities in tropical lows in Indian Ocean easterlies, an 80-year oscillation exists in this central interior region east of the watershed between the Cubango and Quito rivers. If they were correct the increased discharges will reach a maximum in about 2040. This would ameliorate the effects of the scenario described above particularly with regard to the Okavango and Kafue systems.

Under the warmer and drier conditions that are predicted throughout most of the savanna ecozone by 2050, the eutrophic or arid savannas will be better represented at the expense of the dystrophic or moist savannas. The former are typified by various fine-leafed and thorny *Acacia* and *Terminalia* species and the broad-leafed *Colophospermum mopane*, and the latter by *Burkea* and *Brachystegia* (miombo) species. While both host the 'Big Five', the possible eutrophic expansion in Kruger National Park (Figure 8.1) may be 'bad news for tourism because, although mopane woodland supports large numbers of elephants, it does not favour the diversity of large mammals that most tourists visit the park to see' (Midgley *et al.*, 2001: 5).

Savanna vegetation comprises an herbaceous ground layer and an upper woody layer. The degree to which either layer dominates is dependent on a multitude of factors including drought, fire, ungulate grazing and browsing pressure, elephant damage, insect predation of seeds from woody species, and atmospheric CO_2 concentrations. Bond *et al.* (2002) modelled sapling and grass growth at different CO_2 levels and concluded that ambient levels could be responsible for the widespread bush encroachment already pervading. Watson (1995) showed that this trend continued in Hluhluwe-Umfolozi Park in KwaZulu-Natal (Figure 8.1) over the last century despite the implementation of veld burning, ungulate control, destumping and use of arboricides, elephant introduction, and fuelwood harvesting by neighbouring communities. She detailed the detrimental effects that this trend has had on the park's biodiversity as well as on its ecotourist potential due to decreased game visibility. Despite the probable distributional and compositional changes noted above, Desanker and Magadza (2001) are optimistic that a key tourist attraction of the savannas, namely the migration of vast ungulate herds across the Serengeti, will not be detrimentally affected.

Grassland

The grassland ecozone (Figure 8.2) is located at altitudes where frost currently restricts woody plant growth (Hulme, 1996; Midgley *et al.*,

2001). As conditions become warmer, the area of upland grasslands will contract and become less suitable habitats for a number of ungulates including the mountain zebra and bontebok from which national parks in this ecozone take their name.

Given that South Africa's Drakensberg and Lesotho's Maluti mountains (Figure 8.1) are South Africa's most important source of water, the encroachment of woody species into their grass cover will have to be intensely regulated with hot, frequent fires to ensure maximum water yield. Even without such management it is questionable whether bush encroachment would impact on tourism given that the principle attraction in these areas is provided by gazing on the stark, physical grandeur of towering, shear rock faces of the escarpment and hiking the surrounding wilderness. Perhaps more serious would be the impact of rising temperatures on trout. These fish are unable to survive in water temperatures above 21°C and their demise would destroy the contemporary popular leisure activity of flyfishing along streams in the grass-covered mountain foothills.

Karoo

By 2050, the entire Succulent Karoo and most of the Nama Karoo (Figure 8.2) will experience a climate unlike that currently encountered anywhere in South Africa (Midgley *et al.*, 2001). As tourism to the (Nama) Karoo National Park is based on the intrigue of its desolate parched landscape, it may be enhanced by even harsher conditions. Tourists to the Succulent Karoo's West Coast National Park are enticed in spring by the anticipation of seeing the landscape carpeted by a blaze of colour and in summer by the large flocks of paelarctic migrants that congregate in Langebaan Lagoon. The severe detrimental effects of loss of winter rainfall on the park's biota, predicted by Midgley *et al.* (2001), could seriously impair its tourist potential.

Fynbos

Midgely *et al.* (2001) predict that warming and aridification may lead to substantial loss and fragmentation of the Fynbos ecozone (Figure 8.2). This is unlikely to be detrimental to tourism. Although the Fynbos ranks as one of the world's 25 most significant biodiversity hotspots, for most tourists the allure of area has more to do with the excitement of the cable car ride up Table Mountain (Figure 8.1) or the challenge of a mountain climb and the vista from the top, than its Fynbos clothing. More serious would be the predicted decline of winter rainfall on viticulture and its subsequent impact on wine tourism assuming that new cultivars could not be established to accommodate the changing climate.

Aliens, malaria and poverty

Invasive alien plants are already having a very significant detrimental effect on the biodiversity and hence ecotourist potential of most parks in southern Africa. Increased atmospheric CO_2 could favour woody invasives such as *Melia azedarach* and species of Australian *Acacia*. Increased minimum winter temperatures and reduced frost frequency should enable *Prosopis spp.* to invade higher altitudes (Richardson *et al.*, 2000). Nevertheless, it is hoped that the drier conditions will curtail the spread of some of the most aggressive invasives like *Solanum mauritianum* and *Chromolaena odorata*, which prefer more mesic habitats.

Aridification will enable mosquitoes to extend their range southwards and westwards into Namibia and northern South Africa (Hulme, 1996). However, since 1999 there has been a dramatic decrease in the incidence of malaria in northern KwaZulu-Natal, Swaziland and southern Mozambique as a consequence of spraying DDT in and around dwellings. It is hoped that by employing these somewhat controversial eradication strategies throughout the malaria prone area of southern Africa the mosquitoes will be brought under control and that visitors previously dissuaded by the high disease risk will return.

In the wake of the devastating effects of HIV/AIDS, progressive aridification of the region will make rural livelihoods increasingly precarious. This is likely to intensify the slide towards poverty in both urban and rural areas. Under these circumstances poverty-driven behaviour leading to crime and both intra- and inter-country conflict, is likely to be more detrimental to tourism than climate change.

Conclusions

Southern Africa stands to lose specific nature tourism attractions if, by 2050, the predictions are correct regarding increased global warming trends and associated increased aridification. These are the attractions related to the melting of the Mount Kilimanjaro ice cap; the desiccation of the Okavango, Chobe, Zambezi, Kafue and St Lucia hydrological systems; rising temperature in the trout-rich waters of the Drakensberg Mountain foothills and the disappearance of the spring annuals in the Succulent Karoo. However, in general, a warmer and drier climatic scenario should enhance the promotion of popular images of the 'African bush' experience given that the expansion of the eutrophic savannas favours flat-topped thorn trees, large ungulate herds and the 'Big Five'. Even a red hue to the African sunset could be enhanced by increased atmospheric aerosol loading from fires needed to regulate bush encroachment and dust from eroded soils.

While climate change in the region is nothing new and is probably responsible for much of its rich biodiversity, the contemporary pace of environmental change is purported to be occurring several orders of magnitude faster than in the past and will inevitably lead to a decrease in biodiversity. The Peace Parks initiative that is striving to integrate conservation areas in close proximity to one another across national boundaries will substantially enhance the chance of biota adapting to climatic change, and should therefore be actively encouraged. However, if these conservation areas are to survive they must be seen to be beneficial to the increasingly impoverished rural communities surrounding them. At present the major challenge facing southern African governments is to deal with the colonial land legacy, HIV/AIDS and associated poverty so that the region's inhabitants will find themselves in a position to actually benefit from the potentially favourable effects of climatic change on nature tourism.

References

Begg, G. (1978) *Estuaries of Natal*. Pietermaritzburg: Natal Town and Regional Planning Commission.

Beuning, K.R.M., Kelts, K. and Stager, J.C. (1998) Abrupt climatic changes associated with the Younger Dryas interval in Africa. In J.T. Lehman (ed.) *Environmental Change and Response in East African Lakes* (pp. 147–56). Dordrecht: Kluwer Academic Publishers.

Bond, W., Woodward, F.I. and Midgley, G.F. (2002) Does elevated CO_2 play a role in bush encroachment? In A.H.W. Seydach, T. Vorster, W.J. Vermeulen and I.J. van der Merwe (eds) *Multiple Use Management of Natural Forests and Woodlands: Policy Refinements and Scientific Progress* (pp. 202–8). Pretoria: Department of Water Affairs and Forestry.

Desanker, P.V. (2001) Impact of climate change on life in Africa. Available at: http://www.greenpeace.org/climate/climatecountdown/index.htm (accessed on 18 July 2003).

Desanker, P.V. and Magadza, C. (2001) Africa. In Intergovernmental Panel on Climate Change (eds) *Impacts, Adaptation and Vulnerability, Third Assessment Report* (pp. 489–531). Cambridge: Cambridge University Press.

Hall, M. (1984) Prehistoric farming in the Mfolozi and Hluhluwe valleys of southeast Africa: An archaeological survey. *Journal of Archaeological Science* 11, 223–35.

Hernes, H., Dalfelt, A., Berntsen, T., Holtsmark, B., Otto Naess, L. *et al.* (1995) *Climatic strategy for Africa, CICERO Report 1995:3*. Oslo: University of Oslo.

Holmgren, K., Karlen, W., Lauritzen, S.E., Lee-Thorp, J.A., Partridge, T.C. *et al.* (1999) A 3000-year high-resolution stalagmite-based record of palaeoclimate for northeastern South Africa. *Holocene* 9, 295–309.

Huffman, T.N. (1996) Archaeological evidence for climatic change during the last 2000 years in southern Africa. *Quaternary International* 33, 55–60.

Hulme, M. (ed.) (1996) *Climatic Change and Southern Africa: An Exploration of Some Potential Impacts and Implications in the SADC Region*. Norwich: University of East Anglia.

Hulme, M., Doherty, R., Ngara, T., New, M. and Lister, D. (2001) African climate change: 1900–2100. *Climate Research* 17, 145–68.

Johnson, T.C., Scholz, C.A., Talbot, M.R., Kelts, K., Ssemanda, I. *et al.* (1996) Late Pleistocene desiccation of Lake Victoria and rapid evolution of cichlid fishes. *Science* 273, 1091–3.

Joubert, A.M. (1997) Modelling present and future climates over southern Africa. Unpublished Ph.D. thesis. Johannesburg: University of the Witwatersrand.

Joubert, A.M. and Kohler, M.O. (1996) Projected temperature increases over southern Africa due to increased levels of greenhouse gases and sulphate aerosols. *South African Journal of Science* 92, 524–6.

Landman, W.A. and Mason, S.J. (1999) Change in the association between Indian Ocean sea surface temperatures and summer rainfall over South Africa and Namibia. *International Journal of Climatology* 19, 1477–92.

Levy, K. (2001) Global warming: Fact or fiction. *Africa Geographic* 9 (8), 77–81.

Lindesay, J.A. (1987) Relationships between the Southern Oscillation and atmospheric circulation changes over southern Africa, 1957–1982. Unpublished Ph.D. thesis. Johannesburg: University of the Witwatersrand.

Livingstone, D. (1858) *Missionary Travels and Researches in South Africa.* Available at: http://www.ibiscom.com/livingstone.htm (accessed 9 August 2003).

McCarthy, T.S., Cooper, G.R.J., Tyson, P.D. and Ellery, W.N. (2000) Seasonal flooding in the Okavango delta, Botswana: Recent history and future prospects. *South African Journal of Science* 96, 25–33.

Mason, S.J. (1995) Sea-surface temperature-South African rainfall associations, 1910–1989. *International Journal of Climatology* 15, 119–35.

Midgley, G., Rutherford, M.C. and Bond, W. (2001) *The Heat is On: Impacts of Climatic Change on Plant Diversity in South Africa.* Cape Town: National Botanical Research Institute.

Richardson, D.M., Bond, W.J., Dean, W.R.J., Higgins, S.I., Midgley, G.F. *et al.* (2000) Invasive alien species and global change: A South African perspective. In H.A. Mooney and H.A. Hobbs (eds) *Invasive Species in a Changing World* (pp. 303–49). Washington: Island Press.

Ringius, L., Downing, T.E., Hulme, M., Waughray, D. and Selrod, R. (1996) *Climatic change in Africa- Issues and Regional Strategy.* CICERO report. No. 1996:9. Oslo: University of Oslo.

Turnbull, M. (2002) Oceans of life. *Africa Geographic* 10 (4), 32–50.

Tyson, P.D. (1986) *Climatic Change and Variability in Southern Africa.* Cape Town: Oxford University Press.

Tyson, P.D. (1991) Climatic change in southern Africa: Past and present conditions and possible future scenarios. *Climatic Change* 18, 241–58.

Tyson, P.D. and Preston-Whyte, R.A. (2000) *The Weather and Climate of Southern Africa.* Cape Town: Oxford University Press.

Verschuren, D., Laird, K.R. and Cumming, B.F. (2000) Rainfall and drought in equatorial east Africa during the past 1,100 years. *Nature* 403, 410–14.

Vogel, C.H. (1995) People and drought in South Africa: Reaction and mitigation. In T. Binns (ed.) *People and the Environment in Africa* (pp. 249–56). London: Wiley & Sons.

Walker, N.D. (1990) Links between South African summer rainfall and temperature variability of the Agulhas and Benguela: Current systems. *Journal of Geophysical Research* 95, 3297–319.

Watson, H.K. (1995) Management implications of vegetation changes in the Hluhluwe-Umfolozi Park. *South African Geographical Journal* 77 (2), 77–83.

Webster, J.B. (1980) Drought, migration and chronology in the Lake Malawi littoral. *TransAfrican Journal of History* 9, 70–90.

Chapter 9

Changing Snow Cover and Winter Tourism and Recreation in the Scottish Highlands

S.J. HARRISON, S.J. WINTERBOTTOM AND R.C. JOHNSON

Introduction

International reports such as those from the International Panel on Climate Change (IPCC) (Houghton *et al.*, 1996) have concluded that the global climate is currently undergoing spatially and temporally complex changes that are linked to an enhancement of the greenhouse effect in the troposphere. An increase in the average near-surface temperature, or global warming, is the primary manifestation of this, which has implications for the circulation of the atmosphere and of water. In its dependence on both low temperature and water supply, the amount of precipitation falling as ice and accumulating on the ground surface will be affected. At a continental scale, increased temperature reduces the spatial extent, the depth and the persistence of snow cover. At smaller spatial scales in, for example, mountain regions, temperature and precipitation variation is complex and the resulting temporal trend of change is less easy to identify.

Climate change models based on an assumed continuation of global warming, suggest that these trends will continue through the 21st century. This has clear implications for winter tourism and recreational activities, which include a range of snow sports and mountain climbing, and the local services that depend upon them. The questions are what form changes in climate will take and how the tourist industry can adapt to them. Is the future one of steadily failing enterprise or will there be sufficient opportunity for managed adaptation through, for example, diversification?

The Scottish mountains have long been an attraction for those who wish to seek the wild and open spaces but it was not until the 1960s that commercial skiing development arrived. The principal focus was the Cairngorm area, which saw the development of Aviemore as its principal centre. Today, in addition to Cairngorm, there are a further four major skiing areas in Scotland at Glenshee, Nevis Range, Lecht and Glencoe. In a global context these are all at relatively low elevations lying roughly between 700m and 1200m above sea level. Snow cover is provided largely by eastwards-moving cyclonic systems from which precipitation in the mountains frequently falls as snow between October and April. This imparts an ephemeral character to the snow cover. Periods of lying snow are interrupted by melting episodes that can remove the cover from the lower ski slopes, and persistent cover is most probable only in the sheltered corries. Recent changes in winter weather in the Scottish mountains have had a direct effect on recreational and related activities, and there have been concerns expressed regarding future viability, expressed in the typical press headline 'Ski industry is heading downhill' (*Aberdeen Press and Journal*, 1999). A review of the potential impacts of climatic change on tourism and recreation (UK Climate Change Impacts Review Group (CCIRG), 1996) raised the prospect of warmer winters bringing less snow and a declining 'snow confidence'. The picture painted is a gloomy one but it is fortunately incomplete and requires a great deal more brushwork.

In 2000 a study was commissioned by the Scottish Executive the principal objective of which was to develop a more realistic forward projection of changes in snow cover across Scotland and to assess the impacts these may have on a wide range of socio-economic activities. The following is based upon some of the research undertaken (Harrison *et al.*, 2001).

Recent Changes in Snow Cover

The historian's adage that an understanding of the future can be found in the patterns of the past suggests that in the climatic record we can find not only how we respond to ongoing change but also analogues of future changes in climate. With regard to snow cover there is the almost universal problem of shortage of data on which to base any assessment of change. In Scotland, site-specific snow data have been collected by ski-tow companies, avalanche warning services and mountaineering groups while at a larger spatial scale remote sensing imagery has proved useful in recent years. The now defunct Snow Survey of Great Britain also provided information on snow lines on the Scottish mountains. However, the primary database is the record of snow observations from climatological stations at which a routine record is made of days on which snow covers more than 50% of the ground surface in the immediate vicinity of the station. This is a subjective assessment but it provides a long historical record going back to the 19th century. Unfortunately, very few stations lie above an elevation of 250m so extrapolation is required to extend this record into the mountain areas.

The persistence of winter snow cover is extremely variable in Scotland (Manley, 1971; Green, 1973). The long record from Braemar (339m) climatological station (Figure 9.1a) is used as an UK indicator of ongoing climatic change and indicates very clearly the nature of the variability. However, Braemar lies in a deep valley, which makes it untypical of snow conditions across Scotland. An alternative index of snow cover has been developed by Harrison *et al.* (2001). The SNODEX index is based on the long records of days with snow lying at five climatological stations across Scotland, which are expressed in terms of standard deviations from the long-term mean (Figure 9.1b). The resulting values suggest that there has been a decreasing trend since the late 1970s, which is a direct result of increasing winter temperatures.

This is consistent with trends observed by Mordaunt (1997) for sites in the Scottish Highlands, and Wheeler (1999) in his analysis of the records for Durham (UK). Pottie *et al.* (1995) have also produced evidence of the progressively earlier melting of mountain snow-beds in Scotland. Elsewhere, similar trends also appear to be in evidence on the Great Plains of North America (Hughes & Robinson, 1996) and in the Swiss Alps (Laternser & Schneebeli, 2003). However, in identifying a general decrease in the length of the snow season in the Swiss Alps, Beniston *et al.* (2003) found that while a trend could be readily identified below 650m, at elevations above 1300m there was a tendency towards increasing snow cover. The decrease coincides with the period of rapid increase in global temperatures from 1976 inwards identified by Karl *et al.* (2000), but in a European context is linked very strongly to the strength and persistence of the mid-latitude westerly airflow. Maritime air brings not

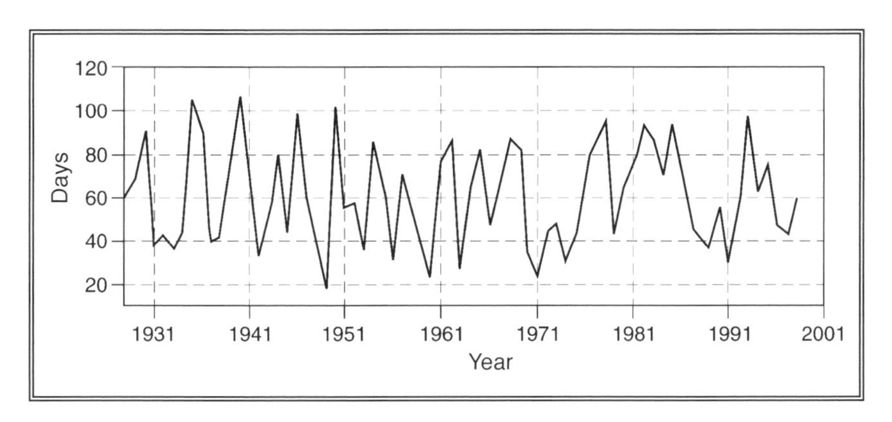

Figure 9.1a The annual number of days with snow lying (09.00 GMT) at Braemar, Scotland, 1927 to 1999

Source: after Harrison *et al.* (2001)

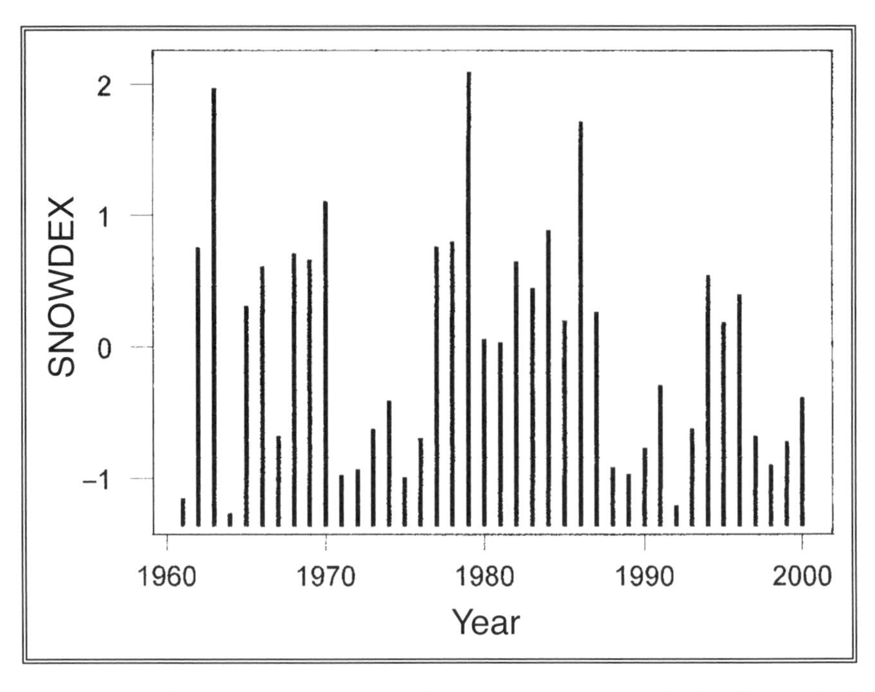

Figure 9.1b Variation in the Scottish winter snow cover index (SNODEX) between winters 1960/1 and 1999/2000

only more equable climatic conditions but also imports moisture. In recent years there has been a tendency towards higher winter rainfalls in Western Europe (Mayes, 1996), which fall as snow at the highest elevations where winter temperatures are likely to be below 0°C. This effect is reflected in an analysis of the number of days Cairngorm ski lifts were operational between 1972 and 1999 (Harrison *et al.*, 1999), which reveals a significant decrease at the lower tows but no significant change above 1000m.

The Effect of Recent Changes in Snow Cover

In order to assess the effects of recent changes in snow cover on a range of related activities, key stakeholders were identified and were asked to complete a short questionnaire that requested responses to the following questions:

- Have changing snowfall patterns affected your organisation, sector or service? (If *Yes* then please provide brief details)
- Has the change been detrimental? (If *Yes* please provide brief details of impacts)

- Have you been able to adapt your operations to accommodate the changes? (If *Yes* please provide brief details)
- Has your organisation any contingency plans for further changes in snowfall? (If *Yes* please provide details)

Contact was made by conventional mail and via a dedicated website. Key stakeholders included a number of snow-sports organisations, ski clubs, ski centres, local tourist boards, mountain guides and climbing instructors. As part of the wider project, representatives in the fields of transport, land management, environmental conservation and management, water, energy supply and construction were also contacted. The responses received from the broad group encompassing snow sports, mountaineering and tourism noted a range of impacts of reduced snow cover.

Snow sports

The snow-sports group noted that snow cover had become more ephemeral in nature with very heavy snowfalls followed by rapid melting, although the very highest areas had experienced more reliable snow cover during some winters. However, on the whole, there had been a steady reduction in the number of skiers visiting the main centres, which was measured in terms of reduced ticket sales for the ski lifts and tows. Palutikof (1999) has noted a strong correlation between lift and tow passes issued and the number of days with snow lying.

The general view was that the snow cover had become unreliable which had led to the cancellation of some key competitive events and visits from ski clubs. Many groups had become unwilling to take the risk of cancellation due to lack of snow and had transferred activities to centres in continental Europe. A side effect of this had been a marked reduction in membership of some ski clubs, to the point of being unable to maintain club facilities. Skiers and snowboarders had become more opportunistic in their behaviour, making last-minute decisions based on, for example, information posted on the Internet. This had impacted directly on service providers who had less flexible operational constraints. On the whole, recruitment of staff had been reduced to a level where a satisfactory compromise could be reached between periods of low and high demand. Accommodation providers were having to deal with a general reduction in demand for holiday accommodation, resulting in a direct loss of income. The result had been a considerable loss of jobs in this sector.

In order to make best use of the available snow cover, increasing grooming and management of snow-lie had become necessary on the lower slopes. Although there had been a more reliable snow cover on slopes above 1000m, skiers were having to travel further to reach it, which had required further investment in the on-site transport infrastructure, such as the Cairngorm Funicular Railway.

Mountaineering

Respondents had, on the whole, come to the conclusion that snow- and ice-climbing had become unreliable. There had been a shift towards mixed buttress climbing and a more general diversification of mountain 'experiences'. A particular casualty had been winter mountain leader training and assessment.

The changing nature of the winter weather had created a number of safety problems. The heavy snows and rapid thaws had increased the avalanche risk in some areas while the absence of snow had tempted many less experienced hill walkers into the mountains. Walkers had occasionally been overtaken by sudden snowstorms and there was the ever-present risk of hypothermia in cold, wet and windy conditions. The presence of increasing numbers of walkers in the winter had begun to result in greater disturbance to, for example, deer and grouse, in addition to the erosion of the fragile arctic-alpine plant communities.

Predicting Future Changes in Snow Cover

It is too simplistic to make the assumption that a warmer winter climate will lead to a general disappearance of snow from the mountain areas of the world and will thereby threaten the future viability of snow-related recreational activities. However, the obvious association between the duration of snow cover and winter temperature cannot be overlooked and the economic risks are real enough (Konig, 1998). Hantel *et al.* (2000) predict that a temperature increase of only 1°C would reduce the snow season in Austria by 73 days over winter and spring. Abegg (1996) suggests that the snow-reliable ski fields at 1500m in Switzerland would be reduced to 63% of current levels if the air temperature increased by 2°C. Whetton *et al.* (1996) write of the total loss of the Australian skiing industry by 2020 in a worst case scenario. Perry and Illgner (2000) maintain that skiing in the Drakensberg Mountains of South Africa will cease to be viable by 2050 if the current rate of temperature increase continues.

In Scotland, Manley's (1971) confident statement that '. . . skiers will certainly find enough cover from late December to early April' had seemed somewhat misplaced by the end of the 20th century. However, there remains the problem of establishing what is going to happen to snow cover in Scotland before its potential consequences can be assessed. Current climatic models are spatially too crude a tool on which to base an assessment of the potential effects upon tourism and recreation in the Scottish mountains. The larger scale of Global Climate Models (GCM) has a spatial resolution measured in 100s of kilometres while Regional Climate Models (RCM) bring this down to approximately 50km. Operationally, a resolution of 1km or less is required in order to relate changes in snow cover to snow-sport activities. This resolution has been achieved using analogue modelling, based on the underlying assumption that the existing climatic record contains winters in the past that were similar in nature to those that are expected to occur as a result of climatic change.

Predictions of winter temperature and precipitation were based on UKCIP98 (Hulme & Jenkins, 1998), which identifies four climatic scenarios, Low, Medium Low, Medium High, and High, the last being effectively the worst-case scenario of unrestrained global warming. The modelling process produces these scenarios for the 30-year periods based on 2020, 2050 and 2080. Four climatological stations having long and relatively homogenous records were selected as being representative of Scotland as a whole. These were Dyce (near Aberdeen), Leuchars (in Fife), Abbotsinch (near Glasgow) and Eskdalemuir (in the Southern Uplands). Average deviations of mean winter temperature and total rainfall from the 1961–90 average were derived for each winter using the three core months of December, January and February. These were crossmatched against the UKCIP98 predictions and suitable analogue years identified for 2020s, 2050s and 2080s Medium Low and Medium High scenarios.

Maps of the number of days with snow lying for each analogue winter were then constructed. The maps were based upon data from 36 climatological stations across Scotland using the three basic spatial co-ordinates of latitude, longitude and altitude (above sea level). The statistical relationship between duration of snow-lie and latitude and longitude was assumed to be linear as a first approximation. However, the relationship between snow-lie and altitude is complex and non-linear, despite it commonly being assumed to be linear. The most appropriate form of the relationship appeared to be best represented by a third-order polynomial function, which explained approximately 98% of the variance in the number of days with snow lying. The performance of the spatial models was tested against site-specific data available in the Snow Survey of Great Britain. A Digital Elevation Model (GTOPO030) developed by the US Geological Survey was then used to generate maps of snow-lie. The model provided topographic data at a spatial resolution of approximately 1km. The snow models were coded into Arc Macro language and processed using Arc/Info's GRID modelling program. Model outputs were projected into the UK National Grid co-ordinate system.

Maps were produced for each of the selected scenarios, which revealed considerable spatial variation in the absolute and relative projected decreases in the duration of snow cover (Harrison *et al.*, 2001). Figure 9.2 indicates the predicted change in the number of days with snow lying expressed as a percentage reduction from the 1961–90 average for a 2080 Medium Low scenario. Regions such as Fife should expect massive reductions in snow-lie while the highest mountain tops experience less than a 25% reduction. The greatest absolute decrease in snow cover has already been shown to be within the altitude range 400–500m (Harrison *et al.* 1999), but the relationship between percentage change and altitude for the same scenario (Figure 9.3) is one of decreasing relative change with altitude. The application of a universal temperature–snowfall relationship would apparently lead to an overestimation of the true nature of the reduction in snow cover on the Scottish mountains, where it is offset by increasing winter precipitation falling as snow.

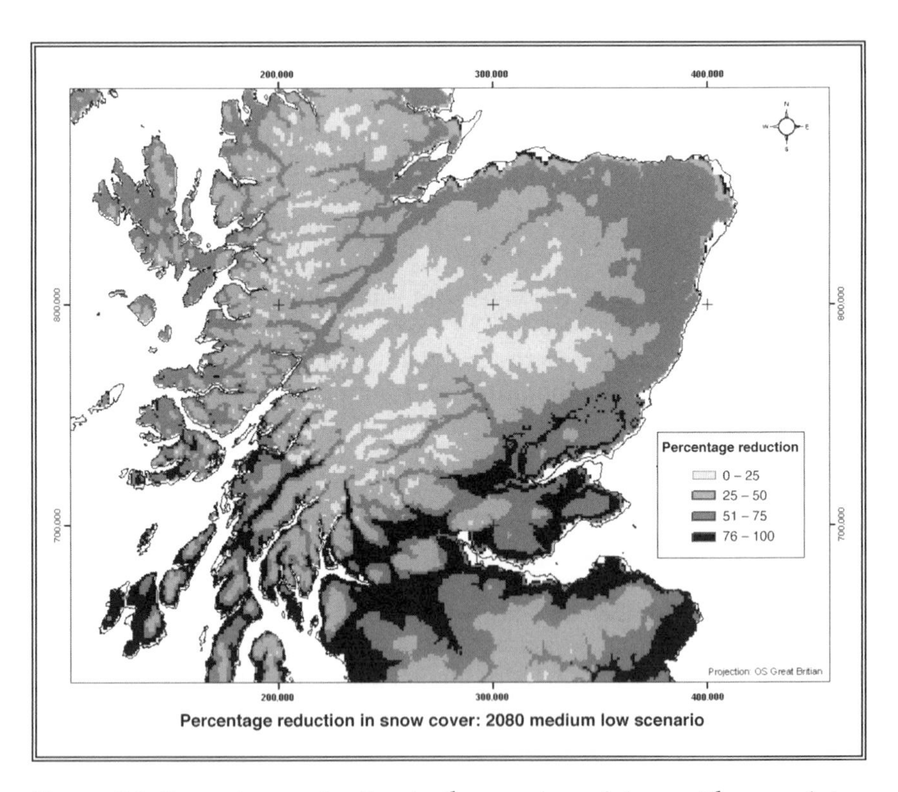

Figure 9.2 Percentage reduction in the number of days with snow lying in Scotland: 2080 Medium Low scenario relative to the 1961–90 long-term average

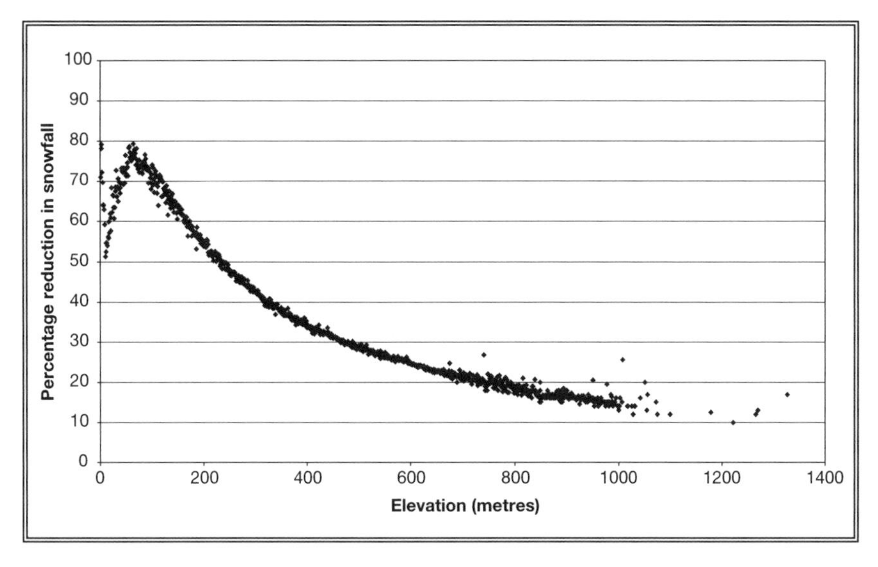

Figure 9.3 Relationship between the predicted percentage reduction in days with snow lying in the Scottish Highlands and altitude (metres) for the 2080 Medium Low scenario

Strategies for Dealing with Future Changes in Snow Cover

Once the maps of predicted change in snow cover had been produced, these were then sent to representative stakeholders, who were asked to assess what effects the changes may have and how they would deal with them. Responses were guided by five questions:

- What potential effects would the changes that we are predicting have on your organisation or area of activity?
- Is it possible to estimate the cost implications (low to high) of such effects on:
  - profitability
  - amenity
  - employment
  - seasonality of activity
  - cost effectiveness of service provision
  - long-term investment strategies.
- What change do you anticipate will be necessary in your organisation or area of activity to accommodate the effect of changes in snow cover?
- Will additional resources be required to effect such changes?

- To what extent does uncertainty in predictive models influence your reaction to the changes that we suggest may take place?

With regard to Scottish winter tourism and recreation the key issues that needed to be addressed were:

(1) Reduced average number of days on which snow can be expected to be lying on the ground surface, which will be more marked on the lower slopes. Greatest reliability will be on the highest slopes, which are less accessible.
(2) Intermittent nature of snow cover with heavy falls separated by periods of rapid melting.
(3) Lower reliability of snow cover resulting in fundamental changes in the behaviour of winter visitors to a more opportunistic pattern.
(4) Greater number of snow-free days on the Scottish mountains resulting in an increase in the numbers of visitors participating in other non-snow activities such as hill walking.

Bürki (2000) has considered a number of response strategies of the skiing industry to changes in climate but the underlying principle is the Darwinian principle of the survival of those most fitted to survive. In the Scottish context, some form of adaptation strategy is required if winter tourism in the Highlands is to survive. Given the importance of winter tourism to the economy of the Scottish Highlands and to Scotland, providing an annual income in excess of £15m, the priority is to find sustainable ways forward. Some changes have already been put in place, one example of which is the Cairngorm Funicular Railway, which provides access to a limited form of mountain experience to a wider range of visitors many of whom are not skiers, climbers or walkers.

The survival of winter tourism in the Scottish Highlands should be based on a long-term strategy to diversify. At its most extreme, diversification could mean a concentration on summer activities, with the possible eventual closure of mountain ski resorts or, alternatively, new winter activity schemes could be introduced such as indoor sports using artificial snow. It is, however, vital that maximum use is made of the snow resource that is available, including snow grooming and the use of snow fences, and access to the more reliable snow on higher slopes needs to be improved using environmentally sensitive technology. Every effort needs to be made to encourage and respond to opportunistic behaviour in winter visitors. This should include appropriate action to ensure availability of short-term (less than three days) residential accommodation. Use should continue to be made of rapid means of communicating with ski and mountaineering clubs, and tourist information centres, using the Internet and e-mail subscriber networks.

The adaptation required of related businesses such as equipment retailers needs to be seen in terms of the range of stock held before each winter. Whereas many used to concentrate on skiing or snow-climbing equipment, there will be a growing market for wet weather protection and hill walking gear, which should be available throughout the year. Other related businesses which could benefit from adapting to changes include food and drink outlets, cycle hire and activity centres, all providing alternatives in case snow sports are not available. As warmer conditions are likely to encourage the less experienced to enter the mountains, there should also be greater investment in visitor education and the provision of warnings, together with greater financial support for mountain rescue organisations.

Conclusions

The models used to provide an indication of future changes in the winter snow cover in the Scottish mountains should be regarded as nothing more than a best estimate of what the tourist industry will face over the course of the 21st century. In adopting UKCIP as the basis of the analogue modelling, the authors have tended to assume that climate change will continue to be driven primarily by global warming, but there are other mechanisms that may result in a very different set of changes. If the westerly winds do continue to dominate the climate of the Scotland, then there is also the real risk that despite the presence of a snow cover, strong wind and low cloud may restrict the use of the ski lifts. Service providers such as the ski centres have experienced more than 20 years of a steadily decreasing snow resource and have already begun to diversify. While the future is clearly not one without a snow cover on the Scottish mountains, the prognosis is that further changes in the tourism and recreation sector of the economy of the Scottish Highlands will be required and it is unlikely that a secure economic future can be based solely on traditional winter activities.

References

Abegg, B. (1996) Klimaanderung und Tourismus. *Klimafolgenforschung am Beispiel des Wintertourismus in den Schweizer Alpen*. Projektbericht NFP 31. Zurich: vdf Hochschulverlag AG ETH.

Aberdeen Press and Journal (1999) Ski industry is heading downhill. *Aberdeen Press and Journal* 16 June.

Beniston, M., Keller, F., Koffi, B. and Goyette, S. (2003) Estimates of snow accumulation and volume in the Swiss Alps under changing climate conditions. *Theoretical and Applied Climatology* 76 (3–4), 125–40.

Bürki, R. (2000) Climate change and winter tourism in the Swiss Alps. In A. Lockwood (ed.) *Tourism and Hospitality in the 21st Century* (pp. 20–3). Guildford: University of Surrey.

Green, F.H.W. (1973) Changing incidence of snow in the Scottish Highlands. *Weather* 28, 386–94.

Hantel, M., Ehrendorfer, M. and Haslinger, A. (2000) Climate sensitivity of snow cover duration in Austria. *International Journal of Climatology* 20, 615–40.
Harrison, S.J., Winterbottom, S.J. and Johnson, R.C. (2001) *Climate Change and Changing Snowfall Patters in Scotland*, Scottish Executive Central Research Unit Report No. 14. Edinburgh: The Stationery Office.
Harrison, S.J., Winterbottom, S.J. and Sheppard, C. (1999) The potential effect of climate change on the Scottish tourist industry. *Tourism Management* 20, 203–11.
Houghton, J.F., Filho, L.G.M., Callander, B.A., Harris, N., Kattenberg, A. *et al.* (1996) *Climate Change 1995: The Science of Climate Change*. Cambridge: Cambridge University Press.
Hulme, M. and Jenkins, G. (1998) *Climate Change Scenarios for the United Kingdom*. UK Climate Impacts Programme Technical Report No. 1. London: DETR.
Hughes, M.G. and Robinson, D.A. (1996) Historical snow cover variability in the Great Plains region of the USA: 1990 through to 1993. *International Journal of Climatology* 16, 1005–18.
Karl, T.R., Knight, R.W. and Baker, B. (2000) The record-breaking global temperatures of 1997 and 1998: Evidence for an increase in the rate of global warming. *Geophysical Research Letters* 27, 719–22.
Konig, U. (1998) *Tourism in a Warmer World*. Wirstschafts-geographie und Raumplannung No. 28. Zurich: University of Zurich.
Laternser, M. and Schneebeli, M. (2003) Long-term snow climate trends of the Swiss Alps 1931–1999. *International Journal of Climatology* 23, 733–50.
Manley, G. (1971) The mountain snows of Britain. *Weather* 26, 192–200.
Mayes, J. (1996) Spatial and temporal fluctuations of monthly rainfall in the British Isles and variation in the mid-latitude westerly circulation. *International Journal of Climatology* 16, 585–96.
Mordaunt, C.H. (1997) Association between weather conditions, snow lie and snowbed vegetation. Unpublished Ph.D. Thesis, University of Stirling.
Palutikof, J.P. (1999) Scottish skiing industry. In M.G.R Cannell, J.P. Palutikof and T.H. Sparks (eds) *Indicators of Climatic Change in the UK*. London: HMSO.
Perry, A.H. and Illgner, P. (2000) Dimensions of winter severity in Southern Africa: Is a skiing industry in the Drakensberg Mountains viable? *Journal of Meteorology* 25, 226–30.
Pottie, J. (1995) Scottish snow bed records from three sites. *Weather* 50, 124–9.
UK Climate Change Impacts Review Group (CCIRG) (1996) *Review of the Potential Effects of Climate Change in the United Kingdom*. London: HMSO.
Whetton, P.H., Haylock, M.R. and Galloway, R. (1996) Climate change and snow cover duration in the Australian Alps. *Climatic Change* 32, 447–79.
Wheeler, D. (1999) Published letter. *Weather* 54, 376–7.

Chapter 10
Climate Change and Tourism in the Swiss Alps

ROLF BÜRKI, HANS ELSASSER, BRUNO ABEGG AND URS KOENIG

Introduction

For many alpine areas in Switzerland winter tourism is the most important source of income, and snow-reliability is one of the key elements of the tourist resource. Skiing and snowboarding, but also snow-related activities such as cross-country skiing or snow hiking depend on there being enough snow. The winters with little snow at the end of the 1980s (1987/8 to 1989/90) caused a stir in the Swiss Alps. The big difference to earlier periods with little snow was that the ski industry had become much more capital intense.

However, there is more to climate change for the alpine tourism industry than snow-deficient winters. Since 1850, Swiss glaciers have lost more than a quarter of their surface. In 2030, 20% to 70% of Swiss glaciers will have disappeared. This is not only a severe loss of mountain aesthetic, but also a problem for glacier skiing in both summer and winter. Furthermore, global warming increases melting of permafrost and makes many mountain areas more vulnerable to landslides. Mountain cableway stations, lift masts and other buildings on permafrost soil will become unstable and it is very expensive to stabilize them. Moreover, warmer temperatures in mountain areas will make hiking and climbing more dangerous due to increasing rock fall. The future climate will not only be warmer, but will also change its pattern. More precipitation or a higher fog level will lead to new conditions for mountain summer tourism such as hiking, trekking or biking. More frequent and more extreme events are another threat for tourism activities and tourism infrastructure.

In the Swiss Alps, a large number of the farmers depend on winter tourism to supplement their income. This is important because government subsidies and the total gross margin could change in the future independent of climate change, whereas additional income from activities in other sectors, such as winter tourism, may change because of climate change. Direct impacts of climate change on the tourism industry may therefore have serious indirect effects on agriculture.

Climate Change and Snow-deficient Winters

If the assumptions of the impacts of climate change hold true, snow cover in the Swiss Alps will diminish which will, in turn, jeopardise the tourism industry. The crucial factor for the long-term survival of mountain cableway companies is the frequency and regularity of winters with good snow conditions, or to put it another way, the number of snow-deficient winters that can be withstood (Abegg *et al.*, 1998). It is challenging to define this number because the economic situation of individual companies varies considerably. The experience acquired by Swiss ski resorts, however, shows that a ski resort can be considered snow-reliable if, in seven out of ten winters, a sufficient snow covering of at least 30 to 50cm is available for ski sport on at least 100 days between 1 December and 15 April.

A temporal and spatial downscaling from two global circulation models (GCMs) (ECHAM: GCM of the Max-Planck Institute for Meteorology; CCC: GCM of the Canadian Climate Center) together with a snow model, simulate the number of days with a certain snow depth in the future (2030–50) (Bürki, 2000). In light of the above concept of snow-reliability, Figure 10.1 shows the results for five ski resorts in the Swiss Alps.

Findings are that, at present, the criteria are matched in Davos-Dorf and Weissfluhjoch-Davos only. Engelberg, Disentis, and the village of Montana located at almost 1500m above sea level, cannot be considered as snow-reliable, because significantly less than 70% of the winters are 'good'. By between 2030 and 2050, only the summit station of Weissfluhjoch will meet the criteria. In Engelberg, there will be no 'good' winters at all. In the case of Disentis and Montana respectively, there will be, depending on the scenario, only a few 'good' winters. Davos, still snow-reliable under current conditions, will no longer be snow-reliable under future climate conditions. For both, the ECHAM and the CCC conditions, the share of 'good' winters accounts for less than 50%. If the threshold value of the snow depth was changed from 30 to 20cm, the 70%-criterion will still be missed, but only slightly.

The results also show that the snow conditions in lower and medium altitudes will be getting worse. This will go along with a rise in the line of snow-reliability. But a comparison between Davos and Montana – almost at the same altitude but with very different snow conditions – also shows that the regional peculiarities, e.g. the regional climate and topography, have to be kept in mind.

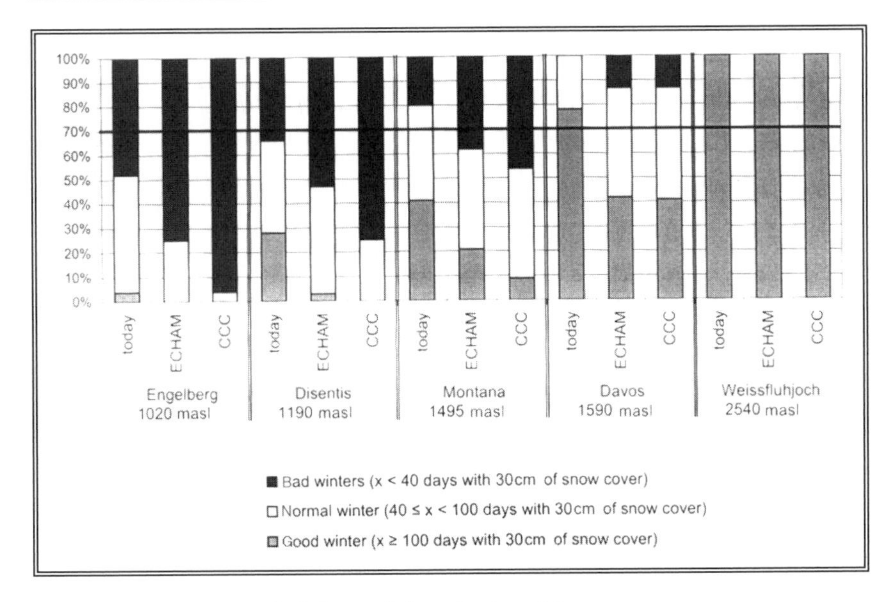

Figure 10.1 Snow-reliability of five Swiss ski resorts

Note: The figure shows the frequency (%) of 'good' (≥ 100 days with a given snow depth) and 'bad' winters (< 40 days with a given snow depth) under different climatic conditions (today: means of the years from at least 1970 to 1995). A ski resort is snow-reliable if at least 70% of the winters are 'good'.

In spite of regional differences, 85% of all 230 Swiss ski resorts can be considered to be snow-reliable today (Table 10.1). However, even today a lot of ski resorts in the lower Alps are not snow-reliable. If the line of snow-reliability were to rise to 1500m, the number of snow-reliable ski resorts would drop to 63%. The Jura, Eastern and Central Switzerland, Ticino, and the Alps in the cantons of Vaud and Fribourg will be particularly jeopardised by global warming. The ski regions of Valais and the Grisons will experience virtually no major problems, since the mean altitude of the cableway terminals in these regions is higher than 2500m above sea level. If the line of snow-reliability were to rise to 1800m, which is a possible scenario according to the results of snow modelling, there would be a further serious deterioration in conditions: only 44% of skiing regions could be designated as snow-reliable. Even in the cantons of Grisons and Valais, approximately a quarter of the ski resorts would no longer be snow-reliable.

Climate change will lead to a new pattern of favoured and disadvantaged ski tourism regions. If all other influencing factors remain the same, ski tourism will concentrate on high-altitude areas, i.e. the cantons of Valais and Grisons. Ski resorts at lower altitudes will withdraw from

Table 10.1 Snow-reliability of Swiss ski resorts

Region	*Number of ski resorts*	*Snow-reliability*					
		1200 masl		*1500 masl*		*1800 masl*	
		No.	*%*	*No.*	*%*	*No.*	*%*
Jura	15	4	27	1	7	0	0
Alps (Vaud + Frib.)	19	16	84	7	37	4	21
Valais	54	54	100	52	96	40	74
Bern (ex. Jura)	35	30	86	20	57	12	34
Central Switzerland	35	26	74	13	37	7	20
Ticino	8	8	100	3	38	2	25
Eastern Switzerland	18	11	61	6	33	3	17
Grisons	46	46	100	42	91	33	72
Switzerland	230	195	85	144	63	101	44

Source: After Abegg (1996); Bürki (2000)

the market sooner or later because of the lack of snow. The only areas with good prospects will be those with transport facilities that provide access to altitudes higher than 2000m. The regions at higher altitudes may experience greater demand, prompting pressure for further expansion. The pressure on ecologically sensitive high-mountain regions will increase. The call for snow-reliable ski resorts constitutes the main reason for the current boom in concept studies and plans for opening up high-mountain regions, or, expressed in different terms, climate change is an argument for opening up high-mountain regions to tourism.

A survey among tourists shows, that skiers will respond flexibly to changing snow conditions. During a period of snow-poor seasons, as expected more often under a changing climate, 49% of the skiers would change to a ski resort that is more snow-reliable and 32 % of the skiers would ski less often. Although only 4% of the respondents would give up skiing, it can be concluded that climate change would have serious impacts on the number of skier days. The most vulnerable ski resorts in the lower regions of the Alps would have to deal with a significant decrease of younger guests, day tourists and novice skiers, which is exactly the target group of these resorts (Bürki, 2000).

Meier (1998) calculated the potential annual costs of climate change in Switzerland at CHF2.3 to CHF3.2 billion (US$1.5 to $2.1 billion) by the year 2050, which is 0.6 to 0.8% of the Swiss gross national product for 1995. CHF1.8 to CHF2.3 billion (US $1.2 to $1.6 billion) would be accounted for by tourism. Despite many reservations voiced against this calculation, it nevertheless shows that tourism is the economic sector that would be affected most by climate change in Switzerland and that the magnitude of this cannot be neglected. The impacts of climate change on winter tourism may be even more severe in Germany and Austria due to the lower altitudes of their ski resorts.

Extreme events may arise more frequently in a warmer future. The winter of 1998/9 for example showed that the possibility of winters with a great deal of snow cannot be excluded in the future. In a study of the 'avalanche winter' of 1999, the direct losses incurred by mountain cableways as a result of avalanches and the large quantities of snow were estimated at CHF17 million (US$12 million). In total, 44 facilities were damaged, including 20 ski lifts, 11 chairlifts, 4 cable railways and 2 funicular railways. The mountain railway companies had to spend an extra 77% on snow clearing, compared with previous years. Roughly 25% more than in normal winters was spent on securing the ski slopes. All in all, the avalanche winter of 1999 probably caused losses of CHF332 million (US$ 240 million), the major portion of these were indirect losses of CHF302 million (US$ 215 million). The main causes for the indirect costs in tourism are unfavourable weather conditions and exaggerated media coverage. It is possible that some tourist companies are able to gain a surplus in revenue after a natural disaster, mainly due to transfer of tourists to areas not affected. However, that surplus cannot possibly make up for the losses (SLF, 2000; Nöthiger, 2003).

Adaptations of Tourism Stakeholders

In Switzerland, the tourism stakeholders at a political, entrepreneurial, operational and organisational level are not sitting back idly contemplating the consequences of a climate change. They are adapting right now in the expectation of climate change. The experiences with snow-deficient winters have shown them that the climate does not determine their economic activities, but, instead, constitutes a key resource and framework condition. The results of a focus group study among tourism representatives in Switzerland can be summed up as follows:

- Climate change has been recognised as a problem for winter tourism. Those responsible for tourism know that what they can offer is highly dependent on snow and that they are at risk from snow-deficient winters. They are familiar with the potential consequences of climate change for winter tourism. While achieving snow-reliability constitutes a central topic, potential climatic change is seen as being only of relatively minor importance.
- Climate change is not regarded as a big threat for winter tourism. Tourism stakeholders think that climatic change is highly

exaggerated by not only the media but also by science and politics. They think that although climate change could intensify the problems that already exist in ski areas at lower altitudes and speed up structural changes in the sector, the majority of ski resorts at medium and high altitudes, however, would scarcely be affected.

- Climate change is already affecting the strategies and plans of the winter sport resorts today. The discussions held in the focus groups clearly revealed an ambivalent relationship to climate change. On the one hand, stakeholders strongly distrust the information disseminated about climate change and play down its potential consequences, but on the other, they use climate change to legitimate forward strategies. Climate change and global warming, together with international competition, have been used as the key arguments for constructing artificial snowmaking facilities, as well as for extending existing ski runs and opening new ones in high-alpine regions (above 3000m).
- Tourism stakeholders agree that winter sports can only survive in the Alps if snow-reliability is guaranteed. The smaller ski fields at lower altitudes either have their hands bound or can scarcely finance the necessary investments (e.g. artificial snowmaking, levelling out ski slopes, opening higher-altitude chambers in skiing areas). They do not have financial resources of their own and banks are (now) only prepared to grant very restrictive loans to ski resorts at altitudes below 1500m that are not particularly profitable. Nevertheless, the representatives believe that smaller ski fields in the alpine foothills play a key role in promoting the importance of skiing. Opinions frequently differ a great deal, however, on whether non-profitable ski regions of this type should be retained and how their financing can be guaranteed. While a number of people are in favour of dismantling non-profitable cableway and ski-lift operations and regard a certain 'healthy shrinkage' of the sector as necessary, others believe that there is an obligation to retain these ski fields for regional economic reasons by providing subsidies.

Strategies

Climate change represents a new challenge for tourism, and particularly for winter tourism in alpine areas. It is not, however, the case that tourism's initial position will undergo a sudden, radical change. Instead, climate change has to be viewed as a catalyst that will reinforce and accelerate the pace of structural change in the tourist industry and more clearly highlight the risks and opportunities inherent in tourist developments even now. The emergence of a two-tier structure in the tourist sector will not be due to climate change alone, but also to other structural changes. On the one hand, we have the top resorts with their already varied and attractive offers and high snow-reliability and, on the other hand, we have the smaller locations with their less-extensive developments, less-refined offers and restricted opportunities for further development.

Since climate change is a relatively long-term development in comparison to other trends in tourism, tourism managers and tourists will have every opportunity to adjust to the different constraints and adopt the corresponding strategies and measures (Figure 10.2). One of the most familiar measures in the struggle against snow-deficient winters is the construction of high-cost artificial snowmaking facilities. Adopting a fatalistic attitude towards climate change and its impacts should not be considered as a true strategy in this respect. Such attitudes are manifested by the fact that neither suppliers nor consumers alter their behaviour. This could also be described by using the term 'business as usual'. Another approach that can be classified under the heading of 'fatalism' is when tourist transport facilities that were used for winter sports are closed down and dismantled without any attempt at promoting and reinforcing other types of tourism – in other words, when withdrawal

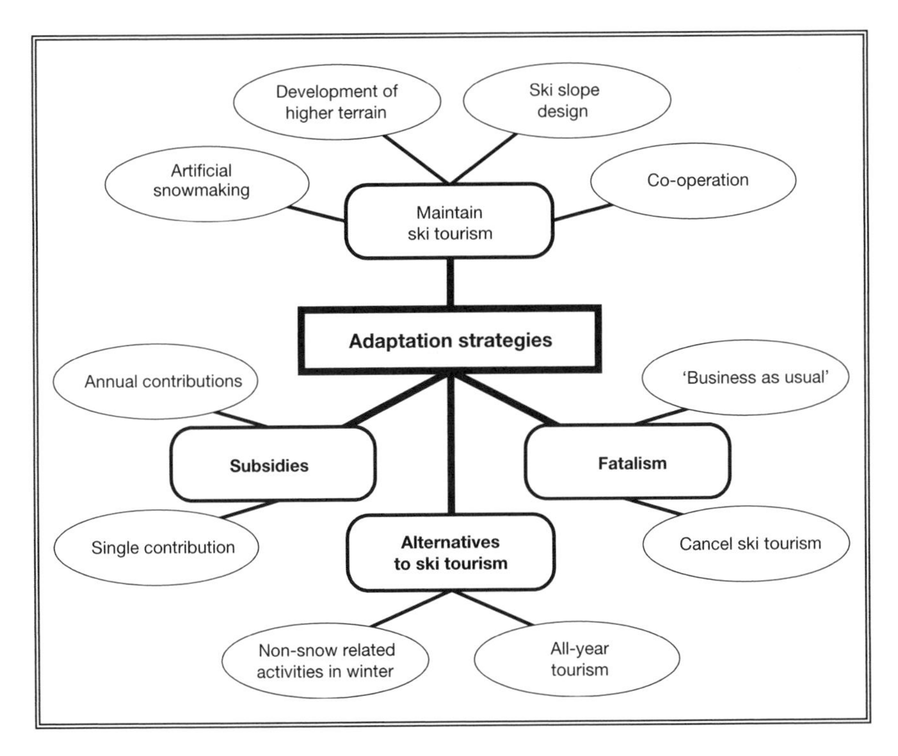

Figure 10.2 Adaptation strategies

from ski tourism is not actively planned. A fatalistic attitude of this type is most readily evident among the operators of small, isolated ski lifts at lower altitudes who experienced severe financial difficulties as a result of the snow-deficient winters.

Conclusions

At first sight, global warming seems to provide an opportunity for the tourism industry in alpine areas. But warmer temperatures and a longer summer season are of minor importance. Overall, climate change is a threat for alpine tourism due to less snow, less glaciers, and more extreme events (e.g. landslides).

Winter tourism depends on good snow conditions and is highly sensitive to snow-deficient winters. Climate research findings show that there will be an increase in the number of winters with little snow on account of climate change. The tourism representatives will not just sit back idly in the face of climate change. They are reacting to the deteriorating snow conditions and the changes in demand. Technical measures to maintain ski tourism, especially artificial snowmaking, rank at the forefront. Tourists demand good snow conditions, and hence, this is what has to be offered by the ski resorts. In any case, the impacts of climate change will involve significant costs for tourism. One of the most important questions will be, how young people would start skiing/snowboarding, if there is only little snow in the big towns and if the little and cheap ski lifts for families at small distances to these towns will be dismantled due to climate change. Although indoor skiing is a growing industry in European towns, it is uncertain that indoor ski domes can replace the role of little ski resorts for beginners in the foothills.

As a sector of the economy that is severely affected by climate change, however, tourism needs to focus more on mitigation strategies in its own best interests. This holds particularly true for the traffic generated by national and international tourism and, above all, for air traffic. Tourist development and tourist projects not only need to be verified and evaluated in terms of their social and environmental compatibility but must also be assessed from the climate-compatibility angle.

References

Abegg, B. (1996) *Klimaaenderung und Tourismus – Klimafolgenforschung am Beispiel des Wintertourismus in den Schweizer Alpen*. Schlussbericht NFP 31. Zürich: vdf Hochschulverlag an der ETH Zürich.

Abegg, B., Koenig, U., Bürki, R. and Elsasser, H. (1998) Climate impact assessment in tourism. *Applied Geography and Development* 51, 81–93.

Bürki, R. (2000) *Klimaaenderung und Anpassungsprozesse im Wintertourismus*. St Gallen: Ostschweizerische Geographische Gesellschaft NF H. 6.

Meier, R. (1998) *Soziooekonomische Aspekte von Klimaaenderungen und Naturkatastrophen in der Schweiz*. Schlussbericht NFP 31. Zürich: vdf Hochschulverlag an der ETH Zürich.
Nöthiger, C. (2003) *Naturgefahren und Tourismus in den Alpen. Untersucht am Lawinenwinter 1999 in der Schweiz*. Davos: SLF (CD-ROM).
SLF (Eidg Institut für Schnee- und Lawinenforschung) (2000) *Der Lawinenwinter 1999 – Ereignisanalyse*. Davos: SLF.

Chapter 11

Effects of Climate Change on Tourism Demand and Benefits in Alpine Areas

ROBERT B. RICHARDSON AND JOHN B. LOOMIS

Introduction

Changes in climate, resulting from higher levels of greenhouse gas concentration, may affect tourism in alpine areas in several ways, including changes in the composition of natural resources, the seasonal availability of particular outdoor recreation opportunities, and the enjoyment of recreation activities, ultimately impacting the number of visitors attracted to an area. Effects to the accessibility of high alpine areas, along with changes in climate patterns, may reduce or eliminate the availability of some recreation activities (e.g. alpine skiing, ice climbing); similarly, opportunities for other types of recreation may be expanded due to a longer summer season (e.g. hiking, camping). Climatic change may impact the visitor's recreation experience in two ways. First, the visitor's utility from his or her recreation experience may be *directly* affected by the weather. Changes in average temperature, precipitation, and snow depth may affect the visitor's enjoyment of outdoor recreation activities, thus affecting choices about the frequency or duration of future visits. Second, changes in climate patterns may impact wildlife populations and the composition of vegetation in the park, and these changes may *indirectly* affect visitation behavior, depending on visitor preferences regarding wildlife viewing and scenery. This chapter discusses the climate effects on tourism demand and recreation benefits in an alpine national park in the United States. Contingent visitation analysis, an application of contingent behavior analysis, is used to measure the effects on tourism demand, and the contingent valuation method is used to measure estimate the effects on recreation benefits (willingness to pay).

A visitor survey in 2001 at Rocky Mountain National Park in Colorado included descriptions of hypothetical climate scenarios that depicted both weather- and natural resource-related variables. Respondents were queried about how the number of trips and length of stay would change contingent upon the scenarios as well as about their willingness to pay higher travel costs for the recreation experience. Survey responses are used to estimate the impact of climate change on park visitation, to test for the relative significance among climate scenarios and resource variables, and to measure the potential effects on recreation visitor benefits (i.e. net willingness to pay). Both direct (weather-related) and indirect (resource-related) climate scenario variables are found to be statistically significant determinants of tourism demand, and daily weather variables were found to be statistically significant determinants of willingness to pay. The results of the two analyses are combined in the measurement of contingent changes in quantity and quality of the recreation good – that is, the number of annual trips and the associated benefits. This allows for the estimation of the recreation demand curve, as well as the vertical and horizontal shifts in the demand curve associated with climate change forecasts. The methodology and findings presented herein have numerous policy implications, including the consideration of strategies that address future changes in park visitation, resource management, infrastructure planning, and the quality of the recreation experience.

Theoretical Framework

Recreation demand

Tourism demand is measured by estimating the number of recreation visits (or trips) taken at various price levels (or travel costs). The economic benefits of recreation are measured by estimating the visitor's willingness to pay higher prices for the trip. In order to estimate the overall economic impact of climatic change on recreation demand and benefits, we first estimate a demand curve for the current level of recreation. This requires data on the quantity of trips, costs of travel, and estimates of utility or satisfaction (represented as willingness to pay). A visitor survey is needed to gather information about park visitors, their preferences toward recreation activities, and how their visitation behavior might change under hypothetical climate scenarios. Contingent visitation analysis can then be used to test for the significance of direct and indirect climate scenario variables.

An individual's utility can be represented by $u(R, I)$, where $u(\cdot)$ is utility, R is an outdoor recreation experience, and I is individual income. The utility function can be restated as $u(x_j, q_j, I)$, where $u(\cdot)$ is utility, x_j is the annual number of trips to recreation site j, q_j represents the quality of site j, and I represents individual income. The individual will maximize his utility subject to his budget constraint, represented by $I = p_j x_j + \mathbf{z}$, where p_j represents the travel cost or implicit price of access to site

j, and **z** represents a vector of all other goods (with prices normalized to one). The resulting Marshallian demand functions $[x(p_j, q_j, I)]$ emerge, with the quantity of trips (x_j) decreasing in price, increasing in quality, and increasing in income (Whitehead *et al.*, 2000).

The theoretical model specifies that the quality of recreation is influenced by several variables, including climate conditions and natural resources. With respect to climate, we posit that climate influences quality (and thus, the number of annual recreation trips) both directly, through the visitor experience, and indirectly, through the enjoyment of the site's plant and wildlife resources, which are directly affected by climate. The model also considers the influences of the visitor's preference for particular recreation activities, travel distance, and the demographic characteristics on total demand for recreation at a site.

The theoretical model can therefore be represented as:

$$V_i = f\,(S^D_{1i}, S^D_{2i}, \ldots, S^D_{ni}, S^I_{1i}, S^I_{2i}, \ldots, S^I_{ni}, A_{1i}, A_{2i}, \ldots, A_{ni}, DIST_i, D_{1i}, D_{2i}, \ldots, D_{ni}) \quad (1)$$

where

$V_i =$	number of annual visits to the recreation site
$S^D_{1i}, S^D_{2i}, \ldots, S^D_{ni} =$	direct climate scenario variables, including temperature, precipitation, snow depth
$S^I_{1i}, S^I_{2i}, \ldots, S^I_{ni} =$	indirect climate scenario variables, including population of wildlife and vegetation
$A_{1i}, A_{2i}, \ldots, A_{ni} =$	activities in which the visitor participated during the visit
$DIST_i =$	distance traveled per visit
$D_{1i}, D_{2i}, \ldots, D_{ni} =$	demographic characteristics of the visitor, including gender, age, level of education, annual income, employment status, and membership in an environmental organization
$i =$	individual respondent to survey.

Following this theoretical model, we test the null hypothesis that the climate scenario variables ($S^D_{1i}, S^D_{2i}, \ldots, S^D_{ni}; S^I_{1i}, S^I_{2i}, \ldots, S^I_{ni}$) have no effect on the number of visits to the site per year. If β is the coefficient on the climate scenario variables, then we suppose:

$$\begin{array}{ll} H_0: \beta^D_1 = 0 & H_A: \beta^D_1 \neq 0 \\ H_0: \beta^D_2 = 0\ldots & H_A: \beta^D_2 \neq 0 \ldots \\ H_0: \beta^D_n = 0 & H_A: \beta^D_n \neq 0 \\ H_0: \beta^I_1 = 0 & H_A: \beta^I_1 \neq 0 \\ H_0: \beta^I_2 = 0 \ldots & H_A: \beta^I_2 \neq 0 \ldots \\ H_0: \beta^I_n = 0 & H_A: \beta^I_n \neq 0 \end{array} \quad (2)$$

where the superscripts D and I represent the coefficients for direct and indirect climate scenario variables, respectively.

Recreation benefits

Recreation benefits are a measure of the utility the consumer obtains from the recreation experience (Loomis & Walsh, 1997). The benefits of a particular recreation visit may be influenced by several variables, including climate conditions and travel costs. In this theoretical model, quality is measured by willingness to pay (WTP). The contingent valuation method (CVM) has been used extensively to measure changes in recreation benefits under varying levels of particular amenities, and is an accepted method of valuing the benefits of recreation, as well as other benefits for which no market exists (Cummings *et al.*, 1986; Loomis, 1987). The premise of CVM is based on a hypothetical market for the use or preservation of a natural resource for which there is no market for the exchange of a good. This hypothetical market includes the description of a good (e.g. recreation experience), a payment vehicle (e.g. travel costs), and a procedure for the elicitation of value (e.g. dichotomous-choice approach). The dichotomous-choice approach for measuring WTP involves asking survey respondents whether or not they would still take their most recent trip if travel costs were $\$X$ higher. The respondent will answer yes if his/her utility from the recreation experience (with the associated loss of $\$X$ in income) is greater than or equal to his/her original utility level without having taken the trip. The 'YES' respondent would hypothetically take the trip ($R = 1$) at the higher travel cost, and the 'NO' respondent would choose not to take the trip ($R = 0$). Therefore, the probability of a YES response is represented as $P(\text{YES} \mid \$X) = P[f(R = 1, I\text{–}\$X) \geq f(R = 0, I)]$ (Hanemann, 1984). With respect to climate, the theoretical model for this measurement of recreation benefits suggests that willingness to pay is influenced by weather conditions on the day of the recreation visit. The model also considers the influences of the visitor's preference for particular recreation activities, the visitor's travel costs, and the demographic characteristics of the visitor on total demand for recreation at the site. The model is represented as:

$$WTP_{it} = f\,(T_t, P_t, W_t, C_t, A_{1i}, A_{2i}, \ldots, A_{ni}, TC_i, D_{1i}, D_{2i}, \ldots, D_{ni}) \quad (3)$$

where

WTP_i = net benefits of (willingness to pay for) recreation experience

T_t = temperature (maximum, minimum)

P_t = precipitation

W_t = average wind speed

C_t =	average cloud cover
$A_{1i}, A_{2i}, \ldots, A_{ni}$ =	activities in which the visitor participated during the visit
TC_i =	travel cost per visit
$D_{1i}, D_{2i}, \ldots, D_{ni}$ =	demographic characteristics of the visitor, including gender, age, level of education, annual income, employment status, and membership in an environmental organization
i =	individual respondent to survey
t =	date.

Following this model, we employ *t*-tests to examine the null hypothesis that daily weather variables (representing temperature, precipitation, wind speed, and cloud cover) have no effect on the respondent's recreation benefits. Thus, from Equation (3), we test:

$$\begin{array}{ll} H_0: \beta_T = 0 & H_A: \beta_T \neq 0 \\ H_0: \beta_P = 0 & H_A: \beta_P \neq 0 \\ H_0: \beta_{WS} = 0 & H_A: \beta_{WS} \neq 0 \\ H_0: \beta_C = 0 & H_A: \beta_C \neq 0 \end{array} \qquad (4)$$

where β is the coefficient on the weather variables. Rejection of this hypothesis implies that weather variables *do* influence the visitor's utility from a recreation experience.

Empirical Analysis and Survey Design

The recreation site for the empirical analysis is Rocky Mountain National Park (RMNP), a 266,000-acre (106,400-hectare) alpine preserve in north-central Colorado, in the Front Range of the Rocky Mountains (see Figure 11.1). The Park protects a large wildlife population, alpine meadows, conifer forests, aspen groves, and several high mountain peaks, including Long's Peak, the Park's tallest. Visitors from around the world travel to RMNP each year to view wildlife, climb mountains, hike on trails, drive the magnificent Trail Ridge Road, and have picnics in a scenic alpine location. RMNP receives over three million visitors annually, with significant seasonal variation (87% of annual visitation occurs between May and October, suggesting an influence of seasonal climate). These characteristics make RMNP an ideal location to conduct the visitation study.

The visitor survey was designed using data for a baseline climate scenario and two hypothetical scenarios as depicted by two global circulation models (Canadian Climate Center (CCC) and Hadley Climate Center (Hadley)), which specified expected temperature levels, precipi-

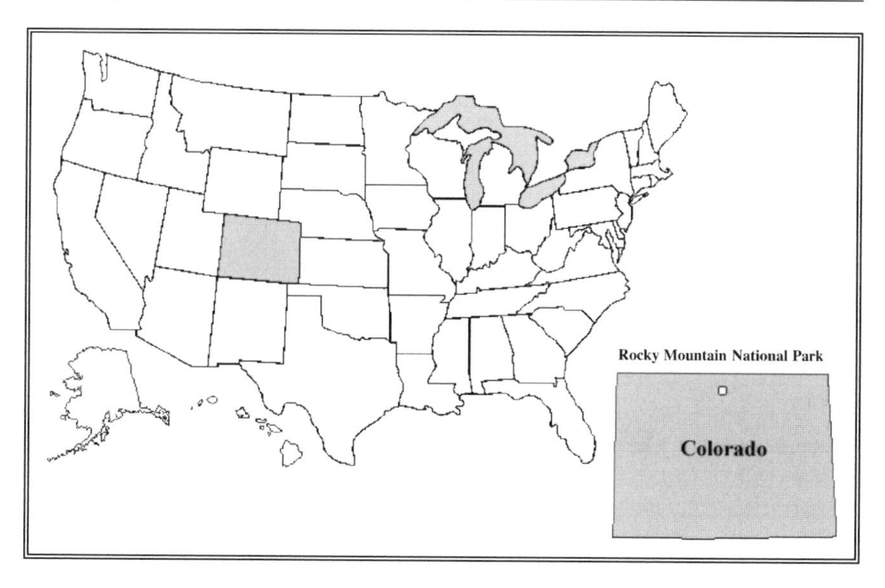

Figure 11.1 Rocky Mountain National Park, Colorado, USA

tation, and snow depth for the year 2020. Population dynamics models were used to estimate the impact of climate on park resources, including wildlife and vegetation composition. Data from the two climate forecasts and the resource estimates were configured as climate scenarios for the survey; four other hypothetical scenarios were created in order to incorporate a wider range of hypothetical climate variation. In total, four survey versions were developed, each with a 'typical day' (baseline) scenario and two hypothetical scenarios. Direct (weather-related) and indirect (resource-related, e.g. elk, ptarmigan, percentage of alpine tundra acreage) climate scenario variables were depicted graphically using icons that proportionally represented hypothetical changes in order to give a more descriptive presentation of climate scenarios (see excerpt of a portion of the survey in Figure 11.2).

The contingent visitation questions asked the respondent to consider if their *number of visits* and *length of stay* would have been affected under the hypothetical climate scenarios; and if so, how many more (or fewer) trips or days would they have visited. The contingent behavior questions are provided as an excerpt from the survey instrument in Figure 11.3 below. Information about travel costs and demographic characteristics was elicited in order to control for their potential effects. The survey was tested in two focus groups for content, clarity, and length, and the design was shortened and refined according to the focus group

	Typical day	Scenario 1	Scenario 2
Temperature Number of days with summer high temperature greater than 80°F	3 days	15 days	20 days
Precipitation Number of summer days with precipitation above 0.25 inches	18 days	15 days	28 days
Elk Each elk symbol represents about 200 elk	1040 elk	1500 elk	600 elk

Figure 11.2 Climate scenarios excerpt from RMNP visitor survey

How would these changes in conditions affect your visitation?

Question:	Scenario 1	Scenario 2
If at the beginning of the year, you knew Rocky Mountain National Park weather and conditions would be as described in Scenarios 1 and 2 rather than the current scenario, would you:	__Visit ***more*** often? No. of additional yearly trips ____ __Visit ***less*** often? No. of fewer yearly trips ____ __ **No change** in no. of trips	__Visit ***more*** often? No. of additional yearly trips ____ __Visit ***less*** often? No. of fewer yearly trips ____ __ **No change** in no. of trips
Would the changes in weather and resources described in Scenarios 1 and 2 affect your length of stay in Rocky Mountain National Park on a typical trip?	Would you stay ___ **Longer**? ____ days longer ___ **Shorter**?____ days fewer ___ **No change**?	Would you stay ___ **Longer**? ____ days longer ___ **Shorter**? ____ days fewer ___ **No change**?

Figure 11.3 Contingent visitation questions

suggestions. The final version was pre-tested with RMNP visitors before distribution during the survey period.

In order to calculate willingness to pay, respondents were also asked if they would have made their trip if travel costs had been higher. Bid payment amounts were randomly chosen, and respondents were asked the following dichotomous-choice contingent valuation question:

> As you know, some of the costs of travel such as gasoline have been increasing. If the travel cost of this most recent visit to Rocky Mountain National Park had been $___ higher, would you have made this visit?
>
> Circle one: YES NO

Bid amounts in the survey ranged from $1 to $495, and this range was chosen based on other recent surveys of willingness to pay for recreation.

During the survey period (June 21–September 12, 2001), visitors were selected randomly in frequently visited areas of RMNP at five specific locations that were selected in order to identify visitors in an array of locations who had been hiking, sightseeing, driving, or engaged in other activities. Survey dates were selected in order to obtain samples from weekdays, weekends, and holidays. On selected sampling dates, visitors were approached randomly at the sampling sites, and surveys were distributed to willing respondents, who took the questionnaire with them to be completed and mailed in at a later date. Mail-returned surveys were chosen because of the complexity of the climate scenarios and the amount of time required to complete the questionnaire. There were 1378 attempts to distribute surveys during the survey period, and 112 were refused. Thus, a total of 1266 surveys were distributed. Following Dillman's Total Design Method (Bailey, 1994), reminder postcards were mailed to survey recipients one week after the day of distribution, and supplementary copies of the survey were mailed three weeks later to non-respondents along with a cover letter. At the end of the survey collection period, 967 surveys were returned, which amounts to a 70% response rate (or a 76% response rate, net of refusals).

Daily Weather Data

In order to test for the effects of actual climate conditions (the day of visit) on willingness to pay, daily weather data were collected from the National Park Service's Weather Information Management System, and included daily observations of the following variables for the entire sample period, provided below in Table 11.1.

The variable *SOW* is a representation of the daily 'state of weather,' which is recorded at 1:00 p.m. and coded as 0–9 (Table 11.2). For the sampling period, there were no observations for the *SOW* variable codes 4, 5, 7, or 8, so these variables were dropped. Dummy variables were created for each of the observations coded 3, 6, and 9 (*OVERCAST*, *RAIN*, and *TSTORM* respectively) for purposes of the statistical analysis. The value of each of these dummy variables is equal to 1 when its respective *SOW* code was recorded as an observation, and equal to 0 when any other code was recorded.

Table 11.1 Daily weather data for RMNP (June–September, 2001)

Variable name	*Definition*	*Mean value*	*Mini-mum*	*Maxi-mum*
SOW	State of weather (see Table 11.2)	n.a.	0	7
TEMP	Temperature (°F) (at 1:00 p.m.)	74.2	39.0	88.0
WS	Wind speed (mph) (at 1:00 p.m.)	5.2	0.0	12.0
PPTAMT	Daily precipitation amount (inches)	0.0	0.0	0.4

Table 11.2 Definitions of state of weather (*SOW*) variables

Variable	*Explanation*
0	Clear (< 10% clouds)
1	Scattered clouds (10–50% clouds)
2	Broken (60–90% clouds)
3	Overcast (> 90% clouds)
6	Raining
9	Thunderstorm in progress (lightning seen or heard)

Data Analysis

Since each survey included contingent visitation questions for two climate scenarios, responses were restructured in such a way that each survey response represents two responses to climate scenarios, thereby doubling the number of observations in the sample. Therefore, although 967 surveys were returned, the number of contingent visitation observations in the sample is 1934.

Statistical analysis is used to test the null hypothesis that climate scenario variables have no effect on visitation (quantity of trips). Ordinary least squares regression is used to estimate a trip response model (change in number of trips) as a function of climate scenario variables, travel cost, and demographic variables; *t*-tests are used to examine the statistical significance of individual variables and of various climate scenarios.

Survey Results

The survey data revealed that most visitors planned their trips well in advance (68 days, on average) and over 66% of respondents indicated that their most recent trip to RMNP was either the 'sole destination' or 'primary purpose' of the trip. More than 70% of respondents indicated that the activities of viewing conifer forests, viewing wild flowers, and driving for pleasure were either 'important' or 'very important' to their decisions to visit RMNP. The average distance traveled to the Park was 643 miles, the average length of stay was more than three days, and over 60% of respondents were from outside of Colorado. These results suggest that summer vacations and the opportunity to view the alpine scenery of RMNP were the main factors in the visitation decision.

We applied a multivariate test involving a qualitative response model that simply distinguishes visitors who would change their behavior (contingent upon the climate scenarios) and those who would not. A binary probit regression analysis on whether or not survey respondents would change their visitation behavior under the hypothetical climate scenarios revealed the following results, presented in Table 11.3. The dependent variable in the binary probit regression is the binary outcome of the contingent behavior question regarding changes to the respondent's visitation behavior. The binary variable is equal to 1 if the

Table 11.3 Binary probit regression results for contingent visitation analysis

Variable	*Coefficient*
Change – number of days with high temperature > 80°F	0.0148 ***
Change – number of days with precipitation > 0.25 inches	–0.0190 ***
Change – number of elk	0.0001 *
Change – percentage of RMNP acres of alpine tundra	0.0254 ***
Distance traveled (in miles)	–0.0004 ***
Gender (1 if male, 0 if female)	0.1446 *
Age (in years)	–0.0128 ***
Retired (1 = yes, 0 = no)	0.2355 *
Member of environmental organization (1 = yes, 0 = no)	0.1229
Education (in years)	–0.0219

Notes: * – significant @ 90%; *** – significant @ 99%; McFadden $R^2 = 0.08$.

respondent indicated that he/she would visit RMNP 'more often' or 'less often' or if he/she would stay 'longer' or 'shorter' (contingent upon the two climate scenarios) and equal to 0 if the respondent indicated 'no change' in the number of trips or in the length of stay. The probit results indicate that the variables representing changes in temperature, precipitation, and the composition of vegetation represented by alpine tundra were significant determinants of the probability of a behavioral change at a level of 99%; the variable representing changes in the elk population was significant at a level of 90%.

Two of the hypothetical climate scenarios were developed using global circulation models. The CCC scenario was included in survey versions A and D, and 8.6% of the 442 respondents to those surveys indicated that their visitation behavior would change under the hypothetical climate scenario. The application of their responses to total RMNP visitation data yields a mean estimate of 1,357,888 *additional* visitor days (Table 11.4).

The Hadley climate scenario was included in survey version B, and 11.1% of the 252 respondents to that survey indicated that their behavior would change under the hypothetical climate scenario. The application of their responses to total visitation data yields a mean estimate of 1,002,080 *additional* visitor days (Table 11.5).

Table 11.4 Survey results – CCC climate scenario

CCC scenario (n = 442)	*Change number of trips*	*Change length of stay*
Number of respondents who would change their visitation behavior	38	51
% of respondents who would change their visitation behavior	8.6%	11.5%
Average additional trips per visitor	+0.10 trips per visitor	+0.13 days per trip
Total Visitation – 1999	3,186,323	
Projected new visitation	3,618,856	
Change in visitation (%)	13.57%	
Change in visitation (number)	432,533	
Average length of stay (days)	3.04	
Mean change in annual visitor days	1,357,588	

Table 11.5 Survey results – Hadley climate scenario

Results: Hadley scenario (n = 252)	*Change number of trips*	*Change length of stay*
Number of respondents who would change their visitation behavior	28	34
% of respondents who would change their visitation behavior	11.1%	13.5%
Average additional trips per visitor	+0.10 trips per visitor	+0.13 days per trip
Total visitation – 1999	3,186,323	
Projected new visitation	3,502,426	
Change in visitation (%)	9.92%	
Change in visitation (number)	316,103	
Average length of stay (days)	3.04	
Mean change in annual visitor days	1,002,080	

Willingness to Pay Results

A binary logit analysis of the responses to the dichotomous choice CVM question of willingness to pay was performed according to the theoretical model presented in Equation 3. The binary logit model results are presented below. Note that the dependent variable in this case is *YPAY*, which is equal to 1 if the respondent indicated that they would pay the bid amount (YES) and 0 if they indicated they would not pay (NO). Note that *t*-statistics are presented in parentheses below their respective coefficients.

$$YPAY = \quad (5)$$

$$\underset{(-1.961)}{-2.68} - \underset{(-6.281)}{0.007BIDAMT} + \underset{(1.688)}{0.028TEMP} + \underset{(2.408)}{4.447PPTAMT} + \underset{(1.950)}{0.411PICNIC}$$

$$+ \underset{(1.682)}{0.371DTRROAD} + \underset{(6.439)}{0.002DIST} - \underset{(-4.390)}{2.62\text{E-}07DISTSQ} + \underset{(3.817)}{9.04\text{E-}06INC}$$

Explanatory variables are *BIDAMT* (bid amount), *TEMP* (temperature [°F] at 1:00 p.m.), *PPTAMT* (daily precipitation amount in inches), *PICNIC* (participated in picnic activities during visit), *DTRROAD* (participated in driving over Trail Ridge Road), *DIST* (one-way travel distance to RMNP), *DISTSQ* (square of DIST variable), and *INC* (household income).

Insignificant variables were eliminated in order to estimate a logit equation with statistically significant coefficients that could be meaningfully reparameterized into a WTP function. This logit equation permits the acceptance of the hypotheses that the coefficients on some of the hypothesized weather variables (e.g. wind speed, cloud cover) are equal to zero. However, we reject the hypothesis that the coefficients on temperature (*TEMP*) and precipitation (*PPTAMT*) are zero, along with that for the coefficient on certain recreation activities (*PICNIC, DTRROAD*), travel cost (represented here as *DIST* and *DISTSQ*), and demographic (*INC*) variables. The coefficient estimates on all of the independent variables are significant above the 90% level.

According to Cameron's (1988) approach, it is possible to calculate an equation that directly relates willingness to pay to weather, activity, and demographic variables. The slope coefficients in Equation 5 are reparameterized by dividing the intercept and all coefficients (other than that on the bid amount) by the coefficient on the absolute value of the bid amount. This conversion for the logit function generates the following equation:

$$WTP = -411.95 + 4.37TEMP + 683.60PPTAMT + 63.18PICNIC \quad (6)$$
$$+ 57.01DTRROAD + 0.30DIST - 4.03E\text{-}05DISTSQ + 0.0014INC$$

The specification in Equation 6 allows that parameters be interpreted in the same manner as ordinary least squares results; a one degree increase in temperature is associated with a increase in willingness to pay of $4.37. Individuals driving over Trail Ridge Road are willing to pay $57.01 more than those that did not. An increase in a visitor's income of $1,000 can be associated with an increase in WTP of $1.40. We expect that the reason for the high coefficient estimate on precipitation is related to the coincidence of the late summer monsoon season (which brings greater levels of rain to the region) and school vacation summer months.

Mean WTP is calculated using the mean values for each of the explanatory variables and is estimated to be $314.95 per trip. Mean travel costs of survey respondents totaled $686.69 per trip. These data, along with average annual visitation, allow for the estimation of a demand curve for recreation at RMNP. Recreation benefits (mean WTP) are represented by the shaded area above the price but below the demand curve in Figure 11.4. Based on survey results that indicated an average group size of 4.3 persons and an average length of stay of three days, we estimate the net WTP per person, per day to be $24.47. This value is within the range of past benefit estimates for hiking and similar recreation activities (Loomis & Walsh, 1997: 187).

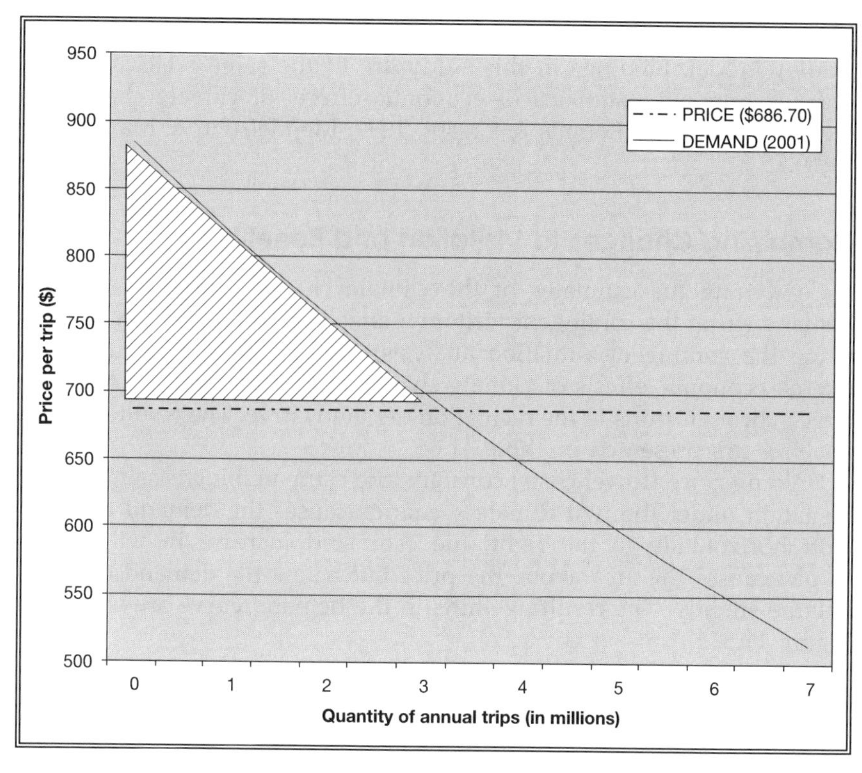

Figure 11.4 Demand curve for recreation at Rocky Mountain National Park, USA

Estimates of Recreation Values with Climate Change Scenarios

Results from two global circulation models are used to estimate potential climate changes for the RMNP area. Both of the scenarios developed by the two models use a baseline time period of 1961 to 1990 for the assessment. The CCC model scenario for 2020 tends to be more than 4°F warmer than the historical baseline period, and predicts a drier overall climate. The Hadley model scenario predicts 2°F warmer and tends to estimate a wetter winter or off-peak months, and drier summer or peak months. In order to estimate the effect of climate change on mean WTP, the 2020 temperature and precipitation forecasts from the CCC and Hadley global circulation models are substituted for mean temperature and precipitation amounts. The results indicate that mean WTP would increase 6.7% to $336.05 per trip under the temperature and precipitation forecast presented by the CCC model; mean WTP would increase

4.9% to $330.38 per trip under the climate forecast presented by the Hadley model. Findings in this study are of the same relative magnitude as previous estimates of economic effects of climate change on recreation benefits (Loomis & Crespi, 1999; Mendelsohn & Markowski, 1999).

Combining Changes to Visitation and Benefits

Combining the estimates of the climate change effects on tourism demand (from the contingent visitation analysis) and recreation benefits (from the contingent valuation analysis), we are able to calculate the overall economic effects of climate change at this site and illustrate the effects through shifts in the recreation demand curve. The results of both analyses are presented in Table 11.6.

Holding price (travel costs) constant, the estimated increase in annual visitation under the two climate scenarios causes the demand curve to shift horizontally to the right; the estimated increase in willingness to pay causes the area above the price but below the demand curve to enlarge slightly. The resulting shifts in the demand curve are shown in Figure 11.5.

Conclusion

Contingent visitation analysis and the contingent valuation method were used to measure the total economic effects of climate change on recreation and tourism in an alpine national park in the United States. The objective of the analysis was to estimate the impact of changes in temperature, precipitation, snow depth, and park resources on the frequency and duration of future visits and the quality of the recreation experience. Visitation behavior was modeled using contingent visitation

Table 11.6 Summary of results

Scenario	*Average annual visitation*	*Price (travel costs/trip)*	*Recreation benefits (WTP/trip)*	*Total expenditures ($ millions)*	*Total recreation benefits ($ millions)*
Current conditions	3,186,323	$686.69	$314.95	$2188	$1004
CCC (2020)	3,618,856	$686.69	$336.05	$2485	$1216
Hadley (2020)	3,502,426	$686.69	$330.38	$2405	$1157

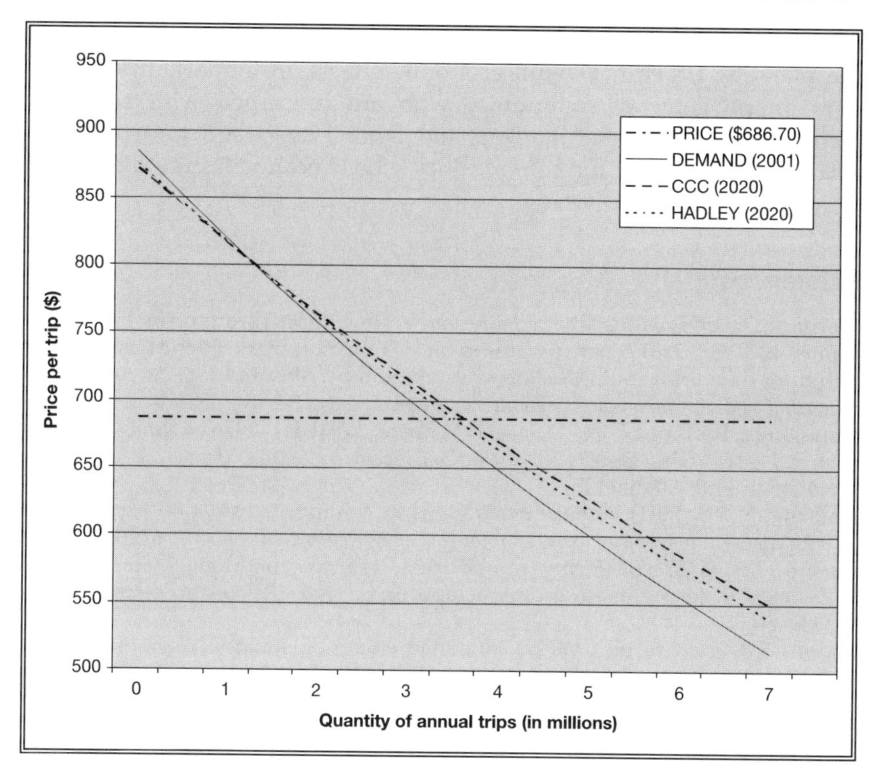

Figure 11.5 Shifts in demand curve associated with climate change scenarios

analysis, where the number of annual visits was analyzed as a function of various climate and natural resource variables. Recreation benefit (willingness to pay) was used as a measure of the quality of the recreation experience, and was analyzed as a function of weather variables on the day of the visit, as well as travel costs and demographic variables. Variables representing both temperature and precipitation were found to be statistically significant determinants of both visitation behavior and recreation benefits. Estimated increases in both the number of annual trips and recreation benefits were illustrated by shifts in the demand curve for recreation.

The framework presented in this chapter for estimating the effects of climate change on both recreation demand and benefits allows recreation planners and resource managers to develop strategies for adapting to future climate change. Increases in visitation have implications for park infrastructure development, environmental impact assessment, and visitor programs and services. Estimated changes in recreation benefits have implications for the efficient allocation of recreation resources in the planning process. Overall economic effects are important for planners in the gateway community who are concerned with economic development, employee housing and labor issues, and transportation planning – especially for communities whose economies are dependent on park recreation visitors.

References

Bailey, K.D. (1994) *Methods of Social Research*. New York, Toronto: The Free Press.

Cameron, T.A. (1988) A new paradigm for valuing non-market goods using referendum data: Maximum likelihood estimation by censored logistic regression. *Journal of Environmental Economics and Management* 15 (3), 355–79.

Cummings, R., Brookshire, D.S., and Schulze, W. (eds) (1986) *Valuing Environmental Goods: An Assessment of the Contingent Valuation Method*. Totowa, NJ: Rowman and Allanheld.

Hanemann, M. (1984) Welfare evaluations in contingent valuation experiments with discrete responses. *American Journal of Agricultural Economics* 67 (3), 332–41.

Loomis, J.B. (1987) Balancing public trust resources of Mono Lake and Los Angeles' water right: An economic approach. *Water Resources Research* 23 (8), 1449–56.

Loomis, J.B. and Crespi, J. (1999) Estimated effects of climate change on selected outdoor recreation activities in the United States. In R. Mendelsohn and J.E. Neumann (eds) *The Impact of Climate Change on the United States Economy* (pp. 289–314). Cambridge: Cambridge University Press.

Loomis, J.B. and Walsh, R.G. (1997) *Recreation Economic Decisions: Comparing Benefits and Costs* (2nd edn). State College: Venture Publishing, Inc.

Mendelsohn, R. and Markowski, M. (1999) The impact of climate change on outdoor recreation. In R. Mendelsohn and J.E. Neumann (eds) *The Impact of Climate Change on the United States Economy* (pp. 267–88). Cambridge: Cambridge University Press.

Whitehead, J.C., Haab, T.C., and Huang, J. (2000) Measuring recreation benefits of quality improvements with revealed and stated behavior data. *Resource and Energy Economics* 22, 339–54.

Chapter 12
Implications of Climate Change on Tourism in Oceania

STEPHEN CRAIG-SMITH AND LISA RUHANEN

The natural environment and climate conditions are very important determinants of the attractiveness of the Oceania region as a holiday destination. The increasing trend towards warmer temperatures could have major consequences for the region's tourism industry, which is heavily dependent on existing climatic and environmental conditions. Impacts likely to affect tourism including such things as coral bleaching, outbreaks of fire, changed migration and breeding patterns of animals and birds, flooding, the spread of vector-borne diseases, shorter skiing seasons, water supply restrictions and urban smog (Agnew & Viner, 2001). Future temperature increases will also be coupled with large increases in atmospheric concentrations and significant changes in rainfall patterns. Rainfall patterns are expected to be more variable from month to month, bringing greater dry-season severity and increased frequency and extremity of events such as storms, cyclones, floods, fires and droughts (Giles & Perry, 1998; Walsh & Ryan, 2000; Williams *et al.*, 2003).

While it is anticipated that most nations will suffer adverse consequences from climate change, small island states, so significant in Oceania, may face the most dire and immediate consequences. Given the massive expanse of the Pacific island region and substantial topographical variations, it is likely that climate change will result in markedly different manifestations across the region. Climate models of Australia indicate that the temperature over the summer months (December to February) may rise by over 1.5 degrees C by 2020, and 3 to 4 degrees C by 2050. Australia will also experience a reduction in cloud cover of 10–15% and a corresponding reduction in seasonal rainfall of 10–20% (Agnew & Viner, 2001). The entire 70,000km of Australian open coast, including the 12,000 islands and extensive estuarine and wetland areas, could be inundated by sea level rise. In addition, storm surges and tropical cyclones continue to pose a large threat across North Australia. Australia's Great Barrier Reef, which alone generates A$1.5 billion from tourism, is predicted to be severely damaged by an increase in sea temperature, seriously reducing the attractiveness of the area as a tourist destination (Viner & Agnew, 1999). Sea level rises in Oceania are likely to increase, coupled with other issues such as increasing temperatures and predictions that the occurrence of tropical typhoons and cyclones could increase by as much as 50–60% and their intensity by 10–20% (Burns, 2000). This does not paint a rosy picture for the islands of Oceania. Already the island of Tuvalu and some islands of Papua New Guinea in the Pacific are experiencing storm over wash and shrinkage of their land area by 20cm per year (Ghina, 2003).

The Climate Institute (2003) has identified a number of climate change effects that are of particular concern to the Oceania region, including:

- *Sea Level Rise* – many of the small island states and territories in the Pacific are no more than a few metres above sea level, therefore, even a small rise in sea level could claim a large percentage of land.
- *El Nino-Southern Oscillation* (ENSO) – the phenomenon will become more frequent and more intense under global warming, which may result in harsher droughts (particularly in Australia) and more violent floods.
- *Water Resources* – the scarcity of resources in interior Australia and the more numerous or intense droughts will exacerbate the existing shortage. Also rising sea levels would contaminate the fresh water supplies for some small islands.
- *Tourism* – the economies of many small island states and territories in the South Pacific are tourism dependent, and a loss of beach to erosion and sea level rises would cripple the islands' economies.
- *Evacuation and resettlement* – extreme sea level rises may result in some low-lying islands literally disappearing under water.

The three main concerns for Oceania of sea level rise, increasing temperatures and decreasing snow cover are discussed in more detail.

One of the most challenging impacts for the Oceania region will be the effects of sea level rise on small island states and coastal areas. To date two islands of Kiribati have already been submerged and the Marshall Islands are under threat (Agnew & Viner, 2001). Sea levels are expected to rise 25cm by the year 2025, and while this may not seem like much, forecasts suggest that it is of considerable concern (Giles & Perry, 1998; World Tourism Organization, 2003). This will potentially lead to a loss of recreational value and carrying capacity of beaches; a loss of property value resulting from decaying amenity value; loss of land value; deterioration of landscape and visual appreciation; and escalating costs for beach and property protection (Jackson, 2002), and will have considerable consequences for the provision of recreational opportunities (Wall, 1998).

Increasing temperatures are likely to have considerable impacts on traditional summer destinations, and the potential loss of summer trade may cripple some areas. If destinations become too hot they are likely to lose high numbers of visitors. Customers' health can be affected by longer heatwaves, which may impact significantly upon the elderly, the sick or children (Perry, 2000; Chapter 5, this volume). Therefore, tourists will increasingly expect holiday accommodation to be air-conditioned. Such accommodation will attract a premium price, while poorer quality self-catering apartments and rooms without air-conditioning will be much less attractive in the summer (Perry, 2000).

Coupled with expected increases in temperature are numerous downstream effects, including an increase in diseases such as malaria and dengue fever for both tropical and temperate Australia, as well as the islands of the Pacific (Agnew & Viner, 2001; Bode *et al.*, 2003). Another potential health risk is the increased likelihood of developing skin cancer, as decreasing cloud cover increases exposure to harmful ultraviolet radiation (Agnew & Viner, 2001). Australia is already known as the skin cancer capital of the world.

Other likely impacts from climate change include coral bleaching, which can be induced by a warming of only 1 or 2 degrees C, as has been evidenced on the Great Barrier Reef and at reefs in the Maldives. Agnew and Viner (2001) cite expectations that in the next 20–40 years Australia's Great Barrier Reef will be severely damaged by increasing sea temperatures, which among other things will considerably reduce its tourist appeal (Agnew & Viner, 2001).

While increasing temperatures may have negative impacts for many tourism destinations, positive benefits may accrue for other destinations (Giles & Perry, 1998; Harrison *et al.*, 1999; Perry, 2000, Chapter 5, this volume). Australia, New Zealand and many of the islands of Oceania may face similar problems to the Mediterranean, as they become too hot for international visitors in the summer months. Australia's tropical destinations such as North Queensland and the Northern Territory may also become less attractive to domestic visitors in the summer months. This could further exacerbate the popularity of the regions' 'high season', which occurs in the more temperate winter months, and stress the carrying capacity of these areas during this period. Alternatively, Australia's 'hot' destinations such as Queensland may lose market share to the more temperate regions of New South Wales and Victoria.

Within the climate change and tourism literature much attention has been given to the impacts of global warming on the skiing industry, where research shows that there are likely to be severe consequences in the event of climate change (Konig & Abegg, 1997). Such changes are increasingly becoming evident in ski resorts around the world with reduced snow cover and shorter seasons. This has been seen in traditional ski resorts in the Swiss Alps (Bürki *et al.*, Chapter 10, this volume).

Although the Oceania region is not globally renowned for its winter sports, the South Island of New Zealand and the New South Wales and Victorian Alps in Australia do have a substantial winter sports industry and these destinations will be affected by climate change due to decreasing snowfall. Konig and Abegg (1997) suggest that although some regions may be able to maintain their winter tourism with suitable adaptation strategies, others would lose their winter tourism industry due to a rising snow line. Low altitude sites are particularly vulnerable due to a greater variability in snow cover (Behringer *et al.*, 2000; Agnew & Viner, 2001). Behringer *et al.* (2000) have suggested that ski resorts in the lower regions of the Swiss Alps, for instance, should plan to cancel ski tourism altogether.

Climate change will also have negative effects on the attractiveness of many ski areas in middle altitudes: a diminished supply of ski runs because parts of the ski areas are no longer snow-reliable. It was also found that skiing close to the ski resorts will become more difficult particularly in the lower altitudes, and will likely lead to the building of resorts at higher locations. Agnew and Viner (2001) anticipate that higher altitude ski resorts will experience increasing pressure if their low-lying counterparts become less commercially viable. To address these impending problems, some have advocated building ski resorts at higher altitudes as an adaptation strategy, however, Konig and Abegg (1997) reject this due to the adverse environmental impacts that would ensue and exacerbate issues such as waste, water and rubbish. Instead they suggest diversification of activities outside conventional ski tourism and suggest such alternatives as theatre and music festivals, which have been successful in destinations, such as Arosa and Gstaad in Switzerland (Konig & Abegg, 1997).

The Australian Alps are facing considerable impacts due their generally lower altitudes. The Australian ski industry contributes in excess of $400 million dollars to the Victorian and New South Wales state economies and creates approximately 12,000 jobs for the winter season (Konig, 1998). However, the commercial viability of winter tourism depends on sufficient snow conditions, namely snow depth and duration and extent of snow cover (Konig, 1998). Currently the Australian resorts have a winter snow cover lasting from a few weeks at the lower elevation sites to up to four months at the higher elevation ski resorts. Global warming will at least shorten the ski season and at worst the average snow cover would be reduced by 66%, suggesting that there will be insufficient natural snow for viable commercial ski operations (Agnew & Viner, 2001). Konig's study found that a significant proportion of skiers (38%) would seek alternative destinations namely New

Zealand and Canada. However, while New Zealand is also facing challenges from reduced snow cover, climate change may positively affect the ski industry in New Zealand because skiing in Australia, which takes place at lower elevations than in New Zealand could be eliminated (Wall & Badke, 1994) thus giving New Zealand a unique position in the region.

Clearly, there are likely to be considerable changes in Oceania over the next few decades. To combat these changes the tourism industry has one of two possible forward strategies: either change the environment to accommodate current tourism activities (e.g. air-conditioning hotel rooms or creating artificial snow for instance) or change the tourism product on offer to cope with the new environmental conditions (abandon skiing and introduce walking tours instead). Air-conditioning, artificial snow creation and cyclone proof structures come at considerable economic cost and most, if not all these environmentally adaptive strategies go away from rather than towards tourism sustainability. Given the trend towards sustainable tourism development and authentic experiences, adapting the environment may not just be expensive, it may also be unacceptable.

Tourism Industry Adaptation Strategies

Climate change poses a very real threat to the global tourism industry. To address some of these issues the first international conference on climate change and tourism took place in Tunisia in April 2003: an initiative of the Tunisian government and the World Tourism Organization (WTO, 2003). It was recognized at the conference that within the current body of scientific analysis, research and prediction, there remain considerable uncertainties about the magnitude of the impact of a changing global climate – for example, the extent of rises in temperatures, changes in precipitation, and the extent and location of extreme events such as floods and droughts. While climate change predictions of necessity cover the long term, the tourism industry tends to have much shorter time horizons, even in the context of physical investment in infrastructure (WTO, 2003). It was felt that, in order to engage the attention of the tourism industry, this gap will need to be bridged, perhaps by emphasizing that climate change is already having an impact on the tourism sector. It was also recognized that in all incidences, the destination's sustainability remains a prime concern, and the conference emphasized that all stakeholders need to redouble their efforts to ensure sustainable solutions.

Many consequences of global climate change discussed in this chapter cannot be influenced by any one individual destination to any great extent. From the standpoint of the tourism industry, it is therefore important to recognize what is likely to happen and to lay plans to mitigate the adverse effects as far as possible (WTO, 2003). However, description, problem-framing and predictions, as well as developing policies must be undertaken under a high degree of uncertainty (Behringer *et al.*, 2000).

The Intergovernmental Panel on Climate Change (IPCC, 2001) identifies a number of issues relevant to Australia and New Zealand. They suggest that adaptation strategies will be needed in water management, land-use practices and policies, engineering standards for infrastructure, and health services. However adaptations will only be viable if they are compatible with the broader ecological and socioeconomic environment, have net social and economic benefits, and are taken up by stakeholders. Responses may be constrained by conflicting short-term and long-term planning horizons, as will be addressed below in relation to two key tourist/geographic regional types commonly found in Oceania, namely Oceania's islands, coasts and beaches and Oceania's mountain areas.

Oceania's Islands, Coasts and Beaches

The effects of climate change on Oceania's islands, coasts and beaches probably produce the most severe repercussions for the tourism industry. The islands of the Oceania region face numerous problems and therefore the need for adaptation to climate change has become increasingly urgent for these small island states. The WTO (2003) has identified small island states and coastal areas in the Pacific as being particularly vulnerable to sea level rise. Many of these island economies are heavily reliant on tourism and if a particular climate is what tourists are ultimately seeking then climate change may have significant consequences for these island economies (Maddison, 2001). The WTO (2003) identifies three main areas of concern:

- Sea level rises cause coast and beach erosion, inundation of flood plains, rising water tables, destruction of coastal eco-systems, salinisation of aquifers, and, at worst, the total submersion of islands or coastal plains.
- Warmer sea temperatures of 1–2 degrees C cause coral bleaching and dying. This leads to a breakdown in the reef protection surrounding most tropical coastlines and an amenity loss for divers and snorkellers. It is predicted that coral bleaching will increase in intensity and frequency to the extent that it will occur annually in the Pacific by 2040.
- Increasing storm frequency, especially in conjunction with rising sea levels, leads to damage to sea defences, protective mangrove swamps and shoreline buildings, and beach erosion. Storm surge also damages coral reefs.

These problems will directly impact upon the tourism industry. There is, therefore, a need for integration of appropriate risk-reduction strategies with other sectoral policy initiatives in areas such as sustainable development planning, disaster prevention and management, integrated coastal zone management, and health care planning. The WTO (2003) find that it is commonly accepted that strategies for adaptation to sea level rises fall into three main categories – delineated as 'protect', 'accommodate' or 'retreat' – and that all need to be considered in the broader context of coastal management in Oceania. The WTO (2003) identifies a number of potential strategies:

- The building of sea wall defences and breakwaters to protect the coast and its hinterland. This has been the traditional response, and may be the only practical option, but it has sometimes been found to create as many problems as it solves and can ultimately destroy a location's natural beauty.
- Enhancement and preservation of natural defences (such as the replanting of mangrove swamps or raising the land level of low-lying islands).
- Adapting to the changed conditions by building tourism infrastructure and resorts further back from the coast (Perry, 2000).
- Importing sand to beaches in order to maintain their amenity value. Beach renourishment, however, may be costly and temporary and may damage the area from which the sand is drawn.
- Impose new building regulations to introduce other types of building materials.
- When dealing with coral bleaching and the death of reefs, alternative man-made protection is unlikely to succeed; ultimately, abandonment may be the only option.

Unfortunately, the IPCC finds that the adaptive capacity of small islands is low because of the physical size of the nations, limited access to capital and technology, shortage of human resource skills, lack of tenure security, overcrowding, and limited access to resources for construction. For many small islands to adapt they will require external financial, technical, and other assistance. It has been suggested that adaptive capacity may be enhanced by regional cooperation and pooling of limited resources, although this is still going to place considerable demands on the local tourism industry.

Oceania's Mountains

Oceania's winter tourism industry is facing considerable challenges. Climate change is affecting mountain regions all around the world and threats to winter sports are already manifesting themselves in Scotland, Switzerland and Austria (WTO, 2003). However, as mountain summers become warmer and drier, opportunities arise for extending the non-ski market. Konig (1999) claims that only those resorts that diversify into all-year tourism will be able to maintain a financially viable tourism industry. Globally, the snow line is receding due to warmer winters (for every 1 degree C increase, the snow line recedes by 150m), and the ski season is becoming shorter (WTO, 2003). These problems signal that urgent adaptation measures are necessary. The WTO's (2003) suggested strategies include:

- increased use of artificial snow can help to extend and supplement natural snow cover as temperatures rise;
- high altitude resorts are likely to become more popular and may have to adapt to greater demand;
- lower altitude resorts, with reduced snow cover, may need to introduce an increased range of alternative attractions to skiing during the winter season;
- as a result of less stable (wetter) snow, greater avalanche protection will be required;
- resorts that are no longer within the reliable winter snow belt may need to reinvent themselves and address alternative markets; and
- as mountain summers become drier and warmer, the summer tourist season may be extended into the shoulder months. Changing demographic patterns, particularly an ageing population with more leisure time, may prove beneficial to attracting this market.

Although urgent action is needed in the winter tourism industry, particularly for resorts in the Australian Alps, which are facing considerable impacts due their generally lower altitudes, Konig's (1999) study of Australia snow resorts found that the majority of resort managers surveyed (nine out of ten) considered the possibility of future climate change due to enhanced greenhouse effect as not important in operating and planning the resort. The surveyed managers claimed that the lack of concern was attributed to: the resorts being concerned with short-term profits; the time-frame of climate scenarios are too long to be of importance for the resorts as they need to know what will happen in the next three to five years; and that the science of future climate change is not good enough and it is not yet proven that global temperature will increase. The managers in Konig's study argue that even if there will be less snow due to climate change, technology may improve so that resorts will be able to produce snow at a much higher temperatures than today. In marked contrast to the Australian situation, Elsasser and Bürki (2002: 255) find that the Swiss Alpine oprerators are not 'sitting back idly contemplating the consequences of climate change'.

General Adaptation Measures for the Tourism Industry

The impending problems from climate change raise a number of more general adaptation measures that must be considered, particularly in terms of sustainability of resources. The WTO (2003) suggests such measures may include:

- Traditional designs may have to be encouraged to deal with alternative methods of cooling buildings in increasingly hot climates to counteract rising energy costs but there is a limit to how this can be achieved to the satisfaction of tourists in extreme temperatures.
- Physical planning issues will require building lines to be moved back from eroding coasts but there is a strong preference for tourists to be able to see the ocean just outside their bedroom windows especially on oceanic islands.
- Coastal infrastructure, such as drainage, waste disposal, electricity, water supply, railways and roads may also have to be moved back from eroding coastal areas.
- Increased insurance costs will have to be factored into resort profitability.

Conclusions

Although the impacts of climate change are not yet known with any certainty, current climatic conditions and scientific predictions suggest that climate change will become a significant issue. Tourism's dependence on stable climatic conditions suggests that it is likely to be affected by climate change sooner than other industries (Bode *et al.*, 2003). The WTO (2003) finds that the effects of changing climatic conditions on the global tourism industry will be influenced at local and sub-regional levels by factors such as:

- the impact of individual climatic characteristics on local destinations;
- the physical environment;
- topographical characteristics;
- local geological factors;
- changing local health risks as a result of climate change;
- the nature of the tourism markets being served; and
- the types of tourism facility and attractions offered.

Oceania faces some very real issues concerning climate change but, unfortunately, many of the small island nations of the region are not in a particularly strong position to counteract many of the likely environmental problems because of their size and level of economic development. Issues of sustainability will need to be addressed as there is a serious possibility that adaptation to the changing climate will proceed with only economic benefits in mind, and short-term economic gains may occur at the expense of increased environmental deterioration (IPCC, 2001). There is a very clear need for the larger countries of the region such as Australia and New Zealand to lead by example and render expertise and financial assistance where needed to the smaller island nations.

References

Agnew, M.D. and Viner, D. (2001) Potential impacts of climate change on international tourism. *Tourism and Hospitality Research* 3 (1), 37–60.

Behringer, J., Bürki, R. and Fuhrer, J. (2000) Participatory integrated assessment of adaptation to climate change in Alpine tourism and mountain agriculture. *Integrated Assessment* 1, 331–8.

Bode, S., Hapke, J. and Zisler, S. (2003) Need and options for a regenerative supply in holiday facilities. *Tourism Management* 24, 257–66.

Burns, W.C.G. (2000) *The Possible Impacts of Climate Change on Pacific Island State Ecosystems.* Occasional paper of the Pacific Institute for Studies in Development, Environment and Security, California.

Climate Institute (2003) *Regional Effects of Climate Change.* Available at: http://www.climate.org/topics/climate/impacts_oc.sthml (accessed 3 November 2003).

Elsasser, H. and Bürki, R. (2002) Climate change as a threat to tourism in the Alps. *Climate Research* 20, 253–7.

Ghina, F. (2003) Sustainable development in small island developing states: The case of the Maldives. *Environment, Development and Sustainabilty* 5, 139–65

Giles, A.R. and Perry, A.H. (1998) The use of a temporal analogue to investigate the possible impact of projected global warming on the UK tourist industry. *Tourism Management* 19 (1), 75–80.

Harrison, S.J., Winterbottom, S.J. and Sheppard, C. (1999) The potential effects of climate change on the Scottish tourist industry. *Tourism Management* 20, 203–11.

IPCC (Intergovernmental Panel on Climate Change) (2001) *Climate Change: Impacts, Adaptation and Vulnerability.* Switzerland: World Meteorological Organization.

Jackson, I. (2002) *Potential Impact of Climate Change on Tourism.* Issues paper prepared for the Organization of American States – Mainstreaming Adaptation to Climate Change (MACC) Project. Available at: http://www.oas.org/macc/Docs/TourismIssues.doc (accessed 3 November 2003).

Konig, U. (1998) Climate change and tourism: Investigation into the decision-making process of skiers in Australian ski fields. *Pacific Tourism Review* 2, 83–90.

Konig, U. (1999) Climate change and snow tourism in Australia. *Geographica Helvetica* 54 (3), 147–57.

Konig, U. and Abegg, B. (1997) Impacts of climate change on winter tourism in the Swiss Alps. *Journal of Sustainable Tourism* 5 (1), 46–58.

Maddison, D. (2001) In search of warmer climates? The impact of climate change on flows of British tourists. *Climatic Change* 49, 193–208.

Perry, A. (2000) *Impacts of Climate Change on Tourism in the Mediterranean: Adaptive Responses.* Available at: http://www.feem.it/NR/rdonlyres/ (accessed 3 November 2003).

Viner, D. and Agnew, M. (1999) Climate change and its impacts on tourism. Report prepared for the World Wildlife Fund, United Kingdom.

Wall, G. (1998) Implications of global climate change for tourism and recreation in wetland areas. *Climatic Change* 70, 371–89.
Wall, G. and Badke, C. (1994) Tourism and climate change: An international perspective. *Journal of Sustainable Tourism* 2 (4), 193–203.
Walsh, K.J.E. and Ryan, B.F. (2000) Tropical cyclone intensity near Australia as a result of climate change. *Journal of Climate* 13, 3029–36.
Williams, S.E., Bolitho, E.E. and Fox, S. (2003) Climate change in Australian tropical rainforests: An impending catastrophe. *Proceedings of the Royal Society of London* 270, 1887–92.
World Tourism Organization (2003) *Climate Change and Tourism.* Proceedings of the 1st International Conference on Climate Change and Tourism, Djerba, Tunisia.

Chapter 13

Tourism, Fossil Fuel Consumption and the Impact on the Global Climate

SUSANNE BECKEN AND DAVID G. SIMMONS

Introduction

In the last few years an increasing body of literature on tourism and energy use has emerged, in which tourism has been described as fossil fuel-dependent industry and a large emitter of greenhouse gases (Becken, 2002a; Dubois, 2003; Gössling, 2000, 2002; Peeters, 2003). Tourism's role in the context of greenhouse gas emissions will increase given that the industry is growing worldwide, and that long-distance travel is increasing in popularity. The diversification from traditional mass tourism into an array of special interest-, nature- and activity-based tourism segments also potentially increases the demand for fossil fuels, and therefore greenhouse gas emissions (Becken *et al.*, 2003). Against this backdrop, countries that capitalise on tourism and at the same time commit to international goals, such as reducing anthropogenic emissions, are challenged to examine carefully costs associated with further tourism development. A first step here is to determine tourism's contribution to a country's consumption of fossil fuels and its greenhouse gas emissions, and also to determine main drivers of this energy use. This chapter discusses the energy use of, and greenhouse gas emissions from, tourism in New Zealand, and identifies methodological and practical issues that are of interest to other countries seeking to undertake an energy inventory of tourism. The involvement of stakeholders in decreasing tourism's dependency on the input of fossil fuels is also outlined.

Factors Determining Tourism Energy Use

The energy use of tourism can be broken down into four major components: (1) travel to the destination, (2) travel at the destination, (3) accommodation, and (4) activities and attractions. Other components, for example restaurants and retailers, also make a contribution to energy use, but have been found to be comparatively minor (Patterson & McDonald, 2002). Different tourist activities can be described by the concept of *energy efficiency*, which means to reduce energy use for a 'given energy service or level of activity. This reduction in the energy consumption is not necessarily associated to technical changes, since it can also result from a better organisation and management' (World Energy Council, no date).

Worldwide, transport is the largest growth sector in terms of energy demand (2.1% per year) (International Energy Agency, 2002), and while the contribution of tourism to transport globally is unknown (Gössling (2002) estimated a minimum energy use of 13,000PJ per annum for global tourism transport), it is likely to constitute an important part of this sector, especially for major tourist destinations. The fundamental importance of transport to tourism and the availability of data explain why it is the one component of tourism supply that has been researched most extensively, compared with other components (Ceron & Dubois, 2003; Høyer, 2000; Müller & Mezzasalma, 1992).

Much tourist travel to destinations is undertaken by private car, but air travel's share has increased considerably in the last decades, with 40% of all international arrivals having been by air in 2000 (Kester, 2002). This trend is adverse given the high energy intensity of air travel and typically long distances travelled compared with other forms of transport, and the increased radiative forcing of emissions into the upper troposphere (Penner *et al.*, 1999). The world's airlines have improved their energy efficiency considerably (about 70% between 1960 and 1990), with an average energy use of 4.8 litres of kerosene per 100 passenger-kilometres in 1998 (International Air Transport Association, 2003); however, it becomes more and more difficult to achieve efficiency gains (Penner *et al.*, 1999), unless zero-emission technologies for aircraft, for example 'hydrogen planes', are developed (NASA, 2001).

Fuel consumption for travel to a destination can be considerable: for example the energy consumed by all international tourists travelling to New Zealand (one way) in 2000 amounted to 27.8PJ and 1900 kilotonnes of emitted carbon dioxide (Becken, 2002a). New Zealand is an example of a 'nature-based' long-haul destination, where 99% of arrivals are by air. When adding emissions from international arrivals to New Zealand's carbon dioxide emissions for the same year, these would increase the total by 6%, not taking into account emissions from New Zealand's considerable outbound tourism. On an individual basis, the energy use associated with travelling to and from the destination will in most cases dominate all other tourist activities, certainly for remote destinations such as New Zealand (Becken, 2001) or the Seychelles (Gössling *et al.*, 2002). Air travel is also important for intra-destination transport, however, other forms of transportation prevail, for example coaches, rental cars, campervans or rail. The energy efficiency of these transport modes differs considerably, being determined both by the fuel efficiency of the vehicle (technological aspects) and load factors (operational aspects). The energy efficiency and carbon dioxide emissions of tourist transport in New Zealand are presented in Table 13.1.

The accommodation sector is even more complex and heterogeneous than transport, comprising different categories of accommodation, business sizes and standards of facilities, all of which impact on energy demand (Deng & Burnett, 2000). The diversity of the sector may explain the scant research undertaken on energy use so far; however, some research indicates that the more service-oriented an accommodation business, the larger the energy use per visitor (Warnken & Bradley, 2002).

Table 13.1 Energy efficiency of different New Zealand transport modes in 2002

Transport mode	*Share (% of international tourists)*[1]	*Energy efficiency (MJ/pkm)*	*CO_2 emissions (g/pkm)*[2]
Domestic air[3]	28.3	2.54	175
Rental car[4]	25.5	0.94	63
Private car[4]	14.6	1.03	69
Coach (tour bus)	22.7	0.32	22
Scheduled coach	5.2	0.51	35
Ferry	11.8	2.63	181
Campervan	3.0	2.39	165
Train	4.3	0.38	26
Shuttle bus, van	n.a. (<1%)	0.56	39
Motorcycle	n.a. (<1%)	0.87	60
Backpacker bus	1.7	0.39	27

Source: Becken and Cavanagh (2003)

Notes:
1. Tourism New Zealand (2002) (year ended June 2002).
2. Applying emission factors of Baines (1993).
3. The energy use per pkm for domestic air travel is comparatively high because of the importance of take-off and climbing phases relative to travel distance, and the older domestic fleet compared with international long-haul fleets.
4. Data for these transport modes were only available for 1998 (Ministry of Transport, 1998). The figures are comparatively low, because of a high occupancy of 2.5 passengers for rental cars and 3.2 for private cars (Becken, 2002b).

Some measure of energy efficiency, for example energy use per visitor-night, per available bed or room, or per square metre needs to be introduced to be able to compare businesses of different sizes. Becken *et al.* (2001) described a two-tier distribution of energy efficiency (energy use per visitor-night) in the New Zealand accommodation sector: 'comfort or service-oriented accommodation', including hotels and bed and breakfasts (B&Bs), with a very large energy use per visitor-night, and 'budget or purpose-oriented accommodation', including backpacker hostels, campgrounds and motels, characterised by lower energy use per visitor-night (Table 13.2).

Alongside the accommodation sector, the recreational component of tourism has rarely been analysed from the perspective of energy use and greenhouse gas emissions (Becken & Simmons, 2002; Stettler, 1997). In New Zealand large energy users (on an annual basis) are the entertainment and experience centres, sporting complexes, large museums and parks, all of which are visited by both tourists and recreationists (Becken & Simmons, 2002). However, the more visitors an attraction or activity receives, the higher the energy efficiency (measured as energy use per visitor), because the potentially large amount of energy required

Table 13.2 Mean energy efficiencies for various accommodation categories in 2000

Category	*Used by (% of international tourists)*[1]	*Energy use per square metre (MJ/m²*year)*	*Energy use per visitor-night (MJ/visitor-night)*	*CO_2 emissions per visitor-night (g)*[2]
Hotel/lodge	56.1	571	155	7895
B&B (inc. yacht and farmstay)	8.3	300	110	4142
Motel	23.9	250	32	1378
Backpacker	12.5	617	39	1619
Campground (inc. cabins, free camping, huts)	10.0	n.a.	25	1364

Source: Becken *et al.* (2001)

Notes:

1. Tourism New Zealand (2002) (year ended June 2002).
2. Applying emission factors of Baines (1993) and accounting for typical fuel mixes.

Table 13.3 Energy use (median) and CO_2 emissions per tourist for eleven attraction and activity categories in New Zealand

Category	*Energy efficiency (MJ/visit)*	*CO_2 per visit (g)*[1]
Building	4	172
Park	7	526
Amusement	22	1507
Industry	8	576
Nature attraction	8	417
Performance	12	589
Other entertainment	6	338
Air activity	424	27,697
Motorised water activity	202	15,312
Adventure recreation	43	2241
Nature recreation	70	1674

Source: Becken and Simmons (2002)
Note: 1. Applying emission factors of Baines (1993) and accounting for typical fuel mixes.

is allocated to a large number of visitors (Table 13.3). In turn, this means that small operations are often comparatively energy intensive on a per visitor basis. Such small businesses are typical for the activity sub-sector and are also often accompanied by a high-service level and on-site transport (e.g. boat cruises at 215MJ/tourist and scenic flights at 442MJ/tourist). 'Viewing wildlife' (152MJ/tourist) is another popular tourist activity in New Zealand that often involves transport with a shuttle bus from the tourist centre to the viewing location, and additional water transport in the case of viewing marine wildlife (e.g. whale watching or swimming with dolphins). Compared with attractions that usually derive their energy from electricity, carbon dioxide emissions are proportionally larger for activities because these often involve transport to, or as part of the activity, and therefore require input from fossil fuels.

Trend Analysis of Energy Use in the New Zealand Tourism Sector

This section summarises the findings of a recent study on trends in energy efficiency of tourism between 1999 and 2001. The study was commissioned by the New Zealand Energy Efficiency and Conservation

Authority (EECA) (Becken & Cavanagh, 2003), which was interested in a detailed analysis of the tourism sector. Tourism had been previously found to contribute about 6.2% to national energy demand (Turney *et al.*, 2002), although its contribution to GDP was only 4.6% in the same year (Statistics New Zealand, 2001). This study focused on energy use within New Zealand and excludes energy use associated with tourists' international transport.

In New Zealand, tourism is characterised by a high level of mobility, because most tourists follow a multi-destination itinerary, with the purpose of visiting the country's geographically dispersed natural attractions. About one quarter of the now two million annual visitors to New Zealand travel as part of an organised coach tour, while the remainder are free and independent travellers (FIT) using various forms of transport such as rental cars, private cars and campervans (Tourism New Zealand, 2002). Domestic tourists outnumber international tourists with over 16.5 million trips undertaken in 2001 (Tourism Research Council, 2002). The diversity of travel choices made by domestic tourists, however, is more limited compared with international tourists, since most domestic tourists travel by car (78%) and domestic air (13%), and stay at private homes (62%) or hotels (17%).

The approach for the trend analysis involved a bottom-up analysis, integrating separate sub-sector and tourist analyses for two reference years (1999 and 2001). (The tourist data sets that were available refer to 1999 and 2001, however, the industry data that are grouped into the 1999 reference year comprise data from between 1998 and 2000, while those linked with the 2001 reference year include data from 2001 and 2002.) Because transport and accommodation were found to be the dominant contributors to the sector's energy use (84% out of 28PJ in 1997/8) (Patterson & McDonald, 2002), the trend analysis focused on these two core sub-sectors. To obtain energy efficiencies in the accommodation sector, two surveys of businesses in different categories were undertaken to collect relevant information on visitation levels and energy consumption by different fuel types. In particular, the 1999 survey (Becken *et al.*, 2001) involved hotels, motels, backpackers, campgrounds and bed and breakfast establishments, while the 2003 survey included hotels, motels and backpackers. Energy use associated with private homes was obtained from EECA (2000) and used for both reference years. The energy use of transport was partly accessed through literature (e.g. Ministry of Transport, 1995, 1998), and additionally required contacting major transport providers in each reference year to provide information on energy efficiency, i.e. energy use per vehicle and passenger-kilometre (MJ/pkm). For the analysis of tourists' travel behaviour, the International Visitor Survey (IVS) and the Domestic Travel Study (DTS) were used for the two reference years. The key step in these analyses was to use cluster analysis to derive tourist types that are characterised by typical transport/accommodation choices, and hence energy use (Becken *et al.*, 2003). The travel behaviour of each tourist type was combined with energy efficiencies obtained in the industry analyses, which then enabled calculation of energy use associated with each tourist type and for all tourists (Becken & Cavanagh, 2003).

Energy use of tourism transport and accommodation amounted to 22PJ in 2001, which represents 4.6% of the energy consumed nationally (Ministry of Economic Development, 2002). The trend analysis revealed that energy use from tourist transport and accommodation decreased between 1999 and 2001 by 7.5% as a result of decreasing energy use by domestic tourism (Table 13.4). This decrease is partly explained by a smaller volume of domestic tourists in 2001 compared with 1999, but also by a decreased energy use per trip (domestic tourists: 1053MJ in 1999 and 950MJ in 2001). Energy use per average trip of an international tourist also decreased from 3385MJ in 1999 to 3082MJ in 2001; however, growth in international visitor arrivals (18%) substantially outweighed per capita gains. The improved energy balance for an individual tourist is primarily a result of improved efficiencies in the transport sector, especially domestic air travel, and to a lesser extent changes in travel behaviour. The main trends in travel behaviour were that domestic tourists travelled for less time (median: 2.9 days in 2001 compared with 3.2 days in 1999 per trip), although they travelled similar distances (mean: 614km in 2001 compared with 623km in 1999). International tourists stayed longer in 2001 (median: 21.1 days compared with 17.7 days in 1999), but travelled equivalent distances in the two reference years (mean: about 1500km per tourist). Hence, daily travel distance decreased for international tourists between 1999 and 2001.

The most important drivers of tourism energy use in New Zealand are internal domestic air travel (34.5% of total energy use in 2001), private

Table 13.4 Summary of energy use by tourism transport and accommodation in New Zealand in 1999 and 2001

Tourists	*Trips 1999*	*Trips 2001*	*Change trips (%)*	*Energy use 1999 (PJ)*	*Energy use 2001 (PJ)*	*Change energy use (%)*
International	1,437,552	1,694,537	17.88	4.87	5.22	7.33
Domestic	16,889,000	16,557,000	–1.97	17.78	15.72	–11.56
Total	18,326,552	18,251,537	–0.41	22.64	20.95	–7.46

Source: Becken and Cavanagh (2003)

cars (30.5%) and private homes (9.6%), however, the role of these drivers differs clearly for domestic and international tourists (Table 13.5). Transport energy use dominates total energy use of tourism, and accordingly the most commonly used fuel sources are petrol and diesel (44.7% in 2001), and aviation fuel (34.5%). All of these fuel sources are associated with the emission of greenhouse gases, in particular carbon dioxide, which highlights the role tourism plays in climate change issues. The third most important fuel source – mostly for accommodation – is electricity (15.1%); greenhouse gas emissions from electricity depend on the type of electricity generation. In New Zealand a large, but varying share from electricity is hydroelectric power (McNicol *et al.* (2002) recommend an average share of 65%), but increasingly it is becoming necessary to generate electricity from gas- and coal-fired power plants, which results in considerable emissions, given the low efficiency of such plants (between 30% and 40%) (Baines, 1993).

Energy use per visitor-night increased between the two reference years for hotels and motels, and only backpacker hostels showed an improvement in energy efficiency. It has to be noted, however, that the period of time between the two reference years may be too short to reveal clear trends in energy efficiency. Nevertheless, there seems to be a need to encourage improvements in energy efficiency, especially for hotels, which make a contribution of over 6% to total energy use (Table 13.5). In contrast, most transport modes improved energy efficiency; domestic air

Table 13.5 Most important drivers of energy use (relative share (%) of 22.6PJ in 1999 and 21.0PJ in 2001)

Key drivers of energy use (%)	*Domestic tourists 1999*	*Domestic tourists 2001*	*International tourists 1999*	*International tourists 2001*	*Overall share 1999*	*Overall share 2001*
Domestic air	39.7	36.1	33.8	29.6	38.4	34.5
Private car	36.3	39.2	4.0	4.4	29.3	30.5
Coach	0.6	0.2	8.7	2.7	2.3	0.8
Rental car	1.2	1.5	11.7	13.8	3.4	4.6
Campervan	1.5	1.7	5.2	6.3	2.3	2.9
'Home'	7.9	8.1	10.9	14.3	8.5	9.6
Hotel	4.5	4.3	12.7	14.9	6.2	6.9
Motel	1.3	1.7	2.1	3.0	1.4	2.0
Backpacker	0.1	0.1	2.3	2.9	0.6	1.0

in particular (over 20%), which is a result of both better technologies, a more modern fleet and higher load factors.

Lessons Learned for Future Energy Analyses

The different studies on energy use presented here all revealed a range of difficulties that need to be overcome when analysing the tourism sector in terms of its contribution to energy consumption and greenhouse gas emissions. One major problem results from tourism's fragmented and diverse nature of its businesses, which require extended data collection to be able to derive meaningful and representative figures for energy efficiencies. Because tourism is not a traditional sector in its own right, involving 'core tourism sub-sectors' alongside incidentally involved sub-sectors, the scope for analysis is not as evident as in other industries, and decisions have to be made as to what is to be included in energy inventories. Air travel is probably the most straightforward, and also the most important component of tourism to be analysed for energy use. There are only few players in the market, and they usually collect data on air movements, passenger numbers and fuel consumption that are publicly available at least in an aggregated form. Also, most destinations collect statistics on tourist arrivals, in particular air arrivals, which are usually accurately recorded for immigration statistics. The inventory of other tourism sub-sectors, however, depends on the availability or existence of energy-efficiency data, as well as tourist databases that include information on where tourists go, using what type of transport, and what they do at the destination. Some countries have International Visitor Surveys similar to that undertaken in New Zealand that could be modified so that energy analyses are possible (e.g. converting tourist itineraries into travel distances). Notwithstanding this, the geography of New Zealand, with only two major points of entry and departure, facilitates the collection of data on international visitors. Using such visitor surveys to aggregate similar tourists into tourist types (e.g. Becken *et al.*, 2003; Becken & Gnoth, 2004) is a useful way to simplify the large number of individual travel patterns into a manageable number of recognisable prototypes.

While there is a large body of literature and data on transport in general, little is known about those aspects of transport that are specific to tourism. For example, buses that are used for tourism rather than for other end-uses (e.g. school buses) are often equipped with additional luxuries, such as air-conditioning, toilets or microwaves. It is therefore possible that energy use associated with tourist coaches differs from the national averages available in official transport statistics. Similarly, little is known about energy-efficiency trends of rental cars, although some anecdotal knowledge exists that there are several competing trends that influence energy use, including more energy-efficient cars, the increasing number of diesel cars, the increasing proportion of cars with larger engines and four-wheel drives, and the application of additional equipment, such as air-conditioning. Another important factor for analysing energy efficiency of transport modes is the occupancy of a vehicle. Again, there is little information on typical load factors for tourist transport (Becken *et al.*, 2003). More research is required to investigate ways to increase occupancy without compromising tourists' experience. Chen (2002), for example, noted that Chinese wholesalers often negotiate with New Zealand suppliers about larger buses, because their customers prefer to have ample space in the coach.

The largest problem in analysing energy use associated with tourist accommodation is the availability and the quality of data, especially for smaller businesses. Many operators are not interested in measures of energy consumption and do not systematically collect this information. It is also difficult to obtain long-term, reliable data, among other factors, because of poor recording practices and high management turnover. What is needed is a consistent way of auditing energy in accommodation, for example by standardised spreadsheets in consultation with management or engineering staff (Warnken & Bradley, 2002). These auditing tools, however, would have to take into account the varying situations for different types of accommodation, for example backpacker hostels versus five-star hotels.

An important step towards tourism energy audits of a high quality would be the recognition of tourism as a sector in its own right. The development of guidelines for Tourism Satellite Accounts by four intergovernmental organisations, the United Nations (UN), the World Tourism Organisation (WTO), the Organisation for Economic Cooperation and Development (OECD) and the European Commission, was an important step towards collecting tourism-relevant statistics and assessing tourism's significance to national economies (WTO, 1999). Similar initiatives should be encouraged in the field of environmental accounting (Patterson & McDonald, 2002), so that the full environmental impacts of tourism – including greenhouse gas emissions – could be assessed, and key contributors challenged to reduce their impacts.

Stakeholders' Involvement in Mitigating Climate Change

The stakeholders immediately concerned with tourism's energy use and climate change are the tourism industry and the tourists themselves. Some industry initiatives already exist to monitor and manage the consumption of fossil fuels. Self-regulation and voluntary initiatives, for example, are one option chosen by hotels that belong to environmental industry groups, such as the International Hotels Environmental

Initiative at an international level or the Environmental Accommodation Providers of Auckland on a national level. Large companies, such as airlines, have long recognised that saving on energy has become an economic (rather than environmental) imperative for the profitability of the company. More formalised ways for improving the environmental performance of individual businesses at all levels include benchmarking and ecolabelling, for example through Green Globe 21, an international environmental benchmarking and certification programme for tourism that includes greenhouse gas emissions as one of nine key performance indicators (Green Globe 21, 2002). Tour operators have a particularly important role in the context of energy use and greenhouse gas emissions, because 'they influence consumer demand, destination development patterns, and their supplier's performance, as well as tourists' behaviour' (Tour Operators' Initiative, 2000). The Tour Operators' Initiative is a response to these responsibilities, and the development of sustainability reporting guidelines is an important part of this initiative.

Tourists exert influence through their consumer behaviour, for example by supporting businesses that seek to achieve best practice in terms of energy efficiency and greenhouse gas emissions. Ecolabels are one means to guide consumers in making their decisions, although there is little evidence to date that tourists recognise tourism ecolabels and base their decisions on such labels (Sharpley, 2001). A large proportion of energy used by tourists could be saved by changing tourists' behaviour, especially within transportation. By making choices concerning transport modes and travel distances, tourists have substantial influence on the energy balance of their trip. Little is known, however, about how educated and aware tourists are of their energy consumption for travelling. An initial study indicated relatively low awareness and knowledge in this area (Becken, 2003); notwithstanding, this study revealed substantial potential for tourists to participate in carbon offsetting schemes that help mitigate their greenhouse gas emissions. The reasons why tourists would participate in such schemes were related to nature protection, tree planting, and 'feeling good', rather than sequestering carbon in particular. This potential should be explored further, building on existing initiatives, such as Trees for Travellers (www.treesfortravellers.co.nz), Future Forests (www.futureforests.com), Climate Care (www.climatecare.org) or Business Enterprises for Sustainable Travel (www.sustainabletravel.org) for business travel.

Tourist destinations increasingly hold an interest in developing sustainable forms of tourism, while at the same time they seek to increase tourists' length of stay. Current marketing and tourism products are often icon-oriented and thereby prompt tourists to travel at a high pace from one primary attraction to the next, which results in high energy use and little economic activity for more regional destinations. By developing alternative regional tourist routes (Briedenhann & Wickens, 2004) and secondary and tertiary attractions, tourists could be encouraged to travel more slowly (i.e. shorter distances per day), consume less energy per day, and to spend more time in the regions. Convenient and price-competitive public transport systems and cycle networks could complement regional touring itineraries. Such initiatives could form part of wider processes of 'greening' a tourist destination. This has been undertaken, for example, in Kaikoura, New Zealand, which obtained Green Globe 21 benchmarking in 2002 (McNicol *et al.*, 2002). Larger scale shifts towards sustainable tourism destinations also require support from marketing agencies at a local, regional and national level, which have the means to turn these efforts into a competitive advantage for the destination, especially for destinations that capitalise on nature or ecotourism.

Finally, governments and intergovernmental institutions, such as the Intergovernmental Panel on Climate Change, World Tourism Organisation and the United Nations Environment Programme, have influencing or even regulating power to mitigate greenhouse gas emissions from tourism. Governments, for example, have the means to introduce taxes on greenhouse gas emissions, which will ineluctably impact on tourism, in particular tourism transport (Turney *et al.*, 2002). Intergovernmental agreements will be necessary to impose such taxes on international travel, for example within the European Union (Royal Commission on Environmental Pollution, 2002). Institutions such as the United Nations or the World Tourism Organisation could use their position to increase awareness of climate change and tourism (as done by the WTO by their First International Conference on Climate Change and Tourism in Djerba, 2003), to provide incentives for increasing energy efficiency, encouraging the uptake of renewable energies by tourism businesses, promoting sustainable tourist behaviour and mitigating impacts by investing in carbon sequestration programmes.

Conclusion

Tourism is a major contributor to energy consumption and climate change. This chapter presented the case of New Zealand, where tourism transport and accommodation alone make up 4.6% of national energy demand. When considering international travel as well, the importance of tourism in the context of climate change becomes even more apparent, and ways need to be negotiated to decrease tourism's greenhouse gas emissions nationally and internationally. Major difficulties associated with energy accounting in tourism are due to tourism's composite nature and the resulting paucity of energy-relevant data, especially for small businesses, for example in the accommodation or attraction sub-sectors. Detailed national tourist travel databases are critical for undertaking a bottom-up energy audit for a destination that includes travel behaviour as an important component. Institutions, such as the World Tourism Organisation, have the potential to encourage data collection that provides a basis of environmental accounts in its member countries. International organisations also exert influence in mitigating greenhouse gas emissions, as do governments, tourist destinations, the industry and tourists themselves.

References

Baines, J.T. (ed.) (1993) *New Zealand Energy Information Handbook.* Christchurch: Taylor Baines and Associates.

Becken, S. (2001) Vergleich der Energieintensität zweier verschiedener Reisestile. *Tourismus Journal* 2, 227–46.

Becken, S. (2002a) Analysing international tourist flows to estimate energy use associated with air travel. *Journal of Sustainable Tourism* 10 (2), 114–31.

Becken, S. (2002b) *Tourism and Transport in New Zealand – Implications for Energy Use.* TRREC Report No. 54, July 2002.

Becken, S. (2003) The perception of climate change and forest carbon sinks by tourists and tourism experts. Unpublished manuscript, Landcare Research, New Zealand.

Becken, S. and Cavanagh, J. (2003) Energy efficiency trend analysis of the tourism sector. Unpublished Landcare Research Contract Report LC0203/293. Prepared for the Energy Efficiency and Conservation Authority.

Becken, S. and Gnoth, J. (2004) Tourist consumption systems among overseas visitors: Reporting on American, German, and Australian visitors to New Zealand. *Tourism Management* 25 (3), 375–85.

Becken, S. and Simmons, D. (2002) Understanding energy consumption patterns of tourist attractions and activities in New Zealand. *Tourism Management* 23 (4), 343–54.

Beckens, S., Frampton, C. and Simmons, D. (2001) Energy consumption patterns in the accommodation sector – the New Zealand case. *Ecological Economics* 39, 371–86.

Becken, S., Simmons, D. and Frampton, C. (2003) Segmenting tourists by their travel pattern for insights into achieving energy efficiency. *Journal of Travel Research* 42 (1), 48–56.

Briedenhann, J. and Wickens, E. (2004) Tourism routes as a tool for the economic development of rural areas – vibrant hope or impossible dream? *Tourism Management* 25 (1), 71–9.

Ceron, J.P. and Dubois, G. (2003) Changes in leisure/tourism mobility patterns facing the stake of global warming: The case of France. Presentation given at '*Global Change and Human Mobility*', International Geographical Union Commission, Palma de Mallorca, Spain, 3–5 April 2003.

Chen, J. (2002) Tour group and independent travel: An analysis of Asian Chinese visitors to New Zealand. Unpublished Masters of Tourism Management thesis, Victoria University, Wellington.

Deng, S. and Burnett, J. (2000) A study of energy performance of hotel buildings in Hong Kong. *Energy and Buildings* 31, 7–12.

Dubois, G. (2003) Le changement climatique. Un enjeu émergent pour le tourisme français. *L'observation du tourisme* 72 (July/August), 7–10.

Energy Efficiency and Conservation Authority (EECA) (2000) Residential sector energy use: Highlights. *Energy Wise Monitoring Quarterly* 15 (June).

Gössling, S. (2000) Sustainable tourism development in developing countries: Some aspects of energy use. *Journal of Sustainable Tourism* 8 (5), 410–25.

Gössling, S. (2002) Global environmental consequences of tourism. *Global Environmental Change* 12 (4), 283–302.

Gössling, S., Borgstrom Hansson, C., Horstmeier, C. and Saggel, S. (2002) Ecological footprint analysis as a tool to assess tourism sustainability. *Ecological Economics* 43, 199–211.

Green Globe 21 (2002) Homepage. Avaiable at: www.greenglobe21.com (accessed 22 August 2003).

Høyer, K.G. (2000) Sustainable tourism or sustainable mobility? The Norwegian case. *Journal of Sustainable Tourism* 8 (2), 147–60.

International Air Transport Association (2003) Aircraft emissions. Available at: http://www.iata.org/soi/environment/aircraftemissions.htm (accessed 12 August 2003).

International Energy Agency (2002) *Word Energy Outlook 2002*. Paris: IEA Publishing. Available at: http://www.iea.org/newsroom/weo2002_highlights.pdf (accessed 25 August 2003).

Kester, J.G.C. (2002) Preliminary results for international tourism in 2002, air transportation after 11 September. *Tourism Economics* 9 (1), 95–110.

McNicol, J., Shone, M. and Horn, C. (2002) *Green Globe 21 Kaikoura Community Benchmarking Pilot Study*. A joint report by Landcare Research and TRREC. Report No.53/2002. Lincoln University, New Zealand.

Ministry of Economic Development (2002) *Energy Data File*, July 2002. Wellington: Ministry of Economic Development.

Ministry of Transport (MoT) (1995) *Greenhouse Gas Emissions from New Zealand Transport*. Wellington: MoT.

Ministry of Transport (MoT) (1998) *Local Air Quality Management: Impacts from the Road Transport Sector. Vehicle Fleet Emissions Control Strategy – Final Report*. Wellington: MoT.

Müller, H.R. and Mezzasalma, R. (1992) Transport energiebilanz: Ein erster schritt zu einer öko-bilanz für reiseveranstalter. In *Jahrbuch der Schweizerischen Tourismuswirtschaft*. St. Gallen, Switzerland.

NASA (2001) *NASA Aerospace Technology News*. Available at: http://www.aerospace.nasa.gov/curevent/news/vol2iss_5/ (accessed 2 August 2003).

Patterson, M.G. and McDonald, G. (2002) How green and clean is New Zealand tourism? Lifecycle and future environmental impacts. Unpublished draft, 15 March 2002. Palmerston North, Massey University.

Peeters, P. (2003) *The Tourist, the Trip and the Earth. Creating a Fascinating World*. Breda: University of Professional Education.

Penner, J., Lister, D., Griggs, D., Dokken, D. and McFarland, M. (eds) (1999) *Aviation and the Global Atmosphere. A Special Report of IPCC Working Groups I and III*. Cambridge: Cambridge University Press.

Royal Commission on Environmental Pollution (2002) *The Environmental Effects of Civil Aircraft in Flight*. London: HMSO.

Sharpley, R. (2001) The consumer behaviour context of ecolabelling. In X. Font and R.C. Buckley (eds) *Tourism Ecolabelling. Certification and Promotion of Sustainable Management* (pp. 41–55). Wallingford: CAB International.

Statistics New Zealand (2001) *Provisional Tourism Satellite Account 1998–2000*. Available at: http://www.stats.govt.nz (accessed 10 November 2001).

Stettler, J. (1997) Sport und Verkehr. Sportmotiviertes Verkehrsverhalten der Schweizer Bevölkerung. *Berner Studien zu Freizeit und Tourismus*, 36.

Tour Operators' Initiative (2000) Homepage. Available at: http://www.toinitiative.org/ (accessed 10 August 2003).

Tourism New Zealand (2002) *International Visitor Survey*. Available at: http://www.tourisminfo.govt.nz (accessed 11 Augst 2002).

Tourism Research Council (2002) *Domestic Travel Survey*. Available at: http://www.trcnz.govt.nz/Surveys/Domestic+Travel+Survey/default.htm (accessed 30 July 2003).

Turney, I., Becken, S., Butcher, G., Patterson, M., Hart. P.*et al.* (2002) *Tourism Industry Association New Zealand (TIANZ): Climate Change response. A Report to Establish the Knowledge Required for a Tianz Response and Policy Formulation with the Government Post Kyoto Protocol Ratification*. Landcare Research Contract Report: LC0102/107. Available at: http://www.tianz.org.nz/Files/Climate Change.pdf (accessed 13 August 2003).

Warnken, J. and Bradley, M. (2002) *Energy Auditing and Estimating Greenhouse Gas Emissions for Australia's Tourist Accommodation Sector*. Gold Coast, Australia: CRC for Sustainable Tourism.

World Energy Council (no date) Available at: http://www.worldenergy.org.

World Tourism Organisation (WTO) (1999) *Tourism Satellite Account (TSA): The Conceptual Framework*. Madrid: WTO.

Part 3: Adaptation and Response: Managing the Relationship Between Tourism, Recreation and Global Climate Change

Chapter 14

Tourism and Climate Change Adaptation: The Norwegian Case

CARLO AALL AND KARL G. HØYER

Introduction

In this chapter we will present examples of how the tourism industry in Norway has adapted to climate change and discuss different ways to strengthen adaptive measures. In doing so it is important to differ between primary and secondary climate change effects on tourism (Richardson & Loomis, 2003). As illustrated in Figure 14.1, the effects on tourism may be *primary* such as effects of changes in temperature on travel patterns. Tourism may also experience *secondary* effects due to changes in nature conditions, which in turn are triggered by climatic changes. In addition to the direct and secondary effects, tourism may also experience *tertiary* effects: that is effects of greenhouse gas (GHG) mitigation policies (Dubois & Ceron, 2003). GHG mitigation policies such as energy taxes and car use restrictions in city centres may have substantial effects on tourism, and hence should lead to adaptive measures. All together we should therefore differ between three dimensions of climate change adaptation in the case of tourism: adaptation to climatic changes in itself, adaptation to changes in nature conditions due to climatic changes, and adaptation to GHG mitigation policies. We will present examples of all three dimensions. But first we will give a more general introduction to the issue of climate change and effects on tourism.

Climate Change and Effects on Tourism

There is surprisingly little research done on the probable effects of climatic change on tourism and how tourism can adapt to such changes (see Chapter 3, this volume). This situation is in spite of the fact that weather and climate often are referred to as having substantial influence on travel flows (see Chapter 2, this volume), and hence the tourism industry has always had to adapt to changes in climate conditions. The scientific literature that does exist indicates however multiple and

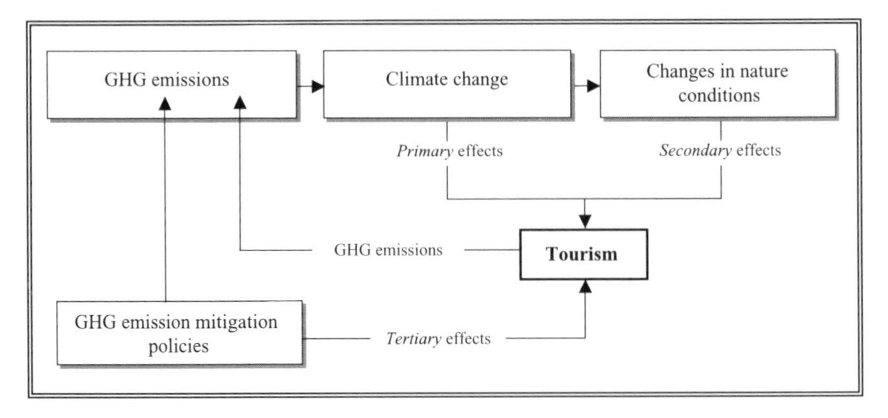

Figure 14.1 A model for analysing the relations between climate change, adaption and tourism

complex causal relationships between man-induced climate change and effects on tourism (Todd, 2003).

It was in April 2003, that the first international conference on climate change and tourism took place, at the invitation of the government of Tunisia and under the sponsorship of the World Tourism Organisation (WTO) (2003). The conference summed up the present status of knowledge on the effects of climate change and possible options for climate change adaptation within the tourism industry. The background paper for the conference states, a priori, that it is *leisure travel* that is most likely to be affected by climate change, whereas travel for business purposes, and, to some extent, travel in order to visit friends and relatives, are less affected (Todd, 2003). This seems a reasonable assumption, at least if we concentrate on the primary effects of climate change. In this respect the weather condition is often the crucial factor. However, if we also take into consideration possible secondary and tertiary effects, it seems reasonable that also travel for business purposes and travel in order to visit friends and relatives may be affected. In this chapter we will concentrate on the effects on leisure travel since this is the largest of the three with respect to the number of travellers and turnover for the tourism industry.

The tourism industry is likely to be more vulnerable to climate change than the tourists (Wall, 1998). The tourism industry has an 'immobility problem' in the sense that only to a very limited degree can it adapt to climate change by moving out of destinations that have become less desirable as a result of climate change, whereas consumers can just change to a new destination. The discussion of negative effects of climate change on tourism and how to adapt to such changes is therefore first of all a discussion relating to the tourism industry.

The amount of, and very nature of, leisure travel can be characterised as driven by a combination of 'push' and 'pull' factors (WTO, 2003). That is, the degree of discomfort at the traveller's residence for a specific time of the year (push factor) combined with the degree of comfort to be experienced at the traveller's destination (pull factor). A changing climate will not only have the potential of changing the attractiveness of destinations, it can also influence the attractiveness of the areas where people have their daily life. So far most research has been on the 'pull factor' side. However, the background paper for the WTO conference gave a general assessment of how both the 'push' and 'pull' factors may be influenced by climate change, and hence what the total impact may be of climate change on the major international tourism flows globally (see Figure 14.2)

We find few examples of climate change adaptation efforts in the context of climate policy within the tourism industry in Norway. Nevertheless, we find many examples of processes that deal with the more general question of how the tourism industry relates to the climate and the quality of nature as a resource base for tourism. Hence, it could be fruitful to differentiate between explicit, implicit and functional climate change adaptation processes (Table 14.1). *Explicit,* in the sense that adaptation to climatic changes or changes in the natural environment

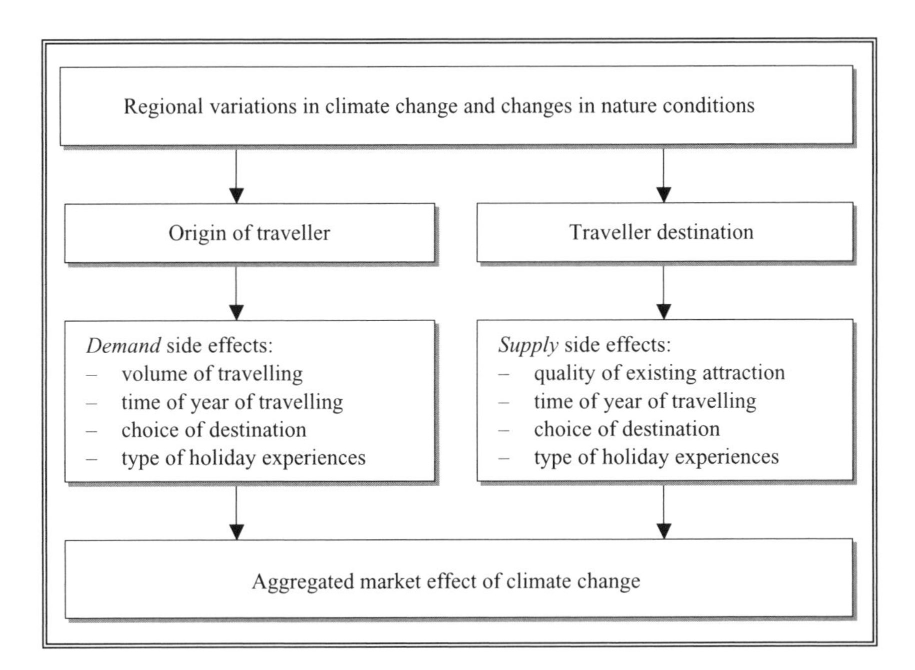

Figure 14.2 Push and pull effects of climate change on tourism

Table 14.1 Four categories of adaptation processes in the case of actual negative changes in the resource base for tourism that *could* be caused of climate change

		Local awareness: Level of reflections on climate change	
		High	*Low*
Nature of negative changes: Caused by climate changes	High	Explicit adaptation	Implicit adaptation
	Low	Symbolic adaptation	Functional adaptation

are carried out within a climate policy context. In the case of such adaptive measures without any links to climate policy, we may speak of an *implicit* climate change adaptation process. Furthermore, we may also speak of a *functional* climate change adaptation process, in the sense that changes within tourism industry relate to changes in nature conditions that in principle *could* have been caused by climatic changes, but *in fact* are caused by other processes. An example of this is natural reforestation due to a decrease of domestic animal grazing, which also could have been caused by a temperature rise.

The Emergence of 'Climate Change Tourism'

Although we have very few examples of explicit climate change adaptation strategies within the tourism industry in Norway, there is one example which in fact could be judged as such, although in a somewhat peculiar manner: the inclusion of climate change as a thematic element in tourism information centres. Pottorff and Neal (1994) have shown that areas struck by hurricanes can experience an *increase* in the number of visitors, at least in limited parts of the affected areas. This conclusion is based on a study of what happened after a hurricane hit Florida. Two weeks after, the number of hotel nights increased by 75% due to the high number of journalists and insurance agents, a relatively high number of curious visitors, and the many local inhabitants that needed temporary housing. This somewhat bizarre positive effect was restricted to the areas that in fact were most severely struck by the hurricane. The surrounding areas experienced a dramatic 95% reduction in the number of visitors. The long-term and sum effect for the whole region struck by the hurricane however is not known.

A Norwegian parallel to the development of 'storm tourism' is the development seen in relation to glacier tourism. From around the end of the 18th century and up to the 1930s, glaciers were even more popular than today. Glacier hiking (on top of the glaciers) was also very popular. However, in the period between the early 1930s and all the way through to the end of the 1960s, the Jostedalen glacier and most of its side glaciers decreased. At the same time so did the interest for visiting the glaciers as well. However, from the end of the 1960s and up to the present, the interest for glacier tourism has gradually increased – as well as the extension of the glaciers themselves. Today there are a number of tourism activities linked in one way or another to the glacier of Jostedalen such as glacier hiking, glacier climbing, horse and boat riding towards the descending glacier falls. During the last ten years we have experienced the establishment of three glacier visitor centres located around the Jostedalen glacier, less than 100km from each other.

Even though glaciers of course have a large tourism potential independent of any link to the issue of climate change, such a link seems to be in action. The climate change discussion has undoubtedly increased the interest for glaciers both as an object for research, for education and – as we have seen in the case of the Jostedalen glacier – for tourism. The increased interest in glaciers relates especially to how the glaciers mirror climatic changes, through increase and decrease in the extension of the glacier itself. So far climate change in the form of increased temperatures has led to an *increase* for most of the glaciers in Norway, whereas other countries have experienced a decrease and even loss of glaciers due to climate change. Since 1850, Swiss glaciers have lost more than a quarter of their surface, and by 2030 it is stipulated that 20% to 70% of Swiss glaciers will have disappeared (Bürki *et al.*, 2003). Another example is from Canada. Glacier National Park in Montana has lost 115 of its 150 glaciers, and scientists predict that the remaining glaciers will have disappeared within 30 years (Scott & Suffling, 2000). As Scott and Suffling (2000: 22) point out: 'The loss of this [Glacier National Park in Montana] park could serve as an important educational tool to illustrate the impacts of climate change'. Using the link between climate change and changes in the glaciers could be a window of opportunity for rising awareness on the issue of climate change. This is being put into practice at the renowned Norwegian Glacier Museum, one of the three glacier visitor centres around the Jostedalen glacier. The Norwegian Glacier Museum is a non-profit foundation established in 1989 by the International Glaciological Society together with several Norwegian governmental and non-governmental organisations and the universities of Oslo and Bergen. Some 53,000 people visited the museum in 2001, making it the most visited built attraction in the county of Sogn og Fjordane, even rating higher than the many famous stave churches in the same region.

The only attractions to rate higher are the natural attractions, such as the Flåm valley, the Sognefjord and the Jostedalen glacier itself – each of them ranging between approximately 200,000 and 300,000 visitors in 2001.

The original aim of the museum was to collect, create and disseminate knowledge about snow, ice and glaciers. However, in 2003 the work started on making a separate 250m^2 climate exhibition. This will be an exhibition with climate change as a separate topic and linked to the issue of glaciers. The basic idea is to utilise the strong pedagogical link between the issues of climate change and glaciers, thus enabling the fact that glaciers increase or decrease to be linked to the issue of climate change, and illustrating how man can influence nature. A Norwegian presentation of the exhibition is available on www.bre.museum.no/klimautstilling.html. In the foreseeable future we may even see effects within the Norwegian tourism industry when it comes to raising awareness and increased interest for the topic of climate change and climate change adaptation.

Norway: The Cradle of Skiing

'Snowy' is for foreigners – as well as for Norwegians – one of the key characteristics of Norway. Many reckon Norway to be the cradle of skiing. The word slalom is originally a Norwegian word for 'a path down a hill'. Cross-country skiing has for many years been on the top-three list of leisure time activities for Norwegians. Norway has, with a population of only four million people, has hosted two winter Olympics (Oslo in 1952 and Lillehammer in 1992). Even so, the number of artificial snowmaking facilities has increased dramatically over the last few years in Norway, as well as in other countries with less favourable climate conditions for ski tourism.

Alpine skiing is not a large tourism sector in Norway, responsible for only 3.2% of the total turnover of tourism in Norway (Norske Fjell, 2000). However, the number of alpine constructions has tripled in the period from 1980 to 1995 (Stølen, 1995). Plans exist for an extensive increase in hotel capacity at renowned and new winter resorts in the range of 38,000 beds with investments ranging up to 1.3 million euros. But even so, a recent study of development possibilities for Norwegian alpine ski resorts gives no analysis whatsoever on the possible effects of climate change (Norske Fjell, 2000).

According to the Norwegian Association of Alpine Resorts, artificial snowmaking facilities have been taken into use in almost *all* of the commercial alpine resorts in Norway, as well as in many of the minor alpine facilities only meant for local use. Such a strategy is known from other winter resort countries such as Austria, France, Switzerland and Canada (Bürki *et al.*, 2003). However, in Norway this strategy is in most cases *not* related to climatic changes, but more so to an adaptation to *market* changes. An increased demand for a longer alpine season combined with heavy investments in hotels etc. has made it profitable (and economically necessary) to invest in artificial snowmaking facilities, thereby being able to prolong the season. The minor local alpine facilities have had to follow up this trend in order not to lose their local customers.

From 1970 up to the present, the number of days with acceptable snow conditions for *cross-country skiing* (snow cover more then 25cm) has been more than halved in 'Nordmarka'. This is the most important skiing area for Norwegians: a 300km^2 area of wooded countryside situated 60 minutes by train from the centre of the capital of Norway, Oslo. During the 1970s the inhabitants of Oslo could experience an average of 150 days of good conditions for cross-country skiing. Twenty years later this is reduced to 75 days (Bjørbæk, 2000).

There are no reports on the situation in Norway as a whole when it comes to development of snow conditions for cross-country skiing, but according to the Norwegian Association of Skiing (NAS) the situation documented in Nordmarka could serve as an indicator for the general development in major parts of lowland Norway. As a response to this situation, the NAS has supported local investments in artificial snow-making facilities up to 0.75 million euros in 1998. However, according to the NAS this strategy has a somewhat different setting to what has been done in alpine ski resorts. First of all, this has been a strategy that in most cases in fact *has* been related to climatic changes, and is even so explicitly related to the climate policy discourse. Even if artificial snowmaking facilities have been taken into use in both major sport arenas, such as Lillehammer and Oslo, as well as minor local cross-country facilities, the total number of artificial snowmaking facilities are still relatively low within cross-country skiing compared to that of alpine skiing. Furthermore, the economic interests vested in cross-country skiing are far less than that of alpine skiing. Hence, the economic losses caused by snow-deficient winters as well as the will to invest in artificial snowmaking facilities are much lower when it comes to cross-country skiing.

The number of Norwegians doing different types of skiing, either cross-country, slalom or the more modern and related activities like snowboarding, has so far *not* been influenced much by the fact that the number of days with good ski conditions in general have decreased during the 1980s and 1990s (Teigland, 2001). No research has been done to explain why Norwegians seem to be so reluctant to give up their skiing. One explanation could be that problems with snow-deficient winters, at least when it comes to cross-country skiing, are most of all a 'demand-side' problem. The way to adapt is simply for the skiers to find alternative locations. Another explanation could of course be that the strategy of investing in artificial snowmaking facilities has been successful. However, *running* the artificial snowmaking facilities is both complicated and expensive. A survey from 1998 showed that 25 out of a total of 800 alpine facilities were run at an economic loss, and an additional 11 were, in practice, bankrupt (Norske Fjell, 2000). If problems with snow-deficient winters will increase, it seems reasonable to predict that non-commercial cross-country ski facilities as well as some of the least favourable commercial alpine resorts will suffer from increased economic pressure in the close foreseeable future.

Natural Reforestation of Mountain Areas in Norway

There is no good English translation of the Norwegian word *friluftsliv*, a word invented by the famous North Pole expeditionary Fridtjof Nansen to encompass the very basic nature of being a Norwegian: the simple life in close contact with nature (Breivik & Løvmo, 1978). The modern version of *friluftsliv* is outdoor recreation, and the Norwegian history of outdoor recreation is closely related to the farming tradition with summer grazing and dairying in the mountains. Derived from this is the deep-rooted Norwegian tradition of hiking in the mountains and spending summer holidays at a 'hut' with very primitive conditions. Norway, a country of only 4.5 million inhabitants, has as many as 400,000 'huts' and summer cottages, many of these located in the mountain areas. In addition there is a large number of hotels located in the alpine and sub-alpine regions, as well as the unique Norwegian system of 390 low standard hotels and 19,000km of marked tracks managed since 1868 by the Norwegian Mountain Touring Association (DNT). Almost 200,000 Norwegians are members of the DNT.

The possibility of experiencing unspoiled nature and 'open land' are important qualities for hikers in the mountain areas. The same goes also for the foreign tourists visiting Norway. A serious threat to both hiking and tourism in the mountain areas of Norway is the rapidly increased rate of natural reforestation. A large proportion of the cultural landscape in Norway is threatened by natural reforestation (Austad *et al.*, 2001). The main reason for this situation is the dramatic reduction in summer grazing of large domestic animals. During the last 40 years the number of cattle on summer grazing has reduced by 70%. No exact figures exist to illustrate the total extent of natural reforestation in Norway, which has mainly taken place in the mountain areas. Still, more limited surveys indicate that the potential for natural forestation is huge. Surveys from parts of the county of Sogn og Fjordane dating back to the 1930s indicate that domestic animal summer grazing in some regions has lowered the actual timber line by as much as 500m compared to the natural timber line (Ve, 1930: 65). Assessments from the late 1990s indicate a

1.2% annual increase of areas covered with forest in the county of Sogn og Fjordane (Austad *et al.*, 2001: 125).

Natural forestation has only recently been recognised as a problem for tourism. And much attention is now being paid by representatives from both the tourism industry and the non-commercial side, and the DNT, in supporting agriculture policies that can enable farmers to keep up with domestic animal summer grazing. In cases where huts for rent and hotels are located in areas heavily affected by natural reforestation, or when an 'open landscape' is an important resource for the local tourism industry, there have been examples of the tourism industry giving financial support to keep up domestic animal summer grazing. For example this is the case in the Flåm Valley, one of the major tourist attractions in Norway with 400,000 visitors each year touring the valley with the steepest railway in Europe. Another example is a programme for using local and traditional food introduced at the hotels run by the DNT, thus at the same time indirectly supporting local farmers and the maintenance of domestic animal summer grazing.

Natural forestation is so far *not* a problem caused by climatic changes. Other processes are far more important. However, a further increase in temperature might boost the process of natural forestation and thus lead to economic losses for the tourism industry as well as negative effects for mountain hikers. The prospects of keeping major parts of the cultural landscape in Norway 'open' seems overwhelming, first of all due to the high pressure on national agriculture policies through the negotiations of the World Trade Organisation (WTO). The suggested heavy reductions in import duty as well as national agricultural subsidies will eventually disfavour small-scale extensive farming, which again will make it much harder to keep up with domestic animal summer grazing in mountain areas. An important explicit climate change adaptation strategy connected to this could therefore be to find ways to keep up, and in some areas even increase, the extent of domestic animal summer grazing.

The Challenge of Thinking Globally When Acting Locally

To be able to discuss possible adaptation strategies to climate change, you most have some sort of qualified information about the probable effects of climate change. However, given that tourism is very dependent on local features, the lack of reliable local scenarios for climate change implies a high uncertainty in forecasting impacts on tourism (Dubois & Ceron, 2003). One way to manage the challenge of forecasting local impacts on tourism of climate change could be through the use of *local vulnerability indicators*. To be able to assess local vulnerability the indicator system should be 'local' in the sense that the indicators are sensitive to local (or regional) variations. As pointed out earlier, climate change will be a challenge for tourism not only in the form of a changing climate, but also in the form of changing regulatory regimes for economic activities as a consequence of GHG emission mitigation policies. In addition to both of these aspects, an adequate indicator system should also include indicators that give information about the institutional capability locally to adapt to climate change. Such a system may thus consist of the following four main indicator categories (Aall & Norland, 2003):

- climate change;
- nature vulnerability;
- socio-economic vulnerability;
- institutional vulnerability.

Climatic change is both an indicator category in itself and an important input for estimating nature vulnerability indicators. Climate change has a direct effect on tourism, in the sense that 'nice' weather could attract new visitors to a given locality, as well as reduce the 'push effect' on the local inhabitants to leave their own locality for vacation.

Nature vulnerability consists of two elements: changes in the physical environment and ecosystem effects of climatic changes. Examples of changes in the physical environment are increased risk of avalanches and landslides and increased risk of floods. Examples of ecosystem effects are natural forestation and the spreading of species to new areas. Vulnerability is generally perceived to be the function of the exposure to external stresses and the difficulty of coping with and adapting to such exposures (Agder, 2001).

Socio-economic vulnerability in this respect is the effects on local economy and social life of climate change both directly and indirectly through changes in the physical environment and ecosystem effects triggered by climate change.

Institutional vulnerability is the capacity of local institutions to carry out climate change adaptation efforts including GHG mitigation efforts. We may differ between a general and a specific institutional vulnerability. The more *general* perspective deals with issues such as local growth prognosis of employment sectors, municipality tax income per capita and municipality governmental transfers per capita (O'Brien *et al.*, 2003a, b). A more specific perspective is directly related to climate policy and climate adaptation. Examples of local institutions that are important in a Norwegian context when it comes to climate policy and climate change adaptation are civil defence, municipal master planning and the environmental authorities (Aall & Groven, 2003). Important aspects of a *specific* institutional vulnerability are the status of local land use planning and the competence on environmental policy and land use planning within the local authority.

Table 14.2 Examples of local vulnerability indicators with relevance for tourism

Indicator category	*Indicators*
Climate change	– Precipitation (total amount and time of year) – Temperature – Duration of winter
Nature vulnerability	– Risk of avalanches – Risk of floods – Share of land covered by forest
Socio-economic vulnerability of climate change	– Employment in primary sector (agriculture, forestry) – Employment in the tourism industry – Share of tourism facilities located in vulnerable areas (e.g. areas with risks of avalanches, river flooding or snow deficiency) – Share of transportation infrastructure located in vulnerable areas (e.g. areas with risks of avalanches, river flooding)
Socio-economic vulnerability of GHG emission mitigation policies	– Share of visitor transportation within and to/from the destination by air and private care – Annual GHG emissions related to turnover (tons/euro) within the local tourism industry
Institutional vulnerability – general perspective	– Growth prognosis of employment sectors – Municipality tax income per capita – Municipality governmental transfers per capita – Percentage of ageing working population – Net percentage migration – Index of social problems
Institutional vulnerability – specific perspective	– Status of local land use planning – The competence on environmental policy and land use planning within the local authority – Status of local civil defence when it comes to risk assessments – The quality of local water pipelines

Source: After O'Brien *et al.* (2003a, b) and Aall and Norland (2003)

The indicators in Table 14.2 are just suggestions, and should of course vary in different local settings. To be able to come up with a user-friendly indicator system that works the suggested indicators should also be tested and adjusted. One possible option would be to test the indicators as part of the local Agenda 21 processes. This would be in line with the recommendations from the seventh session of the United Nation Commission on Sustainable Development (CSD) addressing the issue of sustainable tourism (CSD, 1999). The session was arranged in cooperation with the World Travel and Tourism Council, the International Hotel and Restaurant Association, the International Council of Local Environmental Initiatives and an international organ for voluntary organisations (the Earth Council was established to follow up the United Nation Conference on Sustainable Development in 1992). An important recommendation was to utilise the opportunity to integrate promotion of sustainable tourism into the many ongoing local Agenda 21 processes that in fact are working around the world (CSD, 1999). This would in fact also be in line with the recommendation of the United Nations action plan for sustainable development (Agenda 21) agreed on at the UNCSD conference to focus very strongly on developing sustainability indicators within all sectors an all levels of governance.

References

Aall, C. and Groven, K. (2003) *Institutional Response to Climate Change. Examples from Four Institutional Systems in Norway*. VF-report 3/03. Sogndal: Western Norway Research Institute (in Norwegian with English summary).

Aall, C. and Norland, I. (2003) *Local Climate Vulnerability Indicators*. VF-report. Sogndal: Western Norway Research Institute (in press) (in Norwegian with English summary).

Agder, W.N. (2001) Scales of governance and environmental justice for adaptation and mitigation of climate change. *Journal of International Development* 13, 921–31.

Austad, I., Røysum, M. and Braanaas, A. (2001) Natural forestation – a problem or a resource base? In I. Austad, L.N. Hamre and E. Ådland (eds) *Natural Forestation of the Cultural Landscape*. Communication No. 15 from the Bergen Museum. Bergen: University of Bergen/The Regional College of Sogn og Fjordan (in Norwegian with English summary).

Bjørbæk, G. (2000) Less snow and a warmer climate. *Snø & ski* No. 4 (in Norwegian).

Breivik, G. and Løvmo, H. (1978) *Outdoor Recreation from Fridtjof Nansen and up to Present Days*. Oslo: Universitetsforlaget (in Norwegian).

Bürki, R., Elsasser, H. and Abegg, B. (2003) Climate change – impacts on the tourism industry in mountain areas. Paper presented at *Climate Change and Tourism. Proceedings of the 1st International Conference on Climate Change and Tourism*. Djerba, Tunisia, 9–11 April.

CSD (1999) *Report on the Seventh Session (1 May and 27 July 1998, and 19–30 April 1999)*. Economic and Social Council, Official Records, 1999, Supplement No. 9. E/CN.17/1999/20. New York: United Nations. Available at: www.un.org/esa/sustdev/csddecis.htm (accessed 23 January 2001).

Dubois, G. and Ceron, J.P. (2003) *Draft Proposal for a Research Agenda*. Climate Change, the Environment and Tourism: The Interactions workshop, Milan, 4–6 June. Available at: http://perso.wanadoo.fr/tec-conseil/docsPDF/milan.pdf.

Norske Fjell A.S. (2000) *Perspective Analysis for Norwegian Winter Destinations*. Bergen: Kaizen as (in Norwegian).
O'Brien, K.L., Sygna, L. and Haugen, J.E. (2003a) Vulnerable or resilient? A multi-scale assessment of climate impacts and vulnerability in Norway. Submitted to *Climatic Change* (in press).
O'Brien, K., Aandahl, G. and Sæther, B. (2003b) Vulnerability mapping – an outset for local dialogue on the climate change issue. Submitted to *Plan* (in press) (in Norwegian).
Pottorff, S.M. and Neal, D.M. (1994) Marketing implications for post-disaster tourism destinations. *Journal of Travel & Tourism Marketing* 3 (1), 115–22.
Richardson, R. and Loomis, J.B. (2003) The effects of climate change on mountain tourism – a contingent behaviour methodology. Paper presented at the conference *Climate Change and Tourism. Proceedings of the 1st International Conference on Climate Change and Toruism*. Djerba, Tunisia, 9–11 April 2003. Available at: www.world-tourism.org/sustainable/climate/pres/robert-richardson.pdf.
Scott, D. and Suffling, R. (2000) *Climate Change and Canada's National Park System: A Screening Level Assessment*. Toronto: Adaptation and Impacts Research Group, Environment Canada.
Stølen, A. (1995) *A List of Alpine Resort Facilities*. Report 95–TDH-0017, Taubanetilsynet, Det Norske Veritas Industry AS, Trondheim (in Norwegian).
Teigland, J. (2001) *The Effects on Norwegian Tourism of Mega-trends and Socioeconomic Development 1986–99*. VF-report 14/2001. Sogndal: Western Norway Research Institute (in Norwegian with English summary).
Todd, G. (2003) *WTO Background Paper on Climate Change and Tourism*. Prepared for the World Tourism Organisation. Beaconsfield: Travel Research International Limited. Available at: www.world-tourism.org/sustainable/climate/pres/graham-todd.pdf.
Ve, S. (1930) *An Assessment of the Timber Line in Inner Parts of Sogn*. Communication No. 13. Bergen: Western Norway Forestry Test Station (in Norwegian).
Wall, G. (1998) Implications of global climate change for tourism and recreation in wetland areas. *Climatic Change* 40 (2), 371–89.
WTO (2003) *Climate Change and Tourism. Proceedings of the 1st International Conference on Climate Change and Toruism*. Djerba, Tunisia, 9–11 April 2003. Available at: www.world-tourism.org/sustainable/climate/final-report.pdf.

Chapter 15

Tourism and the Ozone Hole: Varying Perceptions

L. MICHAEL TRAPASSO

Introduction

The ozone hole has been perceived as anything from a serious threat to humankind to a naturally occurring cyclic phenomenon. In past studies (Trapasso, 1982, 1992, 1994) it was found that the perception of people dealing with an environmental issue is as real as the scientific facts regarding that issue. A perception *is* reality to those who take that view. The research question then becomes: How is this phenomenon perceived by the people who live and work in the region, and by others wanting to visit the 'land of the ozone hole'?

Fieldwork was conducted during January and February 1998. It included visits to various locations in Argentina and the Antarctic Peninsula to seek perspectives on this controversial topic. Interviews with scientists, and data in the form of maps, charts, and satellite imagery were obtained from scientific stations, institutes, and university campuses. These primary sources, coupled with recent scientific literature constituted a *scientific perception*. Interviews with public officials comprised a *governmental perception*. Talks with news media professionals helped to form both a *popular*, and to some extent an *economic perception*. Discussions with both tourists and tourism professionals added to the *popular*, and *economic perceptions* as well. As expected, multifaceted views of this environmental problem focused on the interests of those interviewed.

The Scientific Perception

In this author's opinion, it is vital to review 'the scientific facts' first. Thus it is fitting to begin with a *scientific perception*. This perspective is discussed on several different levels: (1) science on the international project level, (2) science on the national or regional level, (3) science at Antarctic research stations, and (4) health-related science.

The ozone hole by its very name reflects a certain perception. For the ozone hole is not a hole at all, but rather an area of ozone depletion resembling a dynamic circular/elliptical expanse surrounding the South Polar Region. As science has revealed so far, both natural atmospheric processes and human-induced pollutants act in concert to diminish the ozone layer in this region during the austral spring. The depleted region reaches its maximum during late September through to mid-October in any given year. As reported by the American Meteorological Society (1998), during the year of the field study, the ozone hole was the largest observed since the early 1980s, measured at more than 26 million km^2, that is, larger than the size of North America.

International project level

To investigate the perceptions of this issue required an extensive stay in Argentina. The southern tip of the country, and the Antarctic Peninsula constitute the 'land of the ozone hole' used in this study. Dr Susana Diaz is a resident scientist at CADIC (Austral Center for Scientific Research) located in the port city of Ushuaia, Argentina. This is the southernmost city in the world and can be found within the boundary of the yearly ozone hole.

Dr Diaz's laboratory at CADIC is an integral part of the US National Science Foundation Polar Network for Monitoring Ultraviolet Radiation, and represents the official UV monitoring station for Ushuaia. Diaz's cooperative research efforts have resulted in an impressive list of publications. Reviewing these and other literature, augmented by in-depth conversations with her, indicated that there is less of a danger to human life than one might assume. 'It all pertains to basic radiation physics; as it relates to sun angles, and cloud cover,' Diaz stated. Her data have shown that the maximum UVB radiation received at Ushuaia does not necessarily coincide with the maximum extent of the ozone hole. She elaborated: 'It is possible to have a greater exposure to UVB during a calm, clear, sunny day in summer, than a cloudy and hazy day during the ozone hole' (Diaz *et al.*, 1994, 1996; Diaz, 1998).

Climatologically, humans at high latitudes find protection by two basic radiation physics laws working in tandem: Lambert's Cosine Law and Beer's Law.

Lambert's Cosine Law

$$I = Io \text{ Cos } V$$

where

I = intensity of solar radiation reaching the surface.

Io = intensity of solar radiation at maximum (i.e. when the sun is 90 degrees).

Cos V = cosine of the zenith angle V (i.e, the angular departure from the vertical).

Thus, the lower the zenith angle the less intense the incoming solar radiation. For example, on the equinox dates, 60 degrees N or S latitude will receive half the solar radiation of the equator.

Beer's Law

$$I = Io\ e^{-(arx)}$$

where

I = intensity of solar radiation reaching the surface.

Io = intensity of solar radiation at maximum (i.e. at the outer edge of the atmosphere).

e = the natural logarithm (2.718 . . .).

–(arx) = extinction coefficient (a and r = absorption and reflection, respectively, by atmospheric gases comprising our atmosphere and x = thickness of the atmosphere).

Sun angles become more oblique with increasing latitude. Thus, solar radiation must pass through more of the atmosphere before striking the surface. Again, weaker radiation reaches Ushuaia than locations at lower latitudes with higher sun angles. These two basic radiation laws work together to help protect the human population from excessive incoming ultraviolet radiation, in essence, it is the geography of the location that is critical here.

The Estacion VAG (Global Atmospheric Watch) Station in Ushuaia is the official ozone monitoring station for the region. This United Nations Environmental Program (UNEP)/World Meteorological Organization (WMO) network contains six locations, Ushuaia being the southernmost. There I met with Ingeniero Sergio Luppo. This former Argentine Air Force meteorologist agreed that he felt no danger of living in Ushuaia. 'The sun angles are too weak,' he said, 'and people cover up with clothing in the early spring.' Luppo echoed a response I had heard at CADIC: 'It is much more dangerous to go to the beaches near Buenos Aires in summer than to be here during the ozone hole' (Luppo, 1998).

National or regional project level

To view the issue from a national standpoint, a number of research facilities were visited. At the headquarters of the Argentine Antarctic

Institute in Buenos Aires, a meeting with Licenciado Alberto Cazeneuve, Chief of the Atmospheric Sciences added another view. According to Cazeneuve (1998), the ozone hole expands to include Ushuaia only a few days per year and therefore is not a constant danger to the people living there. He went on to explain that the scientists working at Argentina's Antarctic research bases are safe because these bases only operate during the summer and are not inhabited during the ozone minimum. Cazeneuve did state that the Argentine researchers in the Antarctic are finding the effects of UV radiation on plankton density and diversity. That is their major concern.

Antarctic research stations

Farther within the land of the ozone hole, three research bases in the Antarctic were visited. Taken chronologically, the first was the Estacion Admirante Brown, an Argentine facility in Paradise Bay. There, marine biologist Dr Rufino Comes granted an interview. He stated that their major research interest concerns marine plankton. Their findings indicate that these tiny sea creatures are indeed showing the effects of increasing UV radiation. The plankton populations have decreased with increasing UV radiation, and their role at the base of the marine food web causes great concern at this time (Comes, 1998).

The next visit was to the American research station at Palmer, which deals with terrestrial plants, phytoplankton, and other marine life forms to a depth of 10m (the depth of penetration into the sea of UVB radiation). These researchers have also found delicate marine fauna and terrestrial flora are adversely affected by increased UVB radiation (Arens & Pineda, 1998). These marine biologists felt they had little to worry about since they are not actively working during the greatest extent of the ozone hole. However, Kevin Bliss, who operated the official UV radiation monitoring station for the US National Science Foundation Polar Network, expressed a different concern. According to Bliss (1998) people experience a distinct difference between doing inside analysis work and outside fieldwork. There is concern about bad sunburn and cataracts among those who work upon the highly reflective snow and ice surfaces. Usually, sun block, hats, and protective clothing become the standard procedure, after the first severe sunburn is experienced.

The last of the Antarctic stations was Port Lockroy, a British Station in Dorian Bay. The two staff members, also seasonal inhabitants, were involved with little else than maintenance duties and monitoring the local penguin population. When asked about the ozone hole they graciously handed me an official pamphlet published by the British Antarctic Survey Office in Cambridge, England, and offered no other comments.

Health-related research

According to the Ministry of Health and Social Action Offices in Ushuaia, there has only been one research study dealing with humans and the incidence of skin disorders in the Tierra Del Fuego (de Calot *et al.*, 1994). When comparing their results to the general Argentine population they found no significant increase in dermatological disorders for the inhabitants of Ushuaia. The authors of the study pointed out however, that the average age of the people in Usuhaia, a relatively new and growing city, is only 27 years old. This is generally too young an age group for melanoma and other tumors to manifest. The youthful sample, coupled with the protection by the cold weather clothing, were reasons suspected for the low incidence of skin tumors.

The Perceptions of the Mass Media

Professionals in the mass media often render valuable insight into environmental issues and how they affect the *perception of the populace.* While in Buenos Aires, interviews were sought with professionals working in television, newspapers, and news magazines.

The Argentine mass media

In trying to establish the stance of the Argentine television media, I visited the offices of Cronica TV, the main all-news cable station in Buenos Aires. One of the news directors, Ms Elena Sambeca, stated that the station would not cover any stories unless it comes directly from the National Meteorological Service (SMN). The SMN headquarters are also located in Buenos Aires. This national television network relies totally on what is given to them by official sources and does not tend to initiate environmental stories (Sambeca, 1998).

During field investigations in Ushuaia, a visit to the local newspaper, *El Diario de Tierra del Fuego,* found the editor, Licenciado Fulvio Baschera, available for an interview. While agreeing that the ozone hole was an important issue, he relies upon CADIC to send any relevant data. He waits for the local scientists to decide whether or not a story concerning the hazards of ultraviolet radiation is newsworthy (Baschera, 1998).

The lack of media involvement was also noticed in other parts of the country. While visiting the National University of the Patagonia in Trelew, Ingeniero Jorge Pedroni, Department Head of Physics in the Faculty of Engineering was available for an interview. Pedroni monitors incoming UV radiation in Trelew (independent of the NSF Polar Monitoring Network), and reports the data to the local newspapers. His department publishes a health advisory informing the public of the danger levels of UV radiation (Pedroni & Massoni, 1996). Pedroni stated that the newspapers in Patagonia do not always print the advisories (Pedroni, 1998).

American mass media in Argentina

At the National Desk of the *Buenos Aires Herald*, the English language newspaper of the capital city, editor Joseph Schneider offered an interesting view. He explained that news stories concerning environmental issues were rarely found in newspapers. 'We used to have an environmental page in our paper, but we had to drop it due to lack of advertising dollars,' Schneider said. Furthermore, he was unaware of any major stories concerning the ozone hole in recent years in Argentine print media (Schneider, 1998).

While at the *Buenos Aires Herald*, I also met with Mr Joe Goldman a correspondent for ABC TV News in Argentina. According to Goldman: 'Few if any, environmental magazines can take-off in Argentina, because they lack the advertising from big business sponsors. Environmental issues are downplayed, rarely can you find articles written about these topics' (Goldman, 1998). Goldman mentioned a story written by his colleague, and correspondent for *Time Magazine*, Uki Goni, as the last article he could remember about the ozone hole.

An interview with Uki Goni (1998) and a subsequent reading of his article (Goni, 1997) revealed a rare occurrence, during mid-May 1997, when an area of ozone depletion was found outside the Antarctic region. This area centered over the southern middle latitudes and encompassed the cities of Buenos Aires, Argentina, and Santiago, Chile. Goni asserted that the National Meteorological Service downplayed the episode, and did not release data until well after the event. The Chilean government in contrast, did release data and warnings to the general public especially in and around Santiago. The Goni article was highly critical of Argentina's failure to warn its people of possible danger.

Economic pressures affecting the media

Discussions with news media professionals, confirmed one of my thoughts at the outset of this study. Argentina's economy relies heavily upon two large industries. The 'ozone hole' issue affects both. The first is cattle ranching and the exportation of beef. This activity involves the process of meatpacking and requires refrigeration, which relies heavily upon ozone-destroying Freon (commercial name for some chlorofluororcarbons). Any issues real or perceived that interfere with the use of this vital chemical coolant is bad for business. The second vital economic activity is tourism. Any fear, real or perceived, strongly affects this highly competitive industry. When people are afraid to visit a country, it's bad for business. It was suggested by these media contacts that, because of these two reasons, the Argentine government and mass media seem reluctant to cover this environmental issue.

The Government Perception

In an attempt to gain the *government perception*, the offices of the Department of Natural Resources and Sustainable Development in Buenos Aires were visited. There, Secretary Roberto Kurtz was available for interview. Secretary Kurtz was the representative for Argentina at the Montreal Protocol, during the summer of 1987. (The Montreal Protocol was an international conference where the curtailment of ozone-destroying chemicals was outlined and negotiated.) Kurtz was able to supply a written copy of the official Argentine stance presented at Montreal. In addition, he also rendered a copy of *Chemical Compound Law 24.040*, which defines national legislation concerning the manufacture, restrictions, and handling of ozone-destroying compounds. Secretary Kurtz stated that the ozone issue is taken seriously by the Argentine government, especially with respect to terrestrial and marine flora and fauna. However, he did not extend that fear to the human population. 'People in places like Ushuaia, run no more of a risk than people smoking a pack of cigarettes every day,' he said (Kurtz, 1998).

A visit to the headquarters of the National Meteorological Service (SMN) in Buenos Aires, found most of the administrators and professional staff attending an out-of-town conference. With access to their library, I spoke with librarian Maria Valez. She directed me to published government pamphlets concerning atmospheric ozone, and the ozone hole (Fuerza Aerea Argentina, 1987, 1993). She stated that no special warnings have ever been published for people living in the south (Valez, 1998).

Perceptions of Tourists and Tourism Professionals

As a visitor to the 'land of the ozone hole' myself, I was particularly interested in the attitude of tourists. The opportunity to speak with people about this issue often came about while on various tours around the country. Two such interviews with tourism professionals seemed to tell an interesting story with remarkable similarity. These professionals live in two different parts of the country and work in two different types of tourism.

Argentine tourists

The first interview took place at the Nahuel Huapi National Park, in the Lake District of the Andes Mountains near San Carlos De Bariloche.

There, National Park Ranger Silvina Arrido was adamant in her comments. 'The sun is killing us! . . . It gets worse each year,' she stated. 'The Park Rangers used to apply crèmes with SPF 45 or 50 as protection; now we use total sun block if we can get it.' When asked about how the tourists react to this environmental issue, she replied: 'They don't! . . . They act as if the sun cannot hurt them. They ignore my warnings and advice' (Arrido, 1998).

Some of these statements were echoed at a later time during a visit to the Valdez Peninsula on the East Coast of the Argentine Patagonia. Rogelio Rhys is a certified guide for NaturaTur of Chubut, a private tour company in Patagonia. He called the ozone hole 'a very, very serious problem, both for the quality and quantity of life . . . I never leave the house without applying sun crèmes with SPF of 45 or greater.' According to Rhys, the incidence of both sunburn and sunstroke among his tour groups has risen in recent years. When asked about the attitudes of the tourists themselves he replied: 'People don't care . . . they don't listen. To them, it is too silly to wear a hat or to put on sun crème. The Wildlife Rangers have only recently started to protect themselves, but the general public does not care' (Rhys, 1998).

Ingeniero Jorge Pedroni of the National University of the Patagonia, who reports UV radiation data to the local newspapers, made similar comments. 'People live for today. They don't care about skin protection. If the problem is not immediately life-threatening, Argentine people will not react,' he asserted (Pedroni, 1998).

Foreign tourists

Foreign visitors to the 'land of the ozone hole' were encountered onboard the *M.S. Explorer* while sailing to the Antarctic Peninsula. This ship is owned by Explorer Shipping Inc., an American company but subcontracted by Abercrombie and Kent Ltd, a British firm. In addition to American and British, there were also French, Italian, and Brazilian tourists on board. Explorer Shipping Inc. allowed me passage onboard as a scientist needing transportation for Antarctic research. Full passage was granted in exchange for guest lectures and help with weather forecasting. The ship's administrators were adamant about interviewing the passengers. I was not to interrogate them, or bother them with written questionnaires.

When learning that I was a climatologist, the tourists spoke rather freely about their views concerning weather and climate, and then comments about the ozone hole could emerge. During casual conversations, I asked three questions. The first, have you ever heard of the ozone hole? Second, if so, do you know what it is? Third, how do you feel about visiting this area of the world? From a total of 34 respondents I identified four categories: Ignorant, Informed, Defiant, and Zealous. The 34 people were composed of couples thus making 17 each male and female. The youngest respondent was 33 years old the oldest was 82. These 34 were chosen from the American, British and other English-speaking passengers so that the conversations could proceed in English.

Total ignorance of the issue was rare, and at times comical. One woman had never heard of the ozone hole. Another thought that the ozone hole occurred in the ocean. The second question was added to determine the level of familiarity with the issue. Other comments included 'I've heard about the ozone hole, does it happen around here?' Fortunately those who lacked of knowledge of the issue constituted only about 18%, or six of the tourists interviewed. The level of defiance was about the same at five people or about 15%. This category was found among senior citizens. Their comments included: 'At my age! I couldn't care less about it'; 'Smoking gets your lungs, drinking kills your liver, now the sun gives you skin cancer'; 'I'm going to live my life the way I want!'

The 'Informed' category increased to around 56% or 19 of the tourists with comments like, 'We're only here for a week or two, I think a short exposure would be OK' and 'I think the ozone hole is not too bad in February.'

Only about four of the tourists (about 12%) were classed as Zealous. Those who appeared on deck, even on warm days totally clothed from head to foot. Others constantly applying sun block and offering protection to anyone they thought was not sufficiently protected.

Conclusions and Recommendations

As with any perception study, one set of conclusions is replaced by several sets of conclusions. Here, people expressed their own views and establish levels of perception. From a *scientific perception,* the ozone hole poses little danger to human life in high latitudes. The low sun angles do not yield high intensities, and cloud cover plays a more important role than originally thought. High latitude geography also produces colder climates where people protect themselves with warm clothing. Keeping warm and protection from UVB radiation both occur with a winter/spring wardrobe. The scientists studying in the Antarctic often conduct their experimentation only during the summer months, after the ozone hole reaches its yearly recovery.

Government agencies are dedicated to writing legislation to deal with the issue. The enforcement of those laws, however, may be a different story.

The news media is split between two perceptions. The *Argentine media,* which relies totally on the initiative of scientists and government officials to report a news story, and *American journalists,* who insist that there is a real resistance to this issue. The *economic pressures* of the meatpacking industry and the travel industry may be the cause of the seemingly disinterested views of the Argentine businesses, and their reluctance to advertise in news stories that focus upon environmental issues.

Tourists in southern Argentina and Antarctica exhibit the widest variety of perceptions of this issue. The *Argentine tourists* seem to take a devil-may-care attitude towards the concept of harmful UVB radiation and seem resistant to self-protection. While Argentine tourism professionals, were adamant about their fears of the ozone hole. *Foreign tourists* span the range from total ignorance of the issue to a very guarded attitude toward self-protection.

Recommendations to the tourism industry are quite simple in this particular case. Unlike some environmental issues where people act against the advice of scientific research, here science permits the defiance towards this perceived fear. It could be recommended to the Argentine tourism industry to create a short, colorful, reader-friendly science report. This document (perhaps a brochure) could state why humans have little to fear in the 'land of the ozone hole.' At the same time this report/brochure may recommend some basic protective measures against exposure to intense solar radiation. Such a report, or colorful brochure can display some of the spectacular sites found around the Tierra Del Fuego and the Antarctic Peninsula. From these sources, a more intensive advertising campaign highlighting the scientific facts, can dispel any fears, and can promote tourism.

Acknowledgments

The author wishes to thank Western Kentucky University for granting the sabbatical leave necessary to complete the fieldwork portion of this study. Appreciation also goes to the Southern Regional Education Board, and Explorer Shipping Inc. for their monetary grants and in-kind support. The author wishes to thank all those who participated in this study. The open access to many Argentine governmental offices was a pleasant surprise. Appearing unannounced to interview various individuals was remarkably successful. The people of Argentina, from average citizens to top-ranking government officials, were generous with their time, and gracious in their demeanor.

References

American Meteorological Society (1998) Antarctic ozone hole sets new record. News and Notes. *Bulletin of the American Meteorological Society* 79, 2570–2571.

Arens, W. and Pineda M. (1998) Marine Biologists, Palmer Research Base, Antarctica. Personal communication.

Arrido, S. (1998) Park Ranger, Nahuel Huapi National Park. Personal communication.

Baschera, F. (1998) Editor, *El Diario de Tierra del Fuego*. Personal communication.

Bliss, K. (1998) US National Science Foundation Polar Network for Monitoring Ultraviolet Radiation. Personal communication.

Cazeneuve, A. (1998) Chief of Atmospheric Sciences, Argentine Antarctic Institute in Buenos Aires. Personal communication.

Comes, R. (1998) Marine Biologist, Estacion Admirante Brown, Antarctica. Personal communication.

de Calot, M. *et al.* (1994) Estudio de las enfermedades dermatologicas en Ushuaia. *Ministry Health and Social Action* 1, 1–7.

Diaz, S. (1998) US National Science Foundation Polar Network for Monitoring Ultraviolet Radiation. Personal communication.

Diaz, S. *et al.* (1994) Effects of ozone depletion on irradiances and biological doses over Ushuaia. *Archives of Hydrobiology and Limnology* 43, 115–122.

Diaz, S. *et al.* (1996) Solar ultraviolet irradiance at Tierra del Fuego: Comparison of measurements and calculations over a full annual cycle. *Geological Research Letters* 23, 355–358.

Fuerza Aerea Argentina (1987) Ozono atmosferico. *Boletin Informativo, Comando de Regiones Aereas, Servicio Meteorologico Nacional* 34, 1–20.

Fuerza Aerea Argentina (1993) El Aagujero de ozono Antártico. *Boletin Informativo, Comando de Regiones Aereas, Servicio Meteorologico Nacional* 56, 1–32.

Goldman, J. (1998) Correspondent for ABC Television News. Personal communication.

Goni, U. (1997) Ozone hole over Buenos Aires. *Time Magazine* 5, 17–19.

Goni, U. (1998) Writer for *Time Magazine*. Personal communication.

Kurtz, R.H. (1998) Secretary, Department of Natural Resources and Sustainable Development in Buenos Aires. Personal communication.

Luppo, S. (1998) Global Atmospheric Watch, United Nations Environmental Program (UNEP)/World Meteorological Organization (WMO). Personal communication.

Pedroni, J. (1998) Department Head of Physics, Faculty of Engineering, National University of the Patagonia. Personal communication.

Pedroni, J. and Massoni, M. (1996) Se viene el tiempo de playa! A disfrutar del sol! Pero. *Universidad Nacional de Patagonia San Juan Bosco, Facultad de Ingenieria and Departamento de Fisica, Informativo Bulletin No. 1*, 1–2.

Rhys, R. (1998) Certified Guide for NaturaTur of Chubut. Personal communication.

Sambeca, E. (1998) Cronica TV, Cable news station in Buenos Aires. Personal communication.

Schneider, J. (1998) Editor of the National Desk of the *Buenos Aires Herald*. Personal communication.

Trapasso, L.M. (1982) Drought: Its regional perceptions and expressions. *Geographical Perspectives* 50, 58–63.

Trapasso, L.M. (1992) Deforestation of the Amazon: A Brazilian perspective. *GeoJournal* 26 (3), 311–322.

Trapasso, L.M. (1994) Indigenous attitudes, ecotourism, and Mennonites: Recent examples in rainforest destruction/preservation. *GeoJournal* 33 (4), 449–452.

Valez, M. (1998) Librarian, Servicio Meteorologico Nacional, Buenos Aires. Personal communication.

Chapter 16

'Everyone Talks About the Weather . . .'

KEITH DEWAR

Mark Twain's famous comment, 'everyone talks about the weather but no one does anything about it', is often held up as a truism. However, as Twain himself might have said, 't'ain't necessarily so'. Humans may not be able to alter the day-to-day weather but they do alter their behaviour to either avoid or take advantage of climatic conditions. These changes in behaviour can be modified, if tourism businesses understand them and their exact cause, for the mutual benefit of both customer and business.

How visitors behave is affected by four weather-related variables:

- on-site weather;
- conditions at the trip origin;
- the forecast;
- conditions anticipated by traveller. (Perry, 1972)

Perry could have added that the same factors influence individual tourist businesses. To understand peoples immediate reaction to weather it is useful to divide their behaviour into two broad categories: direct and mediated.

Direct Weather Effects

Unexpected weather, be it rain or heatwave, has very specific effects on the public and businesses it supports. Information gathered from *Amusement Business,* the US trade publication, shows varying but important changes in attendance and hence income for events such as fairs. Table 16.1 shows the effects of 'bad weather' on major fairs in the US between 2000 and 2002.

Complexity is added to Table 16.1 by synchronicity (discussed later) and the intensity of the weather event. The direct influence of Hurricane Isadore on the North Georgia State Fair was considerable while, on the

Table 16.1 Fairs and poor weather

Event	*Attendance[1] Year 1*	*Attendance Year 2*	*Percentage change*	*Weather condition*
Maryland State Fair	486,216	348,848	–28.3	Rain
Great Allen Town Fair	642,000	547,000	–14.8	Rain
Montana Fair	229,254	212,423	–7.3	Wind
Pensacola Interstate Fair	583,127	479,927	–17.7	Rain
Lake County Fair, Florida	63,701	62,452	–2.0	Rain
South Florida Fair	747,517	659,263	–11.8	Cold
Florida State Fair	559,181	493,919	–11.7	Rain and cool
North Georgia State Fair	282,000	210,000	–25.5	Hurricane Isadore
Siskiyou Fair, Calfornia	59,754	55,279	–7.5	Smoke/haze
South Carolina State Fair	555,101	546,381	–1.6	Rain
Permian Basin Fair, Texas	64,000	63,000	–1.6	Wind
Ozark Empire Fair	217,000	216,703	–4	Hot dry weather (heat index 110)

Note: 1. Attendance in 2000 or 2001, the second year listed is the following year either 2001 or 2002.

other hand, the Mid-South Fair in Memphis was not as affected even when Isadore dumped seven inches of rain on the fair (Burnside, 2002b). Attendance at this second fair actually went up 6600. The prediction of Isadore's arrival combined with lower gate prices had an effect. Just as important was proactive promoting by the fair organisers. The fair manager moved a special 'wrist band discount day promotion' back several days to take advantage of predicted good weather before the storm. It worked, the weather on the special promotion day was sunny and attendance was excellent (Burnside, 2002a).

It is important to remember that not all weather changes are negative. *Amusement Business* also reports increases in attendants at fairs when there is good weather (Table 16.2). However, good weather does not seem to have the same intensity of effect as poor weather.

Water-based recreation is another example central to a great deal of tourism. The Rideau Canal in Eastern Ontario, Canada is a recreational waterway that is weather dependent. This 196km canalised heritage waterway runs between Canada's capital city Ottawa and the city of Kingston on the eastern end of Lake Ontario. The speed of movement and the general behaviour of the 'boaties' are directly affected by weather. Lockmasters will tell you 'in cold weather or very hot weather nobody moves'. An examination of the number of lockages compared to weather conditions specifically temperature, rainfall and humidity over a three-year period shows that 32% of the variance from average daily boat movements is the direct result of temperature. Rain has surprisingly little effect, accounting for only 4% of the variance.

Observation by canal staff also suggests that local boaters are more sensitive to the weather and tend to 'stay home' if weather is bad, while those boaters that have come from considerable distances will brave the conditions to meet timetables or simply not to 'waste time'. Similar observations were noted in other major attractions such as national parks. Domestic tourists and recreationalists will move out if there is poor weather while long-haul travellers are much more likely to be found out-of-doors 'experiencing' the destination, as one middle-aged visitor to the Franz Josef glacier in Westlands National Park, New Zealand pointed out: 'I have come ten thousand miles to see this glacier and I'll be damned if a little rain is going to stop me.' Bad weather as perceived by the visitor also leads to visitors in both of these destinations clustering in specific areas. On the Rideau Canal the urban lockstations are a major focus. In the national parks the closest major town is usually a common 'waiting spot'. Bars, cinemas and related recreational facilities in these areas all note an increase in business during these times. This is particularly true if the weather event is of short duration, usually less than 48 hours. Where a canoe concession may suffer, the local tavern may profit (Folgero, 1993).

Table 16.2 And the good news

Event	*Attendance[1] Year 1*	*Attendance Year 2*	*Percentage change*	*Comments*
California Mid-Winter Fair and Fiesta	95,595	101,027	5.7	Good weather
Nevada State Fair	65,000	73,000	12.3	Excellent
Sunfest 2002, West Palm Beach, Florida	305,000	325,000	6.6	Very hot
Alabama National Fair & Agricultural Show	221,500	227,500	2.7	Good weather
Philidelphia County Fair	83,250	100,000	20.1	Good weather, better promotion
Big Fresno Fair, California	539,607	550,619	2.0	Better weather
New Jersey State Fair	204,672	213,200	4.2	Lower gate as well
North Dakota State Fair	208,626	211,126	1.2	Good weather

Note: 1. Attendance in 2000 or 2001, the second year listed is the following year either 2001 or 2002.

Direct effects of weather on visitor behaviour are relatively straightforward although factors such as intensity of the weather event, promotional activities, distance travelled, age and cultural background may add complexities. More difficult to determine is the indirect or mediated effect of weather.

Mediated Effects

Mediated effects are where there is an indirect response by tourists to a weather event. In these situations it is not the effect of the weather on the individual that causes the behaviour but some intermediary. For example, northern Australia tourists visiting local beaches may find them closed due to the lethal box jellyfish (*Chironex fleckeri*) that invades the area during the period October to April. Onshore breezes can bring them to beach areas in considerable numbers.

Another good supply side example of this mediated or indirect effect was the reaction of a sponsor to conditions during the State Fair of Texas in 2002. Here a lucrative and mutually beneficial sponsorship was worked out between the fair and the '7–Eleven' chain of convenience stores. The promotion involved highlighting 7–Eleven's signature product, the Slurpee, a shaved ice drink made of brand name soft drinks and shaved ice. During the 24 days of the fair it was cooler than normal and 12 inches of rain fell and so did the attendance. Slurpee sales were, as a result, not what the company expected, although they would not disclose exact figures. This resulted in the company 'rethinking' their sponsorship. The impact on the State Fair's earnings could be significant (Burnside, 2003).

Besides direct and indirect aspects of weather there are other variables that add to the complexity of trying to understand visitor behaviour. An incorrect weather forecast that predicts fine weather may lead to a high degree of disappointment remembered long after the event. This weather memory (Perry, 1972) may play a role in future trip decisions to a particular destination both for the visitor who experienced the poor weather as well as those they told of their experience.

Another variable that affects behaviour is synchronicity: the weather conditions at the time of any event can have considerable impact on the visitor's perception of the destination. The 2002 and 2003 Tulip Festival in Ottawa was a classic example of the direct effect of bad weather occurring at the height of the festival. The festival attracts some 500,000 people to the downtown area of the city. In 2002 and 2003 unusually cold, wet weather caused serious problems with the festival closing a day early in 2002. There was a loss of revenue C$100,000. For the two years approximately 7000 fewer visitors came to major events, based on a high of 94,000 at the events in 2001. By 2003 the number stood at approximately 79,000 (Anonymous, 2003). The Tulip Festival also suffers from mediated effects. Unseasonably cool weather can slow the opening of the tulips and the full display falls after the festival is over. People therefore do not visit the downtown area and this can have an effect on the local stores and hotels. Spring festivals, particularly those that depend on flowers, are high-risk ventures.

The simple difference between weekends and weekdays is one of the major factors in determining visitor attendance. A combination of weather and a weekend has a powerful effect on many market segments (Brandenburg & Arnberger, 2001). In the case of the Rideau Canal, weekends account for 50% of the boating activity. Rain on a weekend can therefore have considerable effect on profits. This is particularly true when it is realised that the tourism business in the area is largely summer seasonal, there are only eight to ten 'summer weekends' and poor conditions on even two or three of these can wipe out profit margins.

Destination Response

Tourism businesses must respond to unpredictable weather in a way that minimises risk and optimises profits. Plans can be developed and put in place to mitigate the unwanted weather event when it does occur or to enhance good weather periods. Proactive planning needs to be done to deal with situations at the onset or during the problem event. The planning should include proactive and durational measures, and a relaunch plan is also essential for more long-term and severe weather occurrences (World Tourism Organisation, 1998).

Some of the more important things that the tourism operator can do include:

- Weather-based infrastructure design.
- Proper contingency planning.
- Promotion and advertising.
- Selling the problem.
- Personal adaptation and flexibility related to service quality and management.

Infrastructure design and adaptation

In the 19th century as resorts developed around the world ways had to be found to keep people both cool in hot weather and warm in cool weather. The classic colonial porch or veranda is an excellent example of this form of architecture. Also called a wrap-around or walk-around verandas, these structures offered the visitor the opportunity to sit outside in a warm spring or summer rain and enjoy the view and interact with other guests. In hot weather the veranda similarly acted to protect the guest from the sun and provided a breezy location. Related architectural features such as Virginia breezeways, and sunrooms are similar ideas that do not separate visitors completely from the elements but let them experience the weather conditions with a degree of comfort. Covered entranceways, enclosed walkways, canvas awning along main tourist avenues, gazebos, picnic shelters and similar structures all help visitors deal with unfavourable weather conditions.

Saint John in New Brunswick developed a C$23 million 'Pedway' that runs from the historic city market, a major tourism stop, for approximately one kilometre through two shopping malls, two hotels, a convention centre and ends at the 7000-seat enclosed Harbour Station, a stadium. 'It's the best $23 million the city every spent' remarked the manager of Harbour Station. Saint John is a small maritime city that experiences considerable and fast weather changes from fog to rain and high winds. The walkway allows locals and visitors including some 75,000 cruiseship passengers to sightsee and shop in a large part of the downtown area when outside conditions are poor.

The tourism industry needs to build more appropriate infrastructure as engineer Tony Gibbs points out in relation to the Caribbean:

> In the Commonwealth Caribbean, there are very few legally enforceable infrastructure and building design standards . . . The Caribbean is located in an area of the world exposed to multiple hazards, and if sustainable development is to be achieved, there is no other option than to counteract these natural hazards by designing and constructing resistant facilities. This is feasible. . . . (cited in Miller, 2000)

Contingency planning

Proactive management is central to best practice. Some of the measures that can be taken as part of contingency planning are indicated in Box 16.1. The insurance companies, the event organiser and businesses all understand the cost of poor weather. Weather combined with increasing risk of terrorism are making insurance a much more important controlling variable in travel both for the industry and for the individual (Miller, 2000; TIAC, 2002).

Since 1996 new and creative ways of managing weather-related losses have been devised by insurance companies and other related financial institutions (Guaranteedweather, n.d.). Weather risk insurance includes commodity instruments called weather derivatives (Sytsma & Thompson, 2002; Bloomberg, 2003). These weather derivatives are traded on several exchanges in North America, Europe and Asia. They are all similar to insurance policies but:

- do not require proof of loss to settle the claim;
- do not always require an upfront premium;

Box 16.1 Contingency planning for weather conditions

Proactive elements:

- infrastructure design and adaptation to best suit weather conditions;
- security and storage of goods and data in case of severe or extended weather problems;
- proper insurance coverage;
- promotion and advertising (right expectations, discounts and promotions, rain checks, appropriate refund policy);
- development of emergency plans including a relaunch planning programme for severe weather events;
- selling the problem (Ice Hotel, storm chasing, storm hotels); and
- timing (extending altering sales dates, closures).

Immediately before and during a weather event:

- personal adaptation, provisions of clothing and equipment to reduce visitor discomfort;
- communication, keep customers, staff and the media appraised;
- service quality and flexibility, empower employees to make informed and realistic decisions;
- train staff in emergency procedures and have a working emergency plan.

Recovery and relaunch:

- continue open communication, a cooperative regional toll free phone line and a regularly updated website can be of considerable help;
- cooperation amongst local stakeholders to reopen the destination in the case of severe and long-term disruption of activity, this should be outlined in a proactive planning document or business plan;
- incentives and promotions, cross-selling among local businesses.

- have different accounting and tax treatments. (Koch Industries, 2000)

There are several ways insurance financing can be structured to help companies cope with periods of poor weather. One way is the direct 'swap' in its simplest form: the insurer analyses the weather data for the tourism businesses area of operation and an individual contract is worked out. If for example there are an average of 20 bad days for your tour business then for every day over that 20 the company receives a predetermined amount. For example, a whale watching company loses $10,000 for each bad weather day. If the average number of days per year when conditions prevent the boats from going to sea is under those 20 days then the company pays the insurer $10,000, but for each day over those 20 days the insurer pays the company $10,000.

A second way to insure is a capped contract. Here a direct premium is paid, for example $10,000 for each day beyond the 20 days and a fixed premium of perhaps 25% of calculated risk loss in this case $\$10,000 \times 20$ days $\times 0.25 = \$50,000$. There is a cap placed on how many days can be claimed, based on the calculated risk (Koch Industries, 2000). Many other companies offer flexible contract insurance systems to suit specific businesses. The Weather Risk Management Association's home page (http://www.wrma.org/) provides a starting point for those interested in delving into this complex field of risk capital management.

Promotion and advertising

Where weather is considered a problem, businesses can alter the way they promote their products. For example, hurricanes in the Caribbean result in a considerable disturbance of visitation levels with a disruption in cash flow. Here some airlines and destinations have responded by providing 'hurricane insurance'. Those whose visit is disrupted by a hurricane receive a second flight, nights at the resort free or other incentives. Often the insurance company and the business share the cost.

Store or site promotions can also be used. These can take the form of discounts if the temperature is over a certain temperature or if rainfall occurs during a particular period, there are many other examples of creative ways to insure and assure. It is also essential that there is truth in advertising. Several authors have pointed to the 'blue sky' syndrome (Perry, 1972; Robertson, 1993; Smith, 1993). As Smith (1993: 400–1) points out: 'Whilst it is undoubtedly true that few visitors to Scotland are drawn there by weather alone, their expectations may be raised to an unrealistic level by a dominance of "blue sky" photographs in the brochures issued by the various tourism boards'.

A similar problem was noted by Meyer and Dewar (1999) in examining over 75 brochures and advertisements for the West Coast of the South Island of New Zealand. All scenic photos were blue sky with the exception of two in material produced by Westland's National Park, a government agency. Considering the Westland's area of New Zealand gets between 3.2 and 5 metres of rain each year and the area is one of the most important temperate rainforests in the world the expectations of the visitors may not be met if they come for blue sky. This area of New Zealand is constantly on the defensive trying hard to deny their wet heritage. It is instructive to read the short news article from the Greymouth *Evening Post* to better appreciate the defensive posture of the West Coasters.

> Wet Image Challenged
>
> The phrase the wet West Coast is officially a misnomer, thanks to Grey District councillor Jacquine Grant, who took umbrage with an exhibit at the Museum of New Zealand, Te Papa. During a trip to Wellington Ms. Grant took exception to one of the museum's exhibits entitled Why is the West Coast so Wet? Ms. Grant thought the heading could well be detrimental to the coast's tourist industry, and tried unsuccessfully to get the museum to change its wording. Ms. Grant brought her concerns back to Greymouth, and the Grey District council went to the National Institute of Water and Atmospherics (Niwa). Niwa's figures show that on average Auckland records 190 days rain a year a mere 15 days behind Greymouth's, Hokitika has 209, Kaitaia 194, and Invergargill 220. Niwa's report says the coast's total annual rain days are not much more each year than the country's northern areas of Auckland and Northland, and are less than Invercargill. (*Evening Post*, 1998)

Hartz Mountain National Park Tasmania and the surrounding area also face similar advertising and promotional problems. Robertson (1993: 9) describes the problem this way:

> Unfortunately current promotional information often portrays Hartz Mountains as either a 'fair weather park' with magnificent views or a blizzard-ridden place to avoid. The reality is the grey skies, mountains peeping out from banks of wind-blown clouds and intermittent rain is closer to a typical day.

He goes on to suggest how the National Park is changing it's marketing: 'rather than say "bad weather – go home", the message visitors will be receiving is that wild weather is a part of the mountains. Be prepared and experience some of the raw power of nature that has shaped (and still is shaping) the Southwest wilderness' (Robertson, 1993: 9).

The conditions a visitor will face must be part of proper promotion. The raising of expectations on the hope that tourist will experience a 'good day' is a risk that should be avoided. The intensity of 'weather memory' is directly affected by the variation between expectations and perceptions of the weather event.

Selling the problem

If the tourism business is in an area where particular weather patterns are a problem then it is often possible to sell the conditions. Storm chasing in the Midwest of the United States has become a lucrative tourism business during tornado season (McGraw, 1998). Hurricanes also attract visitors both on business, the media for example, and adventure tourists that are there to experience the conditions.

Another special interest tourism activity is storm watching. Perhaps less of an adrenaline activity than storm chasing it is providing a considerable economic boost to areas of Northwest North America whose business had been decidedly seasonal until the realisation that 'bad weather' can sell. Originally this approach was centred on Pacific Rim National Park and the two small gateway settlements of Tofino and Ucluelet on Vancouver Island, Canada. One of the first hotels to appreciate that wind and waves could make money was the Wickaninnish Inn in Tofino. The hotel has found a niche market for storm watchers. The visitors come to watch the heavy surf churned up by the great Pacific depressions as they roll in off the North Pacific Ocean.

The selling of storm watching in the area has become big business. The activity is best described from the place where the idea was said to have originated:

> What is Storm Watching? Storm watching is a relatively new tourism phenomenon on the West Coast of Vancouver Island, though locals have been watching, bracing against and surviving winter storms here for millennia. Storm watching, on the part of the observer, is predominantly a passive activity. It doesn't involve chasing water spouts or flying over the waves in a Zodiac boat. It doesn't really require anything of the observer but stillness and wonder. Storm Watching is best advantaged from ocean-fronting hotels and B&B's where a good book, a fireplace and maybe a down duvet across your lap are really all you need to experience the fury and the force of the pounding Pacific Ocean on the other side of your rattling, double paned window. (Tofina – Long Beach Chamber of Commerce Home Page, n.d.)

The idea has spread rapidly and storm watching tourism promotions can now be found on the World Wide Web from many places in Canada, Northwest United States and the UK. For example, Lake Huron's Bruce County, Ontario tries to draw visitors in the dreary November early December period by suggesting:

> Beautiful fall days are often intermixed with some of Mother Nature's most wicked weather. Winds quickly shift to the northwest, waves crash into the shoreline and seagulls ride the wind with ease. An awesome sight. The sheer power, and energy that you can experience will take your breath away. Become one with nature, get out and enjoy this spectacular time of year. (Bruce County, 2001)

The Ice Hotel is another attempt to live with and profit from weather conditions. The small town of Jukkasjarvi in Northern Sweden opened the first ice hotel in 1989 (Ice Hotel, n.d.). In 2000 it accommodated 6000 overnight visitors and 60,000 day visits in its 5000 square metre facility. The phenomenon has now spread to Duchensey, Quebec as and offers a stay on ice for prices running from C$250 to C$5000 per night with breakfast and dinner included. This half-million dollar, 32-room, 3000 square metre hotel operates for approximately 80 days from the middle of December to late February depending on the weather. In the two-year period 2001 and 2002 it accommodated 4500 overnight guests and 100,000 day visitors (Ice Hotel, 2000).

Service quality and personal adaptation

Companies can do several very simple things to assist visitors in cases of unexpected weather. Rain and sun can bring out umbrellas, preferably brightly coloured ones with the hotel or company name predominately displayed. Rain can also mean distribution of inexpensive gaily-decorated raincoats, again with the company name clearly visible. Less expensive items such as rain ponchos can be given away or sold at cost.

Being prepared for conditions and having contingency plans is not a new idea. Mr Jarrett owner and manager of the 145-year-old Opinicon Hotel, Chaffey's Lock in Canada, pointed out in 1934: '. . .uncontrollable conditions can necessarily be guarded against by creating a fraternal atmosphere that will eliminate boredom. A lending library is at the disposal of guests, and all open fire places are kept blazing at such times' (Jarrett, 1934).

Jarrett and other hostellers have long understood the importances of alternatives and diversions to scheduled activities. A free drink in the bar, $20 worth of chips redeemable at the local casino, or other forms of impromptu entertainment may make all the difference to the tourists' perception of a visit.

Local and regional tourism authorities can also work with private businesses to provide a diversity of activities to provide diversion in the form of activities not affected by weather or incorporating the event. Perry reports in a government study of Donegal, Ireland that: 'wet-weather facilities should be provided sufficient to entertain half the night visitors and one-eighth of the local population, together with one quarter of the daily average of visitors' (Perry, 1972: 200).

Travellers behave in complex and often unpredictable ways to day-to-day weather conditions. Understanding how they react gives industry the opportunity to develop facilities, activities and marketing strategies that can provide a higher quality experience to the visitor as well as reduce risk and increase profits.

References

Anonymous (2003) *Canadian Tulip Festival Attendance and Admission Analysis* [Table]. Ottawa: Canadian Tulip Festival.
Bloomberg (2003) Betting on the weather. *International Herald Tribune* 3 May.
Brandenburg, C. and Arnberger, A. (2001) The influence of the weather upon recreation activities. In A. Matzarakis and C. de Freitas (eds) *Proceedings of the First International Workshop on Climate, Tourism and Recreation* (p. 123). International Society of Biometeorology. Available at: http://www.mif.uni-freiburg.de/isb/ws/report.htm.
Bruce County (2001) *Storm Watching – Lake Huron and Georgian Bay*. Available at: http://www.naturalretreat.com/storms.htm (accessed 18 May 2003).
Burnside, M.W. (2002a) Mid-south weathers rain, up 6,600 at gate. *Amusement Business* 114 (40), 10.
Burnside, M.W. (2002b) Rain drops numbers for some, State Fair of Texas No. 1 again. *Amusement Business* 114 (51), 22.
Burnside, M.W. (2003) Slurpee takes gulp of fair sponsorship. *Amusement Business* 115 (6), 12.
Evening Post (1998) Wet image challenged. *Evening Post*, 8 September, p. 2.
Folgero, I. (1993) Blame . . . anything!: Causal attribution in the hotel sector. *International Journal of Contemporary Hospitality Management* 5 (5), ii–iii.
Guaranteedweather (n.d.) *Using Conditional Weather Promotions to Increase Sales and Customer Loyalty*. Available at: http://www.guaranteedweather.com/guaranteedweather/industries/retail/casestudies/conditionalweather_promo.html (accessed 5 April 2003).
Ice Hotel (2000) *Ice Hotel Quebec-Canada*. Available at: http://www.icehotel-canada.com/en/hotel.htm (accessed 29 May 2003).
Ice Hotel (n.d.) *Ice Hotel – Above the Arctic Circle, Beneath the Northern Light*. Available at: http://www.icehotel.com/ (accessed 29 May 2003).
Jarrett, D.P. (1934) Service quality. Canadian hotel management. In Opinicon scrapbook, Opinicon Hotel. Chaffey's Lock.
Koch Industries (2000) *Weather Risk Management Products for the Recreational Entertainment Industry*. Available at: www.kochweather.com (accessed 5 December 2003).
McGraw, D. (1998) Whirlwind tourism. *U.S. News & World Report* 124 (22), 58.
Meyer, D. and Dewar, K. (1999) A new tool for investigating the effect of weather on visitor numbers. *Tourism Analysis* 4 (3/4), 145–55.
Miller, D.J. (2000) *Caribbean: Plan Needed to Mitigate Impact of Natural Disasters*. Montego Bay, 15 January (IPS). Available at: http://www.twnside.org.sg/title/mitigate.htm.
Perry, A.H. (1972) Weather, climate and tourism. *Weather* 27, 199–203.
Robertson, S. (1993). Making a great park accessible: Hartz Mountains National Park. *Australian Parks and Recreation* 29 (3), 9–12.
Smith, K. (1993) The influence of weather and climate on recreation and tourism. *Weather* 48 (12), 398–404.
Sytsma, D.L. and Thompson, G.A. (2002) *Weather Risk Management: A Survivor of the Collapse/Demise of U.S. Energy Merchants*. R.J. Ruden Associates, Inc. Available at: http://www.rjrudden.com/rudden/media_articles.htm.
Tourism Industry Association of Canada (TIAC) (2002) *Tourism and Increasing Insurance Costs*. Available at: www.tiapei.pe.ca/TourismResources/TIAC Talk10–07.pdf+Hurricane+Tourism+Insurance+&hl=en&ie=UTF-8 (accessed 5 December 2003).

Tofina – Long Beach Chamber of Commerce Home Page (n.d.) *Storm Watching*. Available at: http://www.island.net/~tofino/stormwatch.html (accessed 18 May 2003).

World Tourism Organisation (1998) *Handbook on Natural Disaster Reduction in Tourist Areas*. Madrid: World Tourism Organisation and World Meteorological Organization.

Chapter 17

Climate Change, Leisure-related Tourism and Global Transport

PAUL PEETERS

Introduction

Environmental concerns in tourism date back four decades. In those days the emphasis was on 'planning/or tourism' (Cohen, 1978) cited by Theuns, 2001)) with a common characteristic of environmental impact analyses for tourism development being their focus on the local and sometimes regional level, without reference to the global impacts of tourist transport. Only recently, have the impacts of transport become recognised as important for the environmental analyses of tourism (see Becken, 2002; Ceron, 2003; Gössling, 2002; Gössling *et al.*, 2002; Høyer, 1999; Peeters, 2003).

Current definitions of tourism are very wide and may hamper environmental analysis. The international definition of tourism as given by UN (2001) includes all travel purposes, as tourism visitors are all travellers staying between one night and one year outside their usual environment. The remainder is defined as 'same-day visitor'. Transport statistics often define the 'tourism' motive as leisure related only. Therefore, in this chapter *leisure*-related transport has been taken as a basis for calculations, thereby excluding travel motives such as business, commuting and shopping, but including 'same-day' visitors. The World Tourism Organisation (WTO) (1998) also makes a distinction between domestic and international tourism arrivals. Though essential for national economic assessments, this parameter is not very useful for environmental analysis, because the number of international tourists will strongly depend on the size of the country, but does not give any clue on the actual distances travelled. The total effects on climate change of tourism transport (or other activities) may be evaluated with the following equation:

$$E = \sum_{m} (\beta_m * V_m)$$

In this form E is the total emission of greenhouse gases (GHGs), β the specific GHG emission (the 'emission factor' per unit of transport production) and V the total transport volume both for transport mode m. Data for β and V may be found in two ways: aggregated and disaggregated. The aggregated (top-down) method uses data on the total amount of fuel or energy use and total vehicle kilometres to find average aggregated values for β. With the specific leisure transport V from transport statistics the equation may be used to find the total environmental effect. The disaggregated (bottom-up) method tries first to find average β's for all different kind of tourism transport vehicles and their average kilometres per year.

Global Transport for Leisure

Passenger transport in 2000

The total volume of passenger transport in 2000 may be in the order of 26,000 billion passenger kilometres (pkm). About 60% of this is served by private cars, 13% by aircraft and the remainder by bus, rail and ship. These numbers have been calculated by examining and extrapolating data from Airbus (2002); Boeing (2003); Gössling (2002); IATA (2002); OECD (2000); Pulles *et al.* (2002); Schafer (1998); and World Business Council for Sustainable Development (WBCSD) (2001). Gössling (2002) assumes that about one-third of world travel is specifically for leisure and tourism (see also Chapter 19, this volume). Adjusting this estimate for air transport using a passenger survey in 2001 of Schiphol (2002), results in 8500 billion pkm for leisure-related purposes, of which 20% is travelled by air, 55% by car and 25% by bus, rail and ship.

Passenger transport in 2020

Schafer assumes 3.7% annual growth for motorised transport between 1990 and 2020 (Schafer, 1998; Schafer & Victor, 2000) giving 53,800 billion pkm total world motorised passenger transport in 2020. Of this total 31,000 billion pkm is realised by car and 12,200 billion pkm by bus, rail and ship, leaving 10,600 billion pkm for air transport. However, recent air transport scenarios by Airbus (2002) and Boeing (2003) suggest a lower value of 8300 billion pkm in 2020. The slightly older detailed air transport model AERO gives 9500 billion pkm for 2020 (Pulles *et al.*, 2002). From these an estimate of 9000 billion pkm in 2020 seems reasonable, reducing the total to 52,200 billion pkm.

The Vision 2020 project of the World Tourism Organisation (1998) concludes that the total number of international tourist arrivals in the world will grow at an average of 4.4% per annum between 2000 and

2020 to 1602 million. The share of long-haul tourism will increase from 18% in 1995 to 24% in 2020, which may translate into a growth rate of leisure-related pkm's of about 4.7% per annum. Within both short-haul and long-haul markets a shift towards longer travel distances seems likely, again increasing the growth rate of leisure-related transport. All these developments will boost air transport to an estimated share of 55%. The development of the less developed countries will raise the proportion of leisure-related travel. Therefore, it has been presumed the leisure-related share of the other two modes may increase to 40% worldwide. With these assumptions a total of 23,200 billion leisure-related pkm in 2020 has been found, with 53% by car, 25% by air and 22% other. The total growth rate of tourism and leisure-related transport will be 5.2% per annum.

The Climate Change Effects of Transport

Emission factors

The human induced greenhouse effect is mainly caused by fossil fuel-related emissions such as carbon dioxide and nitrogen oxides and, specifically for air transport, the forming of contrails and cirrus clouds at cruise flight levels (Houghton *et al.*, 2001; Penner *et al.*, 1999; Williams, 2002). The total effect on climate change can be expressed in the form of carbon dioxide equivalents (CO_2-e). The ratio between CO_2 emissions and CO_2-e emissions is the 'equivalence factor'. Table 17.1 gives the equivalence and emission factors for three transport modes. Eurostat (2000) gives ranges for emissions for European Union transport (second column). The average world-fleet factors for aircraft emissions are based on data from the AERO model (total freight plus passenger emissions and total passenger transport), as described by Pulles *et al.* (2002) (total emissions including freight), IATA (2002) (passengers traffic share of total) and Boeing (2003) (relative fuel efficiency of passenger fleet). The result – 138 gram/pkm – fits well within the given range. The equivalence factor for air transport is 2.7 as given by Penner *et al.* (1999) and recently confirmed by RCEP (2003: 15, 16). An average for cars is not easy to find, due to the diversity of cars and car use. For leisure, larger than average cars are used and caravans, trailers and roof-racks all increasing emission factors, but the average seat occupation may be higher, reducing emission factors per pkm. An average of 100 gram/pkm has been chosen. For other modes an average value from the European range has been chosen of 45 gram/pkm. The equivalence factor for cars and other modes given by Heart and Biringer (2000) for USA and Gugele *et al.* (2002) for Europe averages to 1.05.

Table 17.1 Global emission and equivalence factors per transport mode for leisure-related passenger transport in 2000

Transport mode	*CO_2 (range for Europe) gram/pkm*	*CO_2 (average value) gram/pkm*	*Equivalence factor –*	*CO_2-e (average value) gram/pkm*
Car	25–144	100	1.05	105
Air	82–482	138	2.7	373
Other	13–77	45	1.05	47

Source: see text (p. 249)

Total emissions in 2000

The total 2000 emission of leisure- and tourism-related passenger transport is 0.79 Gtons CO_2 and 1.22 Gtons CO_2-e. This is about 3.1% of the total anthropogenic CO_2 emissions from fossil fuel burn and industrial processes (as given in Appendix II of Houghton *et al.* (2001)). Using global warming potentials as given by IPCC gives a leisure-related transport share of 3.6% for total GHG emissions.

To find the total passenger transport emissions it is not appropriate to use the emission factors from Table 17.1. The emission factor of cars will be slightly higher, mainly due to lower seat occupancy for non-leisure-related travel motives. With some other slight corrections based on Pulles *et al.* (2002), Schafer (1998) and Gugele *et al.* (2003) the total passenger transport emissions found are 2.44 Gton CO_2, and 3.33 Gton-e CO_2-e (see Table 17.2). Leisure-related transport takes a share of 37% of CO_2-e

Table 17.2 Total emissions and relative shares of CO_2 and CO_2-e in 2000

2000	*CO_2*	*CO_2-e*	*% world*		*% target*	
	Gton	*Gton*	*CO_2 %*	*CO_2-e %*	*CO_2 %*	*CO_2-e %*
World	25.32	34.19	100	100	n.a.	n.a.
Passenger transport	2.44	3.33	9.6	9.7	53	55
Leisure-related transport	0.79	1.22	3.1	3.6	17	20
OECD/EST Target	4.62	6.01	18.3	17.6	100	100

emissions of the total passenger transport. Air transport dominates the leisure-related transport CO_2-e emissions in 2000 (see Table 17.3).

Emissions in 2020

To provide an estimate of the climate change impact in 2020, information is needed on the development of environmental efficiency per mode of transport. The final efficiency is the result of trends in technology, performance and operational factors. There are several competing trends that influence the energy efficiency of cars. Technology reduces fuel consumption by increasing engine and transmission efficiency and reducing aerodynamic drag and tyre-rolling resistance. These advantages are largely offset by increases in the average car size, weight and engine power (WBCSD, 2001: 2–12). However, there may be political influences at stake. The European Union has signed voluntary agreements with European, Japanese and Korean automobile industries, aiming at an average CO_2 emission target for new cars of 140g per vehicle kilometre (vkm) in 2008 or 2009 (Gugele *et al.*, 2002). In 2000 the fleet average was 172g/vkm. Though the biggest car travelling nation, the USA, does not have such schemes, and recent history does not show increased fleet efficiencies, it seems likely that progress will be made. The WBCSD (2001) report concluded that car travel growth will be 3.5% per annum and suggested world fuel use for road transport will increase at 2.8% per annum. Apparently a car fuel efficiency increase of 0.7% per annum has been assumed. For air transport, values for operational and technological efficiency increases are provided by Lee *et al.* (2001); Penner *et al.* (1999); and Pulles *et al.* (2002). From these figures an average efficiency increase of 1.3% for the period between 2000 and 2020 seems likely. The worldwide increases in efficiency for busses and rail have been estimated to be 1% per annum. Table 17.4 and Table 17.5 summarise the results for all emissions.

Total 2020 leisure-related transport emissions of CO_2 will be 1.86 Gton, or 2.98 Gton CO_2-e. This is 43% of total passenger transport CO_2 emissions and 49% of CO_2-e emissions. Leisure-related passenger transport

Table 17.3 Modal shares in % of CO_2 and CO_2-e emissions in 2000

2000	*CO_2*			*GHG*		
	Car	*Air*	*Other*	*Car*	*Air*	*Other*
World passenger transport	69	19	12	53	38	9
Leisure-related passenger transport	59	30	11	40	52	8

Table 17.4 Total emissions and relative shares of CO_2 and GHG (CO_2-e) in 2020

2020	CO_2	CO_2-e	*% world*		*% target*	
	Gton	*Gton*	CO_2 %	CO_2-e %	CO_2 %	CO_2-e %
World	41.07	53.39	100	100	n.a.	n.a.
Passenger transport	4.28	6.06	10.4	11.4	92	101
Leisure-related transport	1.86	2.98	4.5	5.6	40	50
OECD/EST Target	4.62	6.01	11.2	11.3	100	100

Table 17.5 Modal shares in % of CO_2 and CO_2-e emissions in 2020

2020	CO_2			*GHG* (CO_2 *eq*)		
	Car	*Air*	*Other*	*Car*	*Air*	*Other*
World passenger transport	68	22	10	50	43	7
Leisure-related passenger transport	57	33	10	38	56	6

CO_2 emissions equal 4.6% of all anthropogenical emissions as given for scenario A1FI in Appendix II by Houghton *et al.* (2001). Air transport will take up 56% of GHG emissions in 2020. The growth rate of CO_2 emissions between 2000 and 2020 is 4.4% per annum and for CO_2-e 4.6%. This means the climate change effects will grow at a slightly faster rate than international tourism arrivals of 4.4% per annum.

Mitigating Climate Change Effects

Sustainable development targets

The total amount of leisure-related transport CO_2 emissions in 2020 contributes 40% to the emissions target for 2030 (and 50% of the CO_2-e emissions target) as set by OECD (for OECD countries 80% reduction with respect to 1990 level in 2030 (OECD, 2000)). This means the 2020 amount of world leisure-related transport will be 50% of the total CO_2 emissions sustainable target for 2030. If the average international tourism volume growth of 4.4% per annum has to be sustainable by improving operational and technical efficiency, an efficiency increase with a factor of 25 between 1990 and 2030 will be required. This is far more than has ever been achieved historically within some previous decades. Therefore, sustainable development of leisure and tourism means volume reduction cannot be ignored. Both '-control' and '*V*-control' will be needed.

Operational 'β-control'

Car transport seat occupancy seems a likely candidate for increased efficiency as it now averages at 2.0 to 2.5 for leisure-related transport. However, the trend of reducing family sizes in the developed world may reduce prospects here. In Europe for example the average household size has decreased steadily from 2.8 in 1980 to 2.4 in 2000 (EEA, 2001). Electronic devices such as car navigation systems may reduce the amount of detouring. This may increase efficiency at a few percentage points. Car-pooling and hitch-hiking schemes have already existed for decades, but so far have had little effect on the seat occupancy of total transport.

Increasing seat occupancy in air transport will be difficult as it is already relatively high (70%) and has been stable during the last decade according to IATA (2002). An increase of ten percentage points may be assumed here. Increasing seat density may have more potential, but of course at the cost of comfort. Another way to increase aircraft fuel efficiency is by adjusting operational parameters such as speed schedules. Most airlines will try within the constraints, given air traffic management and air space, to fly a direct operating cost (DOC) optimum (see Figure 17.1). In this case a higher fuel price ($1/kg instead of $0.25/kg) may reduce fuel consumption by 8%. The same mechanism is at work for cruising altitude. Another operational measure is to optimise flight distance. The Boeing 747–400, for example, has the lowest fuel consumption per seat at the 5000–6000 kilometre range. A refuelling stop on longer flights may reduce fuel consumption at a few percentage points.

Finally, increasing air traffic management efficiency may reduce emissions at a maximum of 10% (see section 4.32 of RCEP (2003)). The seat occupancy of other transport modes differs considerably. Coaches show generally high rates (80–90%), high speed trains intermediate (50–60%) and conventional rail low occupancy rates (30–40%). Some room for increases may be assumed for rail. A general property of public rail systems is an increase in average seat occupancy at busy lines. With growing traffic levels, the allocation of carriages changes from schedule driven (the frequency of service) to demand driven, making it easier to fit supply to demand. Rolling stock management and route optimisation may also increase efficiency for both rail and bus companies. Therefore, it is important to generate large traffic flows when promoting rail

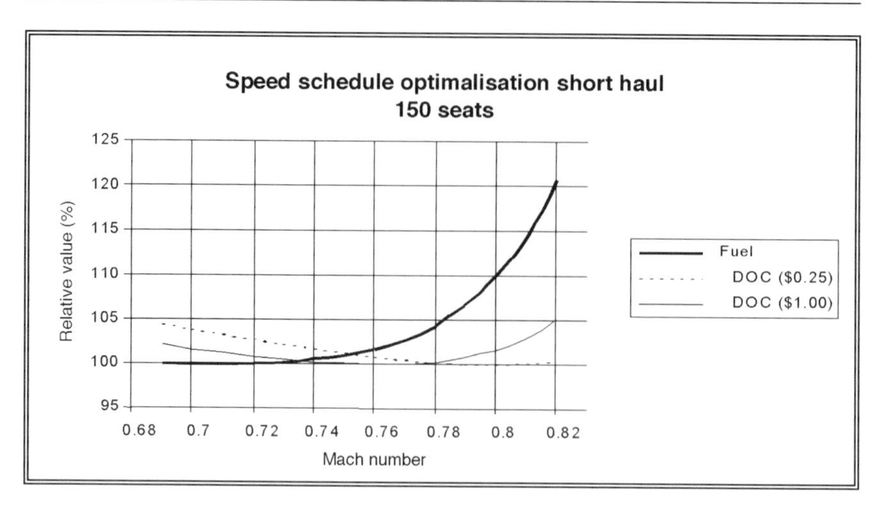

Figure 17.1 Speed optimisation curves for a commonly used short-haul aircraft for both DOC and fuel consumption ($ prices refer to fuel cost per kg)

systems. Ultimately, the effects of increased operational efficiency may be in the order 10% to 40% depending on the mode of transport and the market.

Technological 'β-control'

The technology solution detailed in the EST scenario's focuses on fuel cells and hydrogen vehicles (OECD, 2000). Other reports conclude the combination of hydrogen and fuel cell presents the best opportunities for ground transportation (ICCEPT, 2002; RCEP, 2003). Many experimental versions of fuel cell buses and cars are under development or already operational according to the WBCSD (2001). These technologies seem to become available for all ground and water transport modes. For electric rail transport, power stations are likely candidates for fuel cell technology (ICCEPT, 2002).

Hydrogen technology will only succeed in substantially mitigating climate change if it is produced with renewable energy such as wind, solar, geothermal and, to some extent, biomass. Though large resources are theoretically available, both spatial needs and costs will limit the final amounts to be exploited as shown by van den Brink (2003). These factors may reduce the opportunities for sheer volume growth of transport.

Aircraft may increase fuel efficiency with advanced aerodynamics, increased specific fuel consumption of the propulsion system and reduced construction weights. The operational introduction of these technologies in new aircraft depends on the long-term forecast of kerosene cost. The aircraft designer generally makes a trade-off between the aircraft acquisition cost, operational costs and performance. An example of the effect of high kerosene cost is given in Figure 17.2. The current DOC optimum design point for wing aspect ratio (span squared divided by wing area) is at about eight.

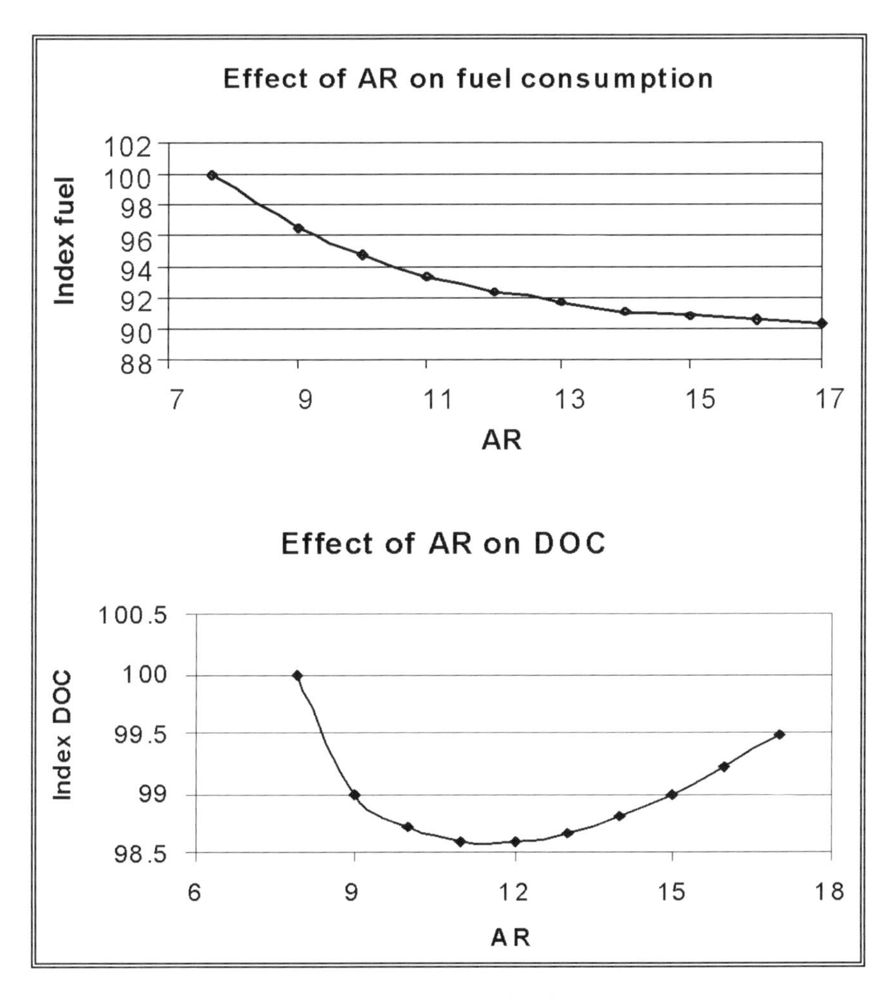

Figure 17.2 Effects of wing aspect ratio (AR) on fuel consumption and DOC for a new 400-passenger aircraft design with a high kerosene price of $1/kg

Source: Peeters (2000)

The average fleet emission factor may be reduced with these technologies at about 20% between 1997 and 2015 (Penner *et al.*, 1999: 220). For further reduction (up to 2050), a trade-off between fuel efficiency and low NO_x emissions will be necessary. Assuming a low fuel consumption scenario, a reduction in fuel consumption of 40–50% can be achieved, while in a low NO_x scenario the reduction will be about 30–40%.

All these figures seem close to the theoretical limits, assuming conventional technology. Including operational -control climate change from air transport may be reduced with a factor two to three. This is way behind the factor 25 required for sustainable development. Clearly, so more has to be done. Hydrogen combined with fuel cells seems the only solution to further reduce climate change effects. However, the use of hydrogen and fuel cells is seen as problematic by Penner *et al.* (1999) and RCEP (2003). On the other hand Snyder (1998) and Peeters (2000) both do not see basic technological or operational objections to fuel cell use in aircraft. The problem may be market introduction as these aircraft will probably have lower cruise speed at reduced cruise levels. Therefore, the aircraft will not be able to compete with current kerosene-based aircraft, unless kerosene prices will be raised very substantially. Legislation may be another route to give incentives to the industry for this sustainable technology revolution. Spatial and cost constraints probably still will limit air transport growth.

V-control

The most important limiting factor in mobility is given by the travel time budget (TTB) (see Schafer, 1998). The average TTB at population level is more or less a constant at 70 minutes per day, and does not vary structurally with parameters as GDP and population density. TTB is a strong parameter: if transport systems become 10% faster, the effect will not be 10% reduction in travel time, but 10% more mobility. Schafer (1998) also describes another limiting factor: travel money budget (TMB). The average TMB is about 12% for most countries, but lower if less than 200 cars per capita are available according to Schafer and Victor (2000). Often a relation is observed between GDP per capita and mobility, but this is probably the combined result of TTB and TMB, whereby TTB has the strongest effect.

Transport models show 'elasticities' for travel cost of between 0.25 and 0.5, while the elasticity for travel time is normally about 1.0. The elasticity for parameter *X* gives the amount of mobility change due to a change in *X* (e.g. an elasticity of –0.3 means 1% increase of transport cost will reduce mobility by 0.3%). The restriction of the TTB (a maximum of 24 hours per day) is much stronger than for TMB. Money budgets may vary considerably and costs may be diverted to third parties, for example employers.

For tourism transport growth a third factor may be involved: predictability. Peters (2003) describes how all efforts of the emerging (leisure) travel industry in the 19th and 20th centuries were aimed at increasing predictability. Examples of increasing predictability are the all-in 'excursions' organised by Thomas Cook and the standardisation of the national road system in the USA, both reducing unpredictability for the traveller. Recent developments in the communication technology make it possible for would-be travellers to virtually 'look at the travel and destinations' in advance. This may increase the number of people to venture on longer journeys, TTB and TMB permitting.

The development of transport technology aimed at increasing speed and reducing cost is therefore the pillar of present-day mass tourism development. Without this technological development, mass tourism is impossible. Air transport is the key to long-haul mass tourism. At least theoretically an almost infinite hidden demand for mobility exists as long as transport systems increase transport speed.

Intercontinental travel causes the largest environmental effects. However, the total number of trips is not very large. Generally, 80% of the environmental problems are caused by only 20% of the market. Therefore, reductions may be reached without a total change of the market and compromising human needs for travel and leisure. The solution will be an uncoupling of tourism growth and transport growth. It is important to bear in mind that it is not the number of leisure days that has to be reduced, but the total kilometres travelled during leisure time. Hence, tourism growth is not restricted, but only transport pkm volume growth. Of course, side effects must be studied, for example, the effects on the emerging tourism industry in developing countries. As long as these are based on intercontinental tourism, doubling the income of the poorest 2.8 billion people would require a volume of air transport of 20 to 50 times total air transport in 2000. This is neither sustainable nor possible.

Conclusions

It has been demonstrated that leisure-related transport plays a major role in GHG emissions and climate change. In almost all business-as-usual scenarios the share of transport will grow during the coming decades. Leisure-related transport's share (pkm based) will increase from 33% in 2000 to 44% in 2020. Leisure-related transport emissions and economic growth are unlikely to uncouple, such as in other sectors. On the contrary, leisure transport emissions still grow faster with the sector volume growth in number of trips and even more in number of leisure and vacation days (due to the decreasing average length of stay).

Unsustainable transport growth will result in a share of more than 50% of GHG emissions in a sustainable development scenario. Sustainable leisure tourism will only be possible if leisure-related transport increases its environmental efficiency with a factor 25 between 1990 and 2030. Technology (including operational technology) may increase the efficiency with a factor two to three within this period. Only a combination of renewable energy sources, hydrogen and fuel cell technology seems to be a viable solution for all modes, including air transport. Both cost and spatial requirements for renewable energy will reduce the opportunities for sheer volume growth of transport. Specifically, the probable 30% lower cruise speed of fuel cell aircraft will reduce growth of long-haul transport, because of constant travel time budget. The number of leisure days does not need to be reduced, but only the total kilometres travelled. This makes the economic sacrifice bearable and may help to reduce the negative impacts of climate change on tourism as described in other chapters in this book.

References

Airbus (2002) *Global Market Forecast 2001–2020*. Toulouse: Airbus S.A.S.

Becken, S. (2002) Making tourism in New Zealand energy-efficient – more than turning off lights. In *Sixth International Conference on Greenhouse Gas Control Technologies, October 2002*, Kyoto.

Boeing (2003) *Current market outlook 2003*. Seattle: Boeing Commercial Airplanes (Marketing).

Ceron, J.-P. (2003) Tourisme et changement climatique. Impacts potentiels du changement climatique en France au XXième siècle. In *First International Conference on Climate Change and Tourism*. Djerba: World Tourism Organisation.

Cohen, E. (1978) The impact of tourism on the physical environment. *Annals of Tourism Research* 5 (2), 215–37.

EEA (2001) *Indicator Factsheet Signals 2001. Household Number and Size*. Brussels: European Environmental Agency.

Eurostat (2000) *Transport and Environment: Statistics for the Transport and Environment Reporting Mechanism (TERM) for the European Union. Data 1980–1998*. Luxembourg: Eurostat.

Gössling, S. (2002) Global environmental consequences of tourism. *Global Environmental Change Part A* 12 (4), 283–302.

Gössling, S., Borgström-Hansson, C.B., Hörstmeier, O. and Saggel, S. (2002) Ecological footprint analysis as a tool to assess tourism sustainability. *Ecological Economics* 43, 199–211.

Gugele, B., Ritter, M. and Marecková, K. (2002) *Greenhouse Gas Emission Trends in Europe, 1990–2000*. Luxembourg: European Environmental Agency.

Gugele, B., Huttunen, K. and Ritter, M. (2003) *Annual European Community Greenhouse Gas Inventory 1990–2001 and Inventory Report 2003 (Final Draft)*. Copenhagen: European Environmental Agency.

Heart, B. and Biringer, J. (2000) *The Smart Growth – Climate Change Connection*. Washington, DC: Conservation Law Foundation.

Houghton, J.T., Ding, Y., Griggs, D.J., Noguer, M., van der Linden, P.J. *et al.* (eds) (2001) *Climate Change 2001: The Scientific Basis. Working Group I, Third Assessment Report*. Cambridge: Intergovernmental Panel on Climate Change (IPCC).

Høyer, K.G. (1999) *Sustainable Mobility – The Concept and its Implications*. Roskilde: Institute of Environment, Technology and Society, Roskilde University Centre.
International Air Transport Association (IATA) (2002) *World Air Transport Statistics*. Montreal, Geneva, London: IATA.
ICCEPT (2002) *Assessment of Technological Options to Address Cimate Change. A Report for the Prime Minister's Strategy Unit*, ICCEPT 101. London: Imperial College Centre for Energy Policy and Technology.
Lee, J.J., Lukachko, S.P., Waitz, I.A. and Schafer, A. (2001) Historical and future trends in aircraft performance, cost and emissions. *Annual Review Energy Environment* 26, 167–200.
OECD (2000) *Synthesis Report on Environmentally Sustainable Transport EST. Futures, Strategies and Best Practices*. Vienna: OECD/Bundesministerium für Land- und Forstwirtschaft, Umwelt und Wasserwirtschaft.
Peeters, P.M. (2000) Annex I: Designing aircraft for low emissions. Technical basis for the ESCAPE project. In *ESCAPE: Economic Screening of Aircraft Preventing Emissions – Background Report* (p. 217). Delft: Centrum voor Energiebesparing en Schone Technologie.
Peeters, P.M. (2003) The tourist, the trip and the earth. In NHTV Marketing and Communication Departments (ed.) *Creating a Fascinating World* (pp. 1–8). Breda: NHTV.
Penner, J.E., Lister, D.H., Griggs, D.J., Dokken, D.J. and McFarland, M. (eds) (1999) *Aviation and the Global Atmosphere: A Special Report of IPCC Working Groups I and III*. Cambridge: Cambridge University Press.
Peters, P. (2003) *De haast van Albertine. Reizen in de technologische cultuur: naar een theorie van passages*. Amsterdam: De Balie.
Pulles, J.W., Baarse, G., Lowe, S., van Drimmelen, R. and McMahon, P. (2002) *AERO Main Report. Aviation Emissions and Evaluation of Reduction Options*. Den Haag: Ministerie van V&W.
RCEP (2003) *The Environmental Effects of Civil Aircraft in Flight*. Special Report. London: Royal Commision on Environmental Pollution.
Schafer, A. (1998) The global demand for motorized mobility. *Transport Research – A* 32 (6), 445–77.
Schafer, A. and Victor, D.G. (2000) The future mobility of the world population. *Transport Research – A* 34, 171–205.
Schiphol (2002) *Statistical Annual Review 2001*. Amsterdam: Amsterdam Airport Schiphol.
Snyder, C. (1998) Zero-emissions aircraft? Scenarios for aviation's growth: Opportunities for advanced technology. In *NASA Environmental Compatibility Research Workshop III*. Monterey, CA: NASA.
Theuns, H.L. (2001) Sustainable long-haul tourism: Does it really exist? *Tourism Recreation Research* 26 (2), 77–80.
UN (2001) *Tourism Satellite Account: Recommended Methodological Framework*. New York: United Nations.
van den Brink, R.M.M. (2003) *Scenario's voor duurzame energie in verkeer en vervoer. Beoordeling op verschillende criteria voor duurzaamheid*. Bilthoven: Rijksinstituut voor Volksgezondheid en Milieu (RIVM).
Williams, M. (2002) *Climate Change Information Sheets*. Geneva: UNEP/UNFCCC.
World Business Council for Sustainable Development (WBCSD) (2001) *Mobility 2001. World Mobility at the End of the Twentieth Century and its Sustainability*. Geneva: WBCSD.
World Tourism Organisation (WTO) (1998) *Tourism Vision 2020 Executive Summary*. Madrid: WTO.

Chapter 18

Sustainable Mobility and Sustainable Tourism

KARL G. HØYER AND CARLO AALL

Introduction

This chapter is about the relationship between movement and change. Changes in means, patterns and levels of human movement are interconnected with changes in tourism. These changes have environmental effects. They are in particular part of the global processes causing climate change. A basic understanding is that there is nothing like a neutral movement, there is no social neutrality and there is no physical or environmental neutrality. Illusions of such movement neutralities have a long history. In almost any traditional culture in the world there are old fairy tales about how long distances easily can be overcome through 'flying carpets' or 'seven-mile boots'. No limits, no change and no impacts. Such adventurous tales are also well known from our own country – Norway. It is of course quite another thing to try to make the adventures become reality, actually quite an appropriate description of the current global development in movements and tourism. However, there are limits to the extent of movements, and there are the interrelated limits to the extent of tourism.

This separation between movement and change has a place in the very foundation of *modernization*. The scholars of the antique age did not draw the line in this way, as notable in the physics and biology of Aristotle. To Aristotle movement and change were one thing; all forms of change were understood as forms of movement. Not only did this apply to movements as such, but also to growth and even to changes in colour in nature. The physics of Galileo and Newton were in contravention to this view. Their programme could not be fulfilled if all changes were seen to be based on one common principle. They had to 'separate out' the physical movements between places and points. This became an important part of the new modern world-view, and the prevailing process of modernization of which we still are part (see, in particular, Hägerstrand, 1993).

The crucial concept in this article is *mobility* and not movement. Without mobility of humans there is no *tourism*. It lays in the very etymological origin of the word 'tour'. The origin is the Latin *tornare* and the Greek *tornos*, which refer to the movement in a circle around a central point or axis. The word tourism, then, actually means taking a round trip. The core lies in the movement itself – away from the starting-point and back again (Høyer, 2000). This is clearly expressed in our own country. In Norway the term applied is *travel life industry*. Tourism is a more limited category within this term. In the Norwegian language a tourist was first of all a rich foreigner. Originally, you found them in upper-class hotels along the fjords of Western Norway, or driving in open seven-seat cars and in first-class trains or coastal ship compartments (Welle-Strand, 1981). Today, however, *tourist class* in travelling is actually low class, and the use of the term is closer to Henry James': 'Tourists are vulgar, vulgar, vulgar' (cited by Pearce and Moscardo (1986: 21) cited in Urry (1995: 129). A rather contradictory view as all Norwegians themselves have become tourists during this century of change.

Like movement, mobility is also very much handled as a category in itself, separated from change. But in other contexts, and not least in our daily language, we are closer to the understanding of antiquity. For example, we apply terms like occupational mobility, population mobility and family mobility: all implying change, but with physical movement as an important precondition (Walzer, 1990).

Leisure Time Mobility

With Norway as an example we shall describe the historical development of mobility connected to transport. The importance of changes and growth in various transport technologies and their infrastructures is highlighted. It may be susceptible to a critique of *technological determinism*. However, it should become clear that we are far from such a position. The historical importance of institutional factors and changes is emphasized. A particular focus is on the relational aspects to tourism development. The thesis is that tourism development can only be fully understood when it is related to developments in transport and mobility. But there is an important reciprocity to this. As transport is necessary to tourism, so tourism is necessary to transport. There is no tourism without mobility, but neither is there mobility without tourism, at least in a modern sense.

Relations between transport and leisure time have always been very close. Even if today we find *leisure time mobility* very dominating, the phenomenon is not at all new. Leisure time activities – sport and tourism – have paved the way for the growth in various transport means. Horse transport only had minor importance in this relationship. The 'new times' were first of all augured by the *bicycle*. A really improved type – with iron frame and pedals on the front wheel – was launched at the World Exhibition in Paris in 1867. It was called a *velocipede* – with connotations to *velocity* – and soon become very popular, particularly after Dunlop introduced the rubber tyre in 1888. Sports and other leisure time uses were crucial in marketing the new transport means, and bicycle sport arrangements had become very extensive by the late 1800s. All users were called *cycle riders*, to remind us of *horse riders*, and of course horse riding was mostly for sport and leisure time with the organised transport function of horses at that time being in the form of coaches. Today we only call them cyclists, but cycle sport and ever more indigenous leisure time use of cycles are just as important. For example, we talk today of cycle tourism, as a separate form of tourism, which some connect to the concept of *sustainable tourism*. But it has changed in very important ways: today we transport the bicycles by cars or we travel by planes to *enjoy* – as we say – the pleasures of cycle tourism and many travel by cars to health studios in order to cycle inside, without going anywhere – a form of *virtual mobility* in late modernity (Urry (2000) differentiates between three different forms of *mobilities*: *corporeal*, *imaginative* and *virtual*).

From Velocipedes to Automobiles

The velocipedes were pure marvels of speed, and all the cycling required higher quality roads than horse transport. Major road improvements were made both in urban and rural areas. The bicycle, so to speak, paved the way for the new transport means – the *automobile*. But at first automobiles did not thrive any more than horses and bicycles. The first automobiles, in Norway in the early 1900s, were met by large opposition, both among politicians and common people. In order to increase their use extensive marketing efforts were needed. One strategy was to change its name. *Automobile* was considered too awkward. A Danish newspaper set up a name-competition in Copenhagen in 1902, and the winner was quite simply *bil*, which since then has been the name for *car* in all three Scandinavian countries (Tengström, 1991).

Stronger efforts were however required. Most people did not need cars in their daily lives; they walked or bicycled to work and nearby shops. New urban rail systems gave the opportunities for longer travels. Thus, cars were neither needed for *production*- nor *reproduction*-related mobilities. Close links were on the other hand made between the car and a third category of mobility: *leisure time mobility*. Car use started as a purely leisure time activity, and this link has later been fairly prominent during the whole car-age history. Early advertisements presented cars as means to go out into the fresh country air and landscapes, and away from the industrialized and polluted cities. This was even marketed as a health measure: while driving in open cars one could breathe in fresh air and thus help to cure tuberculosis caused by city life. *Sport* has also played an important marketing function. Car *races* – with connotations to horse races – were established very early on. In the Scandinavian languages they are called '*løp*', which means 'runs'; there were horse, bicycle and car *runs*. Speed limits were continuously broken, and newspapers had almost daily reports about ever more adventurous car travels, even across whole continents. And the first name – *automobile* – should of course give the impression of a mobility that was automatic: they moved around all by themselves. The word *auto* in automobile is originally Greek and means something that functions by itself. *Mobile* is a Latin word and means something that is movable. *Automobile* is meant to express movements that take place all by themselves. Our use of the term *automobility* is a mobility that in extent and type has the automobile, the car, as a fundamental precondition, and at the same time conveys the historical illusion of a mobility 'all by itself', without any limits or impacts as in the fairy tales with flying carpets and seven-mile boots. Also such automobility plays a major role in structuring the late-modern societies, where leisure time and tourism are important components (Høyer, 1999; Urry, 2000). Similar emphases on the links between cars and leisure time are given in historical works from many countries and continents (Belasco, 1984; Sachs, 1989; Tengström, 1991; Høyer, 1999).

The institutional barriers to cars were however substantial. The first laws from the early 1900s were based on the principle that all use of motor vehicles on roads should, as a point of departure, be prohibited. New restrictions were enforced after the Second World War. In rebuilding the Norwegian economy one could not afford to import large numbers of cars. Only a limited number was imported every year, and concessions to own cars were given by the national authorities based on individual applications. Concessions were given only to individuals who needed cars in their occupations, and they were mostly politicians, doctors, veterinaries, country policemen and civil engineers. The restrictions were not lifted before 1960.

Car Tourism

It was the 1970s before car use among Norwegians really took off. At that time the private car for many households had become a necessity in order to travel to work, and was no longer only a transport means for leisure time. This was an effect of post-war modern, functionalistic urban planning ideology. Basic principles were zoning and separation of various functions in a modern society: dwellings in one area, work places in another, shops and public services in a third, and leisure time activities in another. New distance barriers for the households were thus created, and the private car was the most efficient mean to overcome them without excessive consumption of time (Høyer, 1994).

But the car has still kept its firm grip on leisure time, and vice versa. Not in the least is this due to the development of car-based tourism. By the 1930s farmers located along roads found that they could get some extra income by renting rooms to car tourists. Later this was expanded to camping sites and areas with smaller cabins, solely dedicated to car tourism. The first cars with *caravans* arrived on Norwegian roads in the 1960s. When working-class people in this period got their first cars, they were generally only used for holiday and weekend travel. More recently, this type of tourism has vastly expanded, and taken a variety of new forms: caravans of ever increasing sizes, camping sites with caravans as permanent summer houses, and the later large motor-caravans which travel everywhere, domestically and abroad. In Norwegian they are called '*bobiler*', or 'dwelling-cars'; this is the ultimate form of automobility where you bring everything with you and you are totally disconnected from any anchors to both space and place. In the same way that these examples of tourism are completely formed by the car, other types of tourism have changed with the car and become totally dependent on it in terms of their current form and size.

Railway Tourism at Sunset

Norwegian railways have never played a similar role in forming tourism. This is very different from the situation in England where railways were built to bring people from main industrial cities to beach areas along the coasts. Whole new towns – such as Blackpool – were developed solely to serve this *railway tourism*. It is actually one of the major forms of mass tourism through history. At that time horse-drawn coaches were the main transport means, and the roads were even subject to severe traffic problems during summer holidays (Høyer, 2000). (A similar incident of mass tourism took place some 1900 years before when roads were built to bring middle-class people from Rome to coastal areas during the hot summer season. Some of these towns are still highly valued as tourist resorts today.)

The new towns – connected to *sun and beach* – represented at that time *sunrise* areas, but have later experienced serious problems of *sunset*. In Norway there was only a short period, mostly after the Second World War, where railways had such a role. Middle-class people, mainly from the capital Oslo, used trains to be transported to skiing resorts in the mountains during the Easter holiday.

Trains, the *iron horses*, were first of all a transport means for the earlier forms of modernization, the industrial production society. Railways were primarily a matter of developing the national economy. They did not, as velocipedes and automobiles did, inspire individual mobility. Connections were to work and not to sports and leisure time. There were no train races, no speed limits to break by individual competitors. The engine drivers were anonymous, wore overalls and were deeply rooted in the industrial working class. When car use really took off in the 1970s railways would soon become a symbol of a sunset society. Cars and airplanes were the symbols of the new modern times with expectations of unlimited individual mobilities.

Nevertheless, this changed quite substantially in the early 1990s. Railways were then presented as transport systems for the future sustainable society and crucial means to reduce greenhouse gas emissions from transport. They became synonymous with *sustainable mobility*. But this would soon become more political rhetoric than physical reality. Norwegian transport policies have never managed to give rail transport an important role. The development has been rather the opposite. To the extent that there have been any changes, these changes have strengthened the role of rail as transport means for urban and intercity commuting. This has served to further cement rail transport to the production society. It is our thesis that rail never will become a tool for sustainable mobility before it takes on a completely new role to serve leisure time mobilities, and one of the main problems in this context is that railways never have been able to form their own tourism to the same degree as highways and airways (Høyer, 2003).

From Air Heroes to Airways

Through their whole history Norwegian airways have been tightly connected to leisure time. Just as for velocipedes and automobiles, this is actually where the airplanes started. Sports played a particularly important role. Through several decades airplanes were almost completely a matter of breaking speed limits, breathtaking air acrobatics and adventurous travels across seas and continents. The flying carpets had materialized. Air pilots were called *flyers;* a term we still use even when they have become captains of large air buses, or really ships in the air. They have almost been among our foremost heroes: an impression only strengthened during the Second World War when the *air heroes* achieved a status of quite unlike any other group, and it was some of these same heroes who established the first Norwegian airlines. In Denmark the name of one airline was the same as the secret code name its owner had during the War.

As they became a collective means of transport, airplanes have incredibly managed to keep their association with unlimited individual freedom and mobility. It is for airways as for highways, but very much in contrast to railways. The *institutional system* that has developed to foster individual mobilities is an impressive story. It has been a genuine global system almost from the very start: no national borders or pecularities to substantially inhibit mobility as in the case of rail transport. The first major Norwegian air company was, for instance, inter-Scandinavian for all three countries. All airplanes are similar, manned with crews that all look and behave alike: male air captains with supreme and hero-like control and the most beautiful female air hostesses to attend to comfort. Their language is global, almost like Esperanto at last put into practice. This is a language also spoken by all the international travel agencies that are integrated. All airports and airport hotels are alike, as are the wide surrounding airport landscapes with their same giant poster parades along connecting roads. New high-speed links have even made rail transport a part of the same institutional system, not competing with but instead only making air mobility even more attractive. During the last decade – with its focus on sustainable transport requirements – such a link to the capital Oslo is actually the only success story in Norwegian rail transport (Høyer, 2003).

Tourism is an integral part of the system. Air transport and tourism have grown like Siamese twins. The connections are just as strong whether it is a matter of scheduled or charter flights, or a matter of production-related mobilities in the form of so-called business travels or leisure time mobilities in the form of holiday travel. Air transport is the very backbone of *hotel tourism* in many cities, of *event tourism* related to major music and sport events, of international *cultural* and *air-shopping tourism*, and national and international *conference tourism*. Already in the 1960s the first charter flights started to transport Norwegians to the sun and warm beaches in Southern Europe. Growth has been exceptional in all the years since, not only in numbers but also through a continuous increase in distance. This has made really long-distance travel an opportunity for all social groups, a type of travel that today takes a large share of the total mobilities for many households. As in countries like Greece, Spain and Portugal it has caused fundamental changes in thousands of local communities and their populations, the changes in Norwegians' mobility patterns and extents have been no less. One of the changes is the development of a new form of *dwelling tourism*, in which Norwegians settle in Southern Europe for large parts of the year and travel to and from by plane, in some cases in the form of regular plane commuting. It is the Zygmunt Bauman (2000) *liquid modernity* version of the former road caravans, but with incredible increases in individual mobilities.

Norway has become a society of *aeromobility*, just as it is a society of automobility. This is a global mobility that in extent and type has the aeroplane as a fundamental precondition, and at the same time conveys the historical illusion of movements without limits or impacts as in the fairy tales of flying carpets. It also plays – as automobility – a major role in structuring the late-modern societies, where leisure time and tourism are particularly important components (Høyer 1999).

Mobility and Climate Change

Table 18.1 illustrates the growth in average mobilities for Norwegians between 1850 and the year 2000 (see Høyer (1999) for an account of the data and methods applied to produce these figures). The figures exclude walking and cycling, but cover all other transport means used for passenger transport, including horse transport. While the average daily mobility was about 50 metres per capita in 1850, it has grown to almost a thousand times as much in 2000, and is now above 47 kilometres per day. This includes all mobilities carried out by Norwegians when they travel abroad. The figures demonstrate the importance of automobility, but also the stark growth in aeromobility the last 30 years. An average Norwegian now travels by international airplanes an extent equal to 9 kilometres a day, or more than 3000 kilometres a year. As many Norwegians still do not carry out such travels, the real mobilites for the individuals involved are much larger. One of the authors of this article – a notorious conference tourist – actually has a total mobility of 180 kilometres a day, mostly in order to provide lectures on the problematic relations between transport and climate change! An estimated half of the total mobility – about 25 kilometres per capita per day – is directly linked to leisure time and to a large extent tourism, this includes almost all of international air transport, and about half of the automobility (Høyer, 1999).

In the introduction we emphasized that movements are not environmentally neutral, just as they are not socially neutral. The more extensive mobilities are, the larger are the environmental impacts and as wider roads mean more traffic, longer travel means larger impacts. The aeromobility is more energy intensive than automobility, which again is more energy intensive than ordinary public transport on the ground, and with increased energy intensity follows larger emissions of CO_2 and thus larger contributions to climate change.

This is illustrated in Figures 18.1 and 18.2. Figure 18.1 shows the yearly emissions of CO_2 from all the transport means used by Norwegians, domestically and abroad. They are the emissions connected to the mobilities in Table 18.1. All emission figures are adjusted with an *RFI-multiplier* according to recommendations in the IPCC (1999) report on aviation. RFI stands for Radiative Forcing Index. For all international air transport we have applied an RFI-multiplier of 2.7, which expresses the larger climate change effects of emissions at higher atmospheric levels. For all other transport – domestic air included – the multiplier applied is 1.05.

Table 18.1 Mobilities of Norwegians: 1850–2000 (km/day per capita)

Means of transport	*1850*	*1900*	*1930*	*1950*	*1960*	*1970*	*1980*	*1990*	*2000*
Car	0	0	0.40	1.20	3.80	13.05	21.16	28.60	29.60
Horse, boat, rail and bus	0.05	0.50	2.10	4.00	5.27	5.39	6.03	5.53	6.27
Air: domestic	0	0	0	0	0.07	0.45	0.99	1.72	2.70
Air: international	0	0	0	0	0.15	1.20	4.03	6.99	8.99
All	0.05	0.50	2.50	5.20	9.29	20.09	32.21	42.84	47.56

Source: After Høyer (1999)

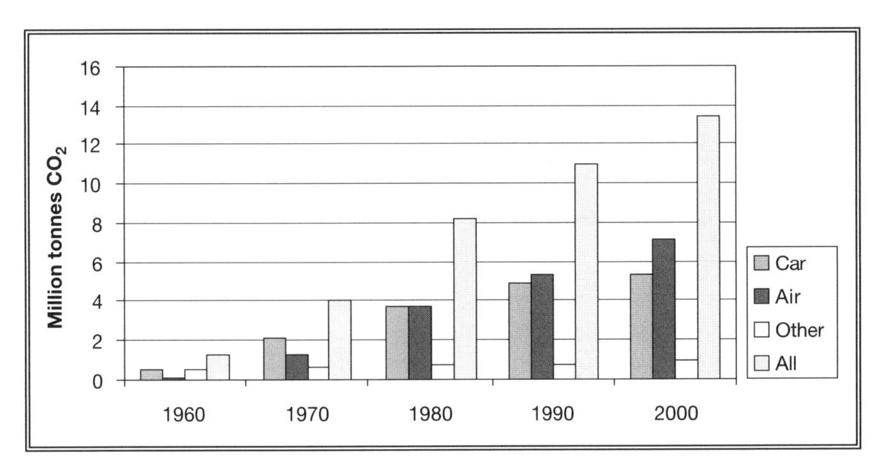

Figure 18.1 Emissions of CO_2 from Norweigan transport (RFI-adjusted figures)

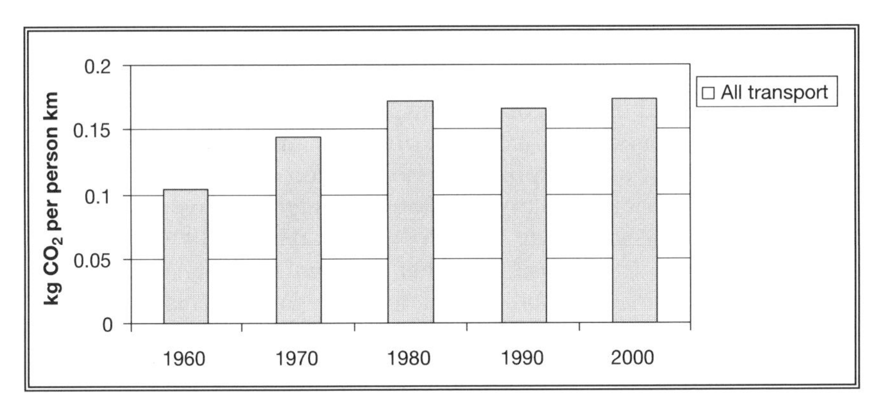

Figure 18.2 Average CO_2 intensity for all Norweigan transport (RFI-adjusted figures)

The growth in emissions contributing to climate change has been very stark, and has increased by more than a factor of three during the last 30 years. Since 1960 the increase has been by a factor ten. It is notable that the climate change effect of the emissions from international air travels by Norwegians now is larger than the effect of their total car travels. Also notable is the relatively minor effect of all other transport on ground level, the rail and bus transport.

Figure 18.2 shows the development in CO_2 intensity of all transport in the same period. Emission factors are adjusted with the same RFI-multipliers as above. During these years we do know that there have been large improvements in energy efficiencies, both for cars and airplanes. But due to the larger share of international air transport, it is worth emphasizing that this has been counteracted by the increased climate change effect of CO_2 emissions.

Towards Sustainable Mobility and Sustainable Tourism

In 2000 one of the authors published an article entitled 'Sustainable tourism or sustainable mobility?' (Høyer, 2000). It was mainly conceptually oriented, and gave an analysis of several theoretical articles published in *Journal of Sustainable Tourism* (*JOST*) during the 1990s. The dominating use of the concept *sustainable tourism* was found to be somewhat of a paradox, in three different ways. Sustainable tourism of course originates in the discourse on sustainable development, basically a global concept, and not least a concept that puts into focus the need to solve environmental problems in the global arena. But the articles mostly conveyed a purely local understanding, applying terms like local carrying capacity. The basic links between tourism and transport were also emphasized, but it was found that most of the articles about sustainable tourism were written as if transport was a non-topic, as if one could have tourism without travel, and then of course transport were not considered to raise any serious environmental issues, surprisingly enough not even locally. It should not come as a surprise that no mention of any relations to the concept of sustainable mobility was found, a highly focused topic both in politics and science during the same period (Høyer, 1999). A conclusion drawn from the analysis was that the two concepts – sustainable tourism and sustainable mobility – needed to be united. This was considered a larger challenge for the tourism side, as tourism and leisure time issues already were integrated in the sustainable mobility discourse. As a background to this article a renewed analysis of the more conceptually oriented contributions published in JOST after our first article was carried out. Included in the analysis are contributions by Sharpley (2000), Hardy and Beeton (2001), Teo (2002), and Hardy *et al.* (2002). It is surprising to find that the above conclusions largely still are valid.

There is no tourism without transport and as we have shown in this chapter neither is there any transport without tourism, at least in a modern sense. Sustainable mobility responds to sustainable development and climate change requirements and implies profound changes in the extents and patterns of transport and mobility. It means severe restrictions on car and air transport. Even though alternative, carbon-free energy sources can help to relieve the problems, it is far from sufficient. The current levels of automobility and aeromobility enjoyed by Norwegians are already well above what can be termed sustainable. A mobility for the future must be based on rail and bus transport as the main transport means for the longer travels, and supplemented with walking and cycling for the shorter ones. All have the potential to take much larger shares of the total mobility than they do today (Høyer, 1999).

These are changes that will deeply affect tourism. Norwegians will again to a large extent have to become tourists in their own country: it is the local, regional and national forms of tourism that first of all need to be developed. New forms of *rail tourism* are crucial, and there are large challenges in the revival of railway stations as nodal points for many forms of tourism. *Intermodality* is a key word in the development of a more sustainable freight transport. It is just as important in creating a sustainable passenger transport, not the least in relation to tourism. A new tourist industry must in particular be based on extensive intermodality – combinations and efficient connections between a multitude of transport means: rail, buses, walking and cycling, and in some cases also boats. There is much to build on.

Acknowledgements

For the part on the historical roots behind the separation of movement and change we are deeply indebted to the works of the late Swedish Professor in cultural geography, Torsten Hägerstrand. He was internationally reknowned as the father of time–space geography, theories and perspectives that are an important background to our article. Our meetings and research cooperation (1989–93) with Hägerstrand have starkly inspired our own works on automobility and environmental change.

References

Bauman, Z. (2000) *Liquid Modernity*. Cambridge: Polity Press.

Belasco, W. (1984) Commercialized nostalgia: The origins of the roadside strip. In B.L. Lewis and L. Goldstein (eds) *The Automobile and American Culture*. Ann Arbor: University of Michigan.

Hardy, A.L. and Beeton, R.J.S. (2001) Sustainable tourism or maintainable tourism: Managing resources for more than average outcomes. *Journal of Sustainable Tourism* 9 (3), 168–93.

Hardy, A.L., Beeton, R.J.S. and Pearson, L. (2002) Sustainable tourism: An overview of the concept and its position in relation to conceptualisations of tourism. *Journal of Sustainable Tourism* 10 (6), 475–97.

Hägerstrand, T. (1993) *Mobilitet* ('Mobility'). Teknikdalen Borlänge, Sweden: Johan Gottlieb Akademien.

Høyer, K.G. (1994) Bilen uten sted – eller stedet uten bil ('Car without Place – or Place without Car'). In T. Forseth (ed.) *Det Gode Sted* ('The Good Place'). Oslo: University Publ./Ministry of Environment.

Høyer, K.G. (1999) *Sustainable Mobility – the Concept and its Implications*. Ph.D. thesis. Roskilde, Denmark: Roskilde University, Department of Environment, Technology and Social Studies.

Høyer, K.G. (2000) Sustainable tourism or sustainable mobility? The Norwegian case. *Journal of Sustainable Tourism* 8 (2), 147–61.
Høyer, K.G. (2003) Den problematiska uthålligheten – Järnvägspolitik och uthålliga transporter ('The problematic sustainability – Railways politics and sustainable transports'). In B.K.Å. Johansson and L.B. Johansson (eds) *Vid vägs ände? Järnvägarna, klimatet och Europas framtida transportpolitik* ('At the End of the Road? Railways, the Climate and Europe's Future Transport Politics'). Stockholm: Banverket.
IPCC (1999) *Aviation and the Global Atmosphere*. A Special Report of IPCC Working Groups I and III. Cambridge: Cambridge University Press.
Pearce, P. and Moscardo, G. (1986) The concept of authenticity in tourist experiences. *Australian and New Zealand Journal of Sociology* 22, 121–32.
Sachs, W. (1989) Die auto-mobile Gesellschaft. Vom Aufstieg und Niedergang einer Utopie. In F.J. Brüggermeister and T. Rommelspacher (eds) *Besiegte Natur. Geschichte der Umwelt im 19. und 20. Jahrhundert*. Munich: Verlag C.H. Beck.
Sharpley, R. (2000) Tourism and sustainable development: Exploring the theoretical divide. *Journal of Sustainable Tourism* 8 (1), 1–20.
Tengström, E. (1991) *Bilismen- i kris?* ('Car Use – in Crisis?'). Stockholm: Rabén & Sjögren.
Teo, P. (2002) Striking a balance for sustainable tourism: Implications of the discourse on globalisation. *Journal of Sustainable Tourism* 10 (6), 459–75.
Urry, J. (1995) *Consuming Places*. London: Routledge.
Urry, J. (2000) *Sociology Beyond Societies. Mobilities for the Twenty-first Century*. London: Routledge.
Walzer, M. (1990) The communitarian critique of liberalism. *Political Theory* 18 (1), 6–23.
Welle-Strand, E. (1981) Bilalderen. Underveis mot nye tider ('The car age. Towards new times'). *Norges Kulturhistorie, Bind 8*. Oslo: Aschehoug.

Chapter 19

Tourism as Victim, Problem or Solution: Story Lines of a Complex Industry–Environment Relation

LOTTA FRÄNDBERG

Introduction

The problem complex of tourism and environmental change involves a multitude of causal relations at different spatial and temporal scales. Radically different understandings of what is the central issue at stake are therefore possible, and dominant arguments and problem formulations on this theme are shifting over time.

In the last few years increasing attention has been given to *mobility* as a crucial aspect of sustainable tourism development. In a number of recent publications, different attempts are made to assess the contribution of tourism transport to climate change and energy use at individual, national and global levels (Frändberg, 1998; Høyer, 2000; Gössling, 2000, 2002; Becken, 2002; Gössling *et al.*, 2002) (see also Chapter 1, this volume). In all of this literature, it is emphatically asserted that proper attention to the environmental consequences of tourism transport has long been due. It is claimed that, in spite of a prolific literature on the topic of sustainable tourism, the issue of tourism transport 'remains unexplored', has been 'neglected',' virtually excluded' or 'largely ignored' in this context (Gössling, 2000; Becken, 2002; Gössling *et al.*, 2002).

This rather conspicuous neglect or marginalisation of transport and climate change issues in relation to sustainable tourism may be approached and explained from several different perspectives. Becken (2002) suggests the complex political situation concerning the allocation of emissions from international travel, as one important reason. However, several authors have identified a more general tendency within sustainable tourism discussions to limit the concern to the destination environment. Hunter (1995) has explained this as a result of 'tourism-centric' perspectives dominating the sustainable tourism paradigm. The ways in which the *general* idea of sustainable tourism is articulated and reproduced by different actors thus seems to be an important key to understanding the neglect of problems generated by tourism mobility thus far.

In the present chapter, the relation between the issues of transport, climate change and sustainable tourism will be analysed using a discourse analytical approach. The text is structured by four 'story lines' on the tourism–environment relationship that has previously been reconstructed as dominant lines of argument organizing the international policy discourse on sustainable tourism in the early to mid-1990s (Frändberg, 1998). These can be captioned as: (1) tourism as *victim* of environmental change; (2) tourism as *generator of problems internal* to the tourism economy; (3) tourism as *generator of problems external* to the tourism economy; and (4) tourism as *solution* to the environment–development dilemma.

The aim of this chapter is to elaborate further on these themes. The scope and character of the study on which the chapter is based, will be outlined in the next section. Thereafter, the four story lines on tourism and the environment will be described in more detail and related to the issues of tourism transport and climate change. The prospect of an ongoing opening of the sustainable tourism policy discourse to include mobility/transport as a central component will be briefly discussed in the final section.

Research Approach

A study of texts on sustainable tourism

'Sustainable tourism' has been discussed extensively both in academic and policy contexts since the early 1990s. Numerous books, reports and policy documents have been published, international conferences have been held and an academic journal specifically concerned with this issue has now been running for a decade (*Journal of Sustainable Tourism*). Although corresponding sectorised sustainability discourses have developed in relation to many other spheres of human activity, there are few areas where the notion has been adopted as widely and with such enthusiasm as in the context of tourism.

This investigation is based on a study in which more than 60 written texts on the topic of sustainable tourism or tourism and the environment were analysed (Frändberg, 1998). Many of the texts presented on this topic in the early to mid-1990s were produced by international organisations and/or for an international readership. For this reason, and for the purpose of including more extensive texts from many different perspectives, interests or actor categories, the study was international in scope. The texts included were produced by national tourism authorities, intergovernmental organisations, industry associations, consumer organisations and non-governmental tourist or environment organisations (for a complete list of the texts used, see Frändberg, 1998: 154–9). The great majority of the texts were published between 1990 and 1995, although these years were not used as strict delimitations in the study. In order to gain some insight also of developments over time, the main study was complemented by an analysis of texts from a small number of Swedish actors over the longer period 1976–97.

Problems connected to transport between home and destination were largely excluded in these texts. There were no clear differences in this respect, between different actor groups. Instead, the material indicates that the weight attributed to the issue of transport as an environmental problem for tourism, depended on the wider discursive context within which the discussion on tourism and the environment took place. This was most evident in the case of texts from the European Union, where sustainable tourism was discussed both in an environmental and a tourism policy context. In the former, 'sustainable mobility' is presented as an obvious and primary concern for achieving sustainable tourism. In the latter, transport is completely excluded as an environmental issue (Frändberg, 1998: 100–1).

These findings indicate that the tourism policy context has been dominated by a logic, which so far has organised out the issue of long-distance mobility from the agenda of sustainable tourism. To understand this situation better, it is necessary to look more closely at what this discourse is actually about.

Story lines

The concept of 'story line' was introduced into the context of environmental discourse analysis by the Dutch sociologist Maarten Hajer in his work on how the policy discourse of acid rain developed in Britain and Holland during the 1970s and 1980s (Hajer, 1995). According to Hajer, story lines are narratives or lines of arguments that combine elements from many different social domains and suggest rather simple causal relationships in the otherwise rather chaotic variety of discursive elements and knowledge claims characterising environmental discourse.

Recognising story lines as important for how environmental policy develops implies an understanding of the relation between actors and discourse as mutually influential and dynamic. According to this view, ways of speaking and writing do not simply convey or mirror actors' understandings, interests and values. Discursive interaction is rather understood as an important generative and constructive part of social life *through which* meaning is given to physical and social realities (Hajer, 1995; Winther Jørgensen & Phillips, 2000). Particular ways of understanding environmental problems and their solutions are 'co-produced' and 'co-refused' in ways that are not always clear if one limits the analysis to interests and beliefs of individual actors or organisations. It is here that the introduction of 'attractive' story lines can be seen to play a role. An important aspect of what makes a specific story line successful or influential is its 'multi interpretability' or its ability to provide many different kinds of actors with a meaningful way of conceiving the environmental problem in question as well as their own role and interest in relation to it (Hajer, 1995).

It could be argued that tourism as an environmental issue is characterised by an additional dimension of complexity compared to many other such issues. Tourism comprises a multitude of different activities and interactions, the nature, extent and geographical scale of which vary tremendously. Yet 'sustainable tourism' or 'tourism and the environment' are evidently seen as issues which can meaningfully be talked and written about at a generalised level. Now if, following Hajer (1995), the sheer complexity of environmental issues renders the role of narratives or story lines significant, then story lines on tourism and the environment could be expected to play a role in organising the discourse on sustainable tourism.

Four story lines on tourism and the environment will be presented in the following. Even though they are reconstructed separately here, they are not to be understood as completely separate or mutually exclusive. When used in discourse they are to some extent entwined and interdependent.

Story Lines on Tourism and the Environment

Particularities of the tourism–environment relationship

The wide and enthusiastic adherence to the idea of sustainability within tourism policy discourses is in part related to an idea of tourism as an economic activity with a very particular relation to the environment – one which differs in crucial ways from that of most other industries. The most common and basic statement in texts about sustainable tourism is that, for its existence and growth, *tourism is fundamentally and uniquely dependent on the environment*. The environment is said to be 'the core of the tourism product' and natural beauty is seen as being tourism's 'raw material'. Statements such as these are mostly made in a general manner without reference to specific types of tourism or certain circumstances or places, and they are made by environmentalist, government and industrial organisations alike.

Tourism as victim

Since tourism is held to be in this way uniquely dependent on a high quality environment, it is sometimes argued that the general environmental degradation hits tourism harder than it does other economic activities. 'Nature' is seen as a factual or potential tourist resource, which is under attack from industrial activity. The relation between tourism and the environment is in other words conceived of in terms of tourism being a *victim* of environmental degradation. The following is a strong statement along these lines, taken from the report *The Tourism Industry and the Environment* published by the Economist Intelligence Unit in 1992:

> Bathing water at a resort, for example, is polluted by raw sewage from large towns, often including dangerous industrial effluents. Coral reefs, attractive to sports divers and less intrepid holidaymakers in glass-bottomed boats, are destroyed by pesticide run-off from farmland, by soil released by clear-cut logging in the watersheds above the coast, and by blast fishing. Historic sites are being eroded by acid rain. Slash and burn farming is destroying habitat that, as national parks, could generate huge tourism revenues. Uncontrolled poaching is taking to the brink of extinction the very species that tourists are most anxious to see. Global warming is believed to be changing the world climate in ways no one can predict, yet no good businessman should willingly accept a future for which he cannot plan. Industry investment in plant such as aircraft is being depreciated by acid rain damage while beachfront concrete foundations are being eaten away by pollution-encouraged burrowing. (Jenner & Smith, 1992: 13)

Such one-sided images of tourism as victim of environmental degradation are more common in very early texts on tourism and the environment. In the Swedish case, the development over time can partially be described in terms of a succession of dominating story lines. From an emphasis on tourism as victim in the late 1980s, tourism was increasingly problematised as a resource consumptive and environmentally degrading activity during the 1990s. The argument as such is still present as a significant element also in more recent texts, however. In these, the emphasis is not as much as earlier on threats towards tourism from non-tourist activities in the destination area, but rather on the effect that environmental problems of a global or regional character may have on individual tourist resorts and on tourism demand in general. An important case in point here is that the relation between climate change and tourism has thus far mostly been approached and discussed in terms of tourism as victim. A future rise in temperature is often presented as a potential threat to the attractiveness of ski resorts and coastal areas.

When tourism is in this way understood as a victim of environmental degradation, the interests of tourism are seen to coincide with that of the environment. This is sometimes taken to imply that the tourism industry should 'lobby' for the environment. Clearly, environmental problems connected to tourism transport – such as its possibly substantial contribution of greenhouse gas emissions – are along with all other problems generated by tourism, marginalised or excluded when this story line constitutes the dominant perspective.

However, towards the mid-1990s, the discourse on sustainable tourism had become predominantly concerned with problems *caused* by tourism, how these are best remedied and, above all, prevented in the future. Two main story lines about the nature, extent and implications of these problems were found to compete in the discourse studied.

Tourism as generator of problems internal to the tourism economy

One of these is, just as 'tourism as victim', based on the assertion that tourism is uniquely and fundamentally dependent on a high quality environment. The central argument is that, since tourists travel in order to enjoy 'high quality' environments, environmental problems caused by tourism activity will negatively affect the tourism business. Remedying or preventing such problems is therefore *inherently* in the self-interest of tourism entrepreneurs.

In the texts studied, this narrative is mostly signalled by expressions such as tourism 'destroying the product on which it depends' or 'ruining its own main asset'. More elaborate explanations of how this dynamic is meant to work has been provided by different tourism scholars (see Briassoulis, 1995; Cohen & Richardson, 1995). Since Cohen and Richardson's argumentation is particularly explicit, it will be quoted here at length:

> There are important respects in which nature tourism (and indeed all tourism) differs from m/re [manufacturing and resource extracting] industries. In m/re industries, resources are typically extracted at one site but consumed at another. In nature tourism, tourist attractions must be 'consumed' on site. These differences have interesting implications for ecological and economic sustainability. In m/re industries, techniques that ecologically damage the site of operation do not literally degrade the product offered . . . Hence, for inherently unsustainable m/re industries, there is no basic economic motive to behave in an ecologically sensitive manner . . .
>
> In contrast, since the product offered by nature tourism is the natural site itself, ecologically insensitive development will indeed literally degrade this product, especially when development devolves towards hard resort tourism. A site can be abandoned only at great cost; a new site may be difficult to find. Furthermore, the new site may not attract tourists. Unlike m/re industries, where the product stays essentially the same despite changes in sourcing, the unique benefits once offered by the abandoned tourist site cannot simply be transferred to the new site. (Cohen & Richardson, 1995: 109–10)

An important aspect of this 'internality narrative' is that it implies the existence within the tourism economy of a negative feedback mechanism (potential or factual) working in the direction of protecting the quality of the environment. The image of the relation between tourism and the environment, which is therefore sometimes evoked, is one of a *self-regulating system*.

Lines of argument such as these have not escaped criticism in the academic literature on tourism and the environment. Butler (1993) has critically examined arguments concerning inherent self-regulation and feedback in the context of tourism and the environment, mainly from the perspective of destination dynamics and temporal change in tourism. From a somewhat different perspective, the British environmental scientist Colin Hunter has emphatically criticised the dominant 'paradigm' of sustainable tourism development for being too 'tourism-centric' and 'parochial' (Hunter, 1995; 1997). According to Hunter (1995: 155), one of the most serious consequences of this tourism-centric sustainability paradigm is that: 'it encourages inappropriate and inconsistent consideration of the scope and geographical scale of tourism's resource base'. The use of non-renewable energy resources for travel and the contribution to climate change from this and other forms of energy use in tourism are obviously part of what is being thus excluded from consideration.

Tourism as generator of problems external to the tourism economy

The image of internality and self-regulation is also challenged by more inclusive perspectives within the policy discourses on tourism and the environment studied here. These perspectives emphasise the fact that tourism also significantly affects resources, systems or values that are in fact *external* to the tourism economy.

Tourism is here considered an environmental problem comparable to other large-scale economic activities. Although the dependency of tourism on 'high quality environments' may be recognised, the argumentation is not occupied with the environment as tourist resource. Environmental degradation is mostly seen as one dimension in a problem complex involving social as well as cultural changes in communities that are being transformed into mass tourism destinations. It is argued that the needs of residents and visitors may in fact be in direct conflict over limited resources, and the problems brought upon the local communities may be completely or partly invisible or irrelevant to tourists. Remedying such problems is therefore not necessarily in the interest of the tourism industry.

This understanding is most commonly expressed in the form of demands that local communities must be involved in decision-making processes concerning tourism development if sustainability is to be achieved. It is a perspective put forward most strongly in texts from tourism and/or environmental networks or organisations such as Tourism Concern/WWF, Ecotrans and Friends of the Earth, but it is also stressed in documents from some other organisations. (Participation of the local population in tourism development is moreover a highly prioritised item in most definitions or lists of criteria concerning eco-tourism.) In the Friends of the Earth Mednet report *Sustainable Tourism in the Mediterranean*, the negative consequences of tourism are described as follows:

> Inadequate management of tourism has been one of the causes of environmental destruction throughout the whole of the Mediterranean region. This is mainly caused by over development exceeding the carrying capacity of destinations and is most obvious in the massive use of local resources by visitors. This has an inevitable effect on access to these resources by the local communities and most often causes changes in the traditional cultures of the local people. The main negative consequences of tourism are seen in:
>
> - the destruction of wildlife habitats and the loss of traditional livelihood;
> - the mis-use of scarce hydric resources;
> - the proliferation of solid wastes;
> - inadequate waste water management;
> - the degradation of the natural surrounding;
> - the loss of cultural identity and traditional values of the host populations. (Friends of the Earth Mednet, n.d.)

Although the welfare of the resident population is mostly considered an end in itself, it is also common to refer to the meeting of the interests of the local community as a necessary precondition for a long-term success of tourism in the area in question. In the European Community report *Taking Account of the Environment in Tourism Development* (1993) the authors argue at length for the importance of including problems in the destination area, which are generated by tourism but are invisible or irrelevant to tourists, when analysing tourism–environment relationships. By reference to issues such as 'social tension' and 'property and land values', consequences that were previously seen as externalities are constructed as long-term internalities. The story lines of environmental problems being internal vs. external to the tourism economy are thus not completely separate in discourse.

Still, the main conflict over the meaning and objectives of 'sustainable tourism' is to be found between 'internalistic' or 'tourism-centric' approaches on the one hand, and perspectives that emphasise the interests and subsistence of the resident populations on the other. In other words, to the extent that the internality story line has been contested within the policy discourse on sustainable tourism, it has been done so from the point of view of tourism places and local populations in destinations. The important point here is that although the issue of long-distance tourism transport and its contribution to climate change can be seen to potentially challenge any internalistic lines of argumentation, it has not been used in this way in the sustainable tourism discourse studied here. This may be taken to imply that at the time, there was no group of actors of significance in the tourism–environment policy field, who could meaningfully interpret the notion of sustainable tourism and their own role or interest in relation to it, in terms of such a story line.

Tourism as solution

The last story line that will be outlined here, can be initially captured by the title of a large international conference held in Montreal in 1994: 'Building a Sustainable World Through Tourism' (IIPT, 1994). In the narrative underlying this phrase, the relation between tourism and the environment is characterised by notions such as 'mutual dependency' or 'reciprocity'. Not only is tourism fundamentally dependent on unspoilt nature or high quality environments, but these *environments also depend on tourism* for their future existence. Tourism is claimed to represent a 'perfect fit', a 'coincidence' or to 'lie at the point of convergence' between the needs for environmental sustainability on the one hand and economic development on the other. The more moderate claim of compatibility between tourism development and protection of the environment is thus here raised to a notion of tourism as an answer, or *solution*, to the dilemma of sustainable development.

In discussions concerning *ecotourism* the argument of (potential) mutuality or symbiosis between tourism and the environment has been present for a long time (Valentine, 1993). Many ecotourism proponents would argue that the combination of environmental conservation and economic development in fact constitutes the core idea of ecotourism. In an article critically evaluating the practice and discourse of ecotourism, Bandy (1996: 542) describes how many different kinds of organisations all have come to promote ecotourism as an attractive option for sustainable development and that: 'Ecotourism has come to represent not only environmentalist and tourist desires, but also a potential solution to the political-economic problems of environmental decay throughout the world'. Because of the potential of certain forms of tourism to provide incentives for conservation of resources, endangered species or ecosystems, environmental organisations with conservation issues high on their agenda may have come to approach the issue of sustainable tourism *through* the concept and ideas of ecotourism. As a consequence, the relation between tourism and the environment may have been conceived of primarily in terms of tourism as a possibility, a potential means for conservation, rather than as a current generator of environmental problems and resource consumption.

As explained above however, the idea of tourism as a force for conservation, or of nature as dependent on tourism, is not limited to discussions about ecotourism. An important case in point is the *Agenda 21 of the Travel and Tourism Industry*, which is a joint publication by the World Tourism Organisation, the World Travel and Tourism Council and the Earth Council (WTO, 1995). In this document, the vague notion of tourism and environment as mutually beneficiary is developed into a complete story line of tourism as solution to the environment–development dilemma. Arguments about tourism as an incentive for conservation and 'environmental enhancement' are here combined with ideas such as tourism being: a leading industry in terms of environmentally sound technology; a provider of essential infrastructure in destination areas; and a provider of sound growth alternative for developing countries (WTO, 1995: 35–6).

Since it is not only the big tourism-promoting organisations, but also many governments and non-governmental conservation and/or development organisations that advocate tourism as a strong option for sustainable development, the narrative of 'tourism as solution' should be considered also as part of the wider contemporary environmental discourse. 'Internalisation' of environmental costs, 'dematerialisation' of production and 'decoupling' of economic growth from increase in the use of energy and materials are central notions in the narrative of ecological modernisation that currently dominates environmental policy discourses in Europe (see for example Cohen, 1999; Hajer, 1995; Murphy, 2000). Again, understood as an ideal 'tourism-destination-environment system', tourism may appear as a welcome means to replace material and energy intensive forms of production with the production of services and the selling of experiences. Furthermore, nature-oriented tourism may be regarded as the perfect means with which to internalise caretaking of the environment.

However, the contribution of greenhouse gas emissions from tourism transport cannot easily be integrated into this image of 'perfect fit' between tourism development and sustainability. If the growth in the tourism economy is to be 'decoupled' from an increase in energy use and material turnover connected to tourism transport, the time-spatial characteristics of tourism – such as modal shift to slower modes of transport and shorter distances travelled – need to be brought into discussion besides the more uncontroversial measures of minimising pollution and energy use from taken for granted volumes of travel.

Concluding discussion

As the preceding analysis shows, the discourse on sustainable tourism of the early to mid-1990s was occupied with arguments about the potentially or factually benign character of the interaction between the tourists, the tourism business and the destination environment. This situation effectively marginalised or excluded problems caused by tourism mobility – such as its contribution to climate change – as either *irrelevant* or *undermining* in relation to the central ideas promoted. No group of actors of significance in the tourism policy field could meaningfully interpret the tourism–environment relationship and their own role or interest in relation to it, in terms of a story line emphasising long-distance transport as a major problem.

What do these findings then imply in terms of possible future developments of sustainable tourism discourses and policies? Are the environmental problems connected to mobility/flow on the one hand and destination dynamics on the other, so disparate in terms of possible solution strategies and actor groups involved, that an integration of these within a tourism policy context will be extremely hard to achieve also in the years ahead? Or should instead the above-described discourse of the mid-1990s be understood as representing a *temporary closure*, which is open to disturbance if only the right story line is launched by the right group of actors? Is such a shift perhaps already on its way?

The fact that the present publication focuses on the interface of tourism and climate change certainly indicates such an opening. The geographical scope and scale of consideration is here inherently inclusive. Several contributions are furthermore specifically concerned with environmental problems connected to tourism transport, reflecting the increasing academic publication on this topic in the last few years. A particularly interesting example of new thinking in this area is found in an article by Becken (2002) on energy use and carbon dioxide emissions from air travel by visitors to New Zealand. Becken here ventures into a discussion of how to resolve the seemingly contradictory goals of an expanding tourism business on the one hand and reductions of greenhouse gas emissions from transport on the other. In this context she presents a number of 'options to improve the environmental record of international travel' which are: increasing the average length of tourist stay in order to potentially decrease the frequency of long-haul journeys; promoting domestic tourism; increasing promotion efforts in countries that are geographically close to the destination (Becken, 2002: 128). All of these options are directed towards *changing the time-spatial characteristics of tourism travel.* Although such ideas have previously figured as part of what an environmentally aware traveller should take into consideration, they are here being presented as recommendations for policy-makers confronted with the challenge to 'decouple' a growing tourism economy from increases in resource use and emissions.

But what will it take for such an understanding to be influential in the tourism policy discourse? Which actors could be expected to promote it as a central aspect of sustainable tourism? One possible scenario is a direct or indirect cooperation between environmental organisations and the domestic tourism industry of sending countries, in highlighting the connection between greenhouse gas emissions and long-distance travel. As the evidence of, and attention given to, aviation's impact on the global climate has increased substantially in the last few years (IPCC, 1999; Krüger-Nielsen, 2001; Olsthoorn, 2001; Upham *et al.*, 2003), it will probably become more difficult for environmental non-governmental organisations engaged in tourism to limit their concern to issues of conservation or resource consumption in the destination area. To the extent that emissions from international transport will somehow be included in climate change negotiations and a system for allocating emissions from international air transport put in place, national environment and tourism authorities will probably also be mobilised in this direction. For many actors it may however prove difficult to even indirectly question the extent and speed of personal mobility – in particular long-distance leisure mobility – as this would challenge presently highly influential narratives concerning the benefits of globalisation on the one hand and sustainability through ecological modernisation on the other.

References

Bandy, J. (1996) Managing the other of nature: Sustainability, spectacle and global regimes of capital in ecotourism. *Public Culture* 8 (3), 539–66.

Becken, S. (2002) Analysing international tourist flows to estimate energy use associated with air travel. *Journal of Sustainable Tourism* 10 (2), 114–31.

Briassoulis, H. (1995) The environmental internalities of tourism: Theoretical analysis and policy implications. In H. Coccossis and P. Nijkamp (eds) *Sustainable Tourism Development* (pp. 25–39). Aldershot: Avebury.

Butler, R. (1993) Tourism: An evolutionary perspective. In J. Nelson, R. Butler and G. Wall (eds) *Tourism and Sustainable Development: Monitoring, Planning, Managing* (pp. 27–43). Waterloo: Department of Geography, University of Waterloo.

Cohen, J. and Richardson, J. (1995) Nature tourism vs. incompatible industries: Megamarketing the ecological environment to ensure the economic future of nature tourism. *Journal of Travel and Tourism Marketing* 4 (2), 107–16.

Cohen, M. (1999) Sustainable development and ecological modernisation: National capacity for rigorous environmental reform. In D. Requier-Desjardins, C. Spash and J. van der Straaten (eds) *Environmental Politics and Societal Aims* (pp. 103–28). Dordrecht: Kluwer Press.

European Community (1993) *Taking Account of the Environment in Tourism Development*. Brussels: ECONSTAT, European Commission DG XXIII.

Friends of the Earth Mednet (n.d.) *Sustainable Tourism in the Mediterranean.* Brussels: Friends of the Earth Europe.

Frändberg, L. (1998) *Distance Matters: An Inquiry into the Relation between Transport and Environmental Sustainability in Tourism.* Humanekologiska skrifter No.15. Göteborg: Section of Human Ecology, Göteborg University.

Gössling, S. (2000) Sustainable tourism development in developing countries: Some aspects of energy use. *Journal of Sustainable Tourism* 8 (5), 410–25.

Gössling, S. (2002) Global environmental consequences of tourism. *Global Environmental Change* 12, 283–302.

Gössling, S., Borgström Hansson, C., Hörstmeier, O. and Saggel, S. (2002) Ecological footprint analysis as a tool to assess tourism sustainability. *Ecological Economics* 43, 199–211.

Hajer, M. (1995) *The Politics of Environmental Discourse: Ecological Modernization and the Policy Process.* Oxford: Claredon Press.

Hunter, C. (1995) On the need to re-conceptualise sustainable tourism development. *Journal of Sustainable Tourism* 3 (3), 155–65.

Hunter, C. (1997) Sustainable tourism as an adaptive paradigm. *Annals of Tourism Research* 24 (4), 850–67.

Høyer, K. (2000) Sustainable tourism or sustainable mobility? The Norwegian case. *Journal of Sustainable Tourism* 8 (2), 147–60.

International Institute for Peace Through Tourism (IIPT) (1994) *Building a Sustainable World Through Tourism.* Conference Invitation. Vermont: IIPT.

IPCC (1999) *Aviation and the Global Atmosphere.* Cambridge: Cambridge University Press.

Jenner, P. and Smith, C. (1992) *The Tourism Industry and the Environment.* London: Economist Intelligence Unit.

Krüger Nielsen, S. (2001) *Air Travel, Life-style, Energy Use and Environmental Impact.* Lyngby: Technical University of Denmark.

Murphy, J. (2000) Ecological modernisation. *Geoforum* 31 (1), 1–8.

Olsthoorn, X. (2001) Carbon dioxide emissions from international aviation: 1950–2050. *Journal of Air Transport Management* 7, 87–93.

Upham, P., Maughan, J., Raper, D. and Thomas, C. (2003) *Towards Sustainable Aviation.* London: Earthscan.

Valentine, P. (1993) Ecotourism and nature conservation: A definition with some recent developments in Micronesia. *Tourism Management* 14 (4), 107–15.

Winther Jørgensen, M. and Phillips, L. (2000) *Diskursanalys som teori och metod.* Lund: Studentlitteratur.

World Tourism Organisation (WTO) (1995) *Agenda 21 of the Travel and Tourism Industry.* Madrid: WTO.

Chapter 20

Tourism's Contribution to Global Environmental Change: Space, Energy, Disease, Water

STEFAN GÖSSLING

Introduction

As demonstrated by many chapters in this book the implications of global environmental change (GEC) have started to be felt by the tourist industry. However, there is a two-way relationship between tourism and the environment. Tourism and travel, as widespread human activities, are important drivers of GEC (cf. chapters by Becken & Simmons (Chapter 13), Peeters (Chapter 17), and Dubois & Ceron (Chapter 6), all this volume), which has gone largely unnoticed because virtually all analyses of the environmental impacts of tourism have been local in scale. As this chapter will stress, these local impacts add up to global phenomena. In particular, this contribution seeks to assess four of the most important fields of GEC to which tourism contributes and which are interrelated with issues of climate change. These include (1) changes in land cover and land use, (2) energy use and emissions of greenhouse gases, (3) exchange and dispersion of diseases, such as HIV and SARS, and (4) the use of fresh water.

Tourism and Physical Environment

Land use and land cover change

Human land use and land cover change have transformed 30% to 50% of the earth's ice-free surface, and are seen as the single most important component of GEC affecting ecological systems (Vitousek *et al.*, 1997). The use and conversion of lands is central to tourism, but is difficult to calculate. Land is converted for the construction of accommodation establishments, airports, roads, railways, paths, trails, pedestrian walks, shopping areas, parking sites, picnicking areas, campsites, summer-houses, vacation homes, golf courses, ports, marinas, ski areas and lifts, as well as for the production of food to supply hotels and restaurants, to bury solid wastes, to treat wastewater, and for the production of items needed by this industry (e.g. computers, TVs and beds). Obviously, the area influenced by tourist activities, such as beaches, is even larger.

A precise calculation of the total area needed for tourism is virtually impossible, due to the lack of data, the large number of assumptions that have to be made, and the difficulty of including certain tourism-related activities. In order to provide a rough estimate of tourism-related land requirements, data are provided for accommodation establishments, golf courses and traffic infrastructure (calculation/method in Gössling, 2002). As Table 20.1 shows, accommodation accounts for about 1450km^2 of land worldwide, as compared to 13,500km^2 for golf courses, and 500,000km^2 for traffic infrastructure.

In total, tourism-related land use may account for roughly 515,000km^2, even though this calculation obviously excludes a great number of tourist activities and indirect land uses. The figure of 515,000km^2 represents 0.34% of the terrestrial surface of the earth or 0.5% of its biologically productive area. This seems to be a minor share of total land use, but leisure-related land alteration is often concentrated in relatively small areas with ecologically sensitive or, with respect to its biological productivity, ecologically valuable regions. The World Wide Fund for Nature (WWF, 2001) reports, for example, that about 54% of the Mediterranean coastline is now urbanized, mainly with construction related to tourism (e.g. hotels, airports and roads).

Use of energy and emission of greenhouse gases

The use of energy for tourism can be divided in transport-related purposes (travel to, from and at the destination) and destination-related purposes excluding transports (accommodation and tourist activities) (Becken & Simmons, 2002).

Table 20.1 Tourism-related land use (1999)

Land alteration	*Area (km^2)*	*Level of certainty*
Accommodation	1450	Fair
Golf courses	13,500	Good
Traffic infrastructure	500,000	Very poor
Total	> 514,950	

Source: Gössling (2002)

Travel

Evidence from travel surveys suggests that leisure-related travel accounts for about half of all travel in industrialized countries; this is about 20pkm (passenger kilometre) per day (Gössling, 2002). In transitional and developing countries, daily travel distances are shorter, with a lower share of leisure-related travel and a greater proportion of public transport used (cf. Schafer & Victor, 1999, Schafer, 2000). Table 20.2 shows the distribution of leisure distances travelled in industrialized, reforming and developing countries.

According to this estimate, leisure-related travel may account for roughly 8 billion pkm, which compares to 23,970 billion pkm in 2001 travelled globally for all purposes (Gössling 2002). The table also shows that leisure-related transport is unequally distributed: the industrialized countries, which constitute only 15% of the world's population, account for 82% of the leisure distances travelled. This figure is even more skewed with respect to air travel, the environmentally most harmful form of travel. With respect to this means of transport, the industrialized countries account for 97.5% of the distances covered for leisure-related purposes (Gössling, 2003).

In order to calculate the energy use associated with leisure-related transport, it is necessary to multiply the passenger kilometres travelled in industrialized, reforming and developing countries with a factor for energy use (Table 20.3). An estimate for water-borne traffic, which does not appear in travel surveys, was added to the calculations. As shown in Table 20.3, global energy use associated with leisure-related transport is in the order of 13,223PJ, entailing 1,263Mt of CO_2-equivalent emissions (in 2001). For another calculation including development scenarios see Peeters (Chapter 17, this volume).

Accommodation and activities

Energy use in accommodation hotels varies considerably, both with respect to the sources of energy used as well as the amount of energy consumed (Becken & Simmons, Chapter 13, this volume). Based on the estimate that 5.2 billion nights were spent in different accommodation establishments in 2001, and that the average amount of energy used per bed night was 97.5MJ, the resulting energy use is in the order of 508PJ, entailing 81Mt of CO_2-emissions (including leisure and business tourists) (Gössling, 2002).

There is comparably little information on the energy intensity of tourist activities at their destination. Energy intensity of different tourist activities varies widely, and it seems difficult to allocate an average amount of energy for such purposes. Assuming that an average tourist uses 250MJ of energy for 'activities' during a longer vacation, which seems to be a conservative estimate for leisure tourists (cf. Becken & Simmons,

Table 20.2 Passenger kilometre by means of transport: leisure-related purposes (2001)

Means of transport	*Industrial countries*[1] *(population: 900 million)*			*Reforming countries*[2] *(population: 400 million)*			*Developing countries*[3] *(population: 4750 million)*			*Total pkm (billion)*
	%[4]	*pkm/cap/day*	*pkm (billion)*	*%*[4]	*pkm/cap/day*	*pkm (billion)*	*%*[4]	*pkm/cap/day*	*pkm (billion)*	
Car	70–5	14.5	4763	40	1.5	219	20	0.1	173	5155
Air travel	15–20	3.5	1150	5	0.2	29	0	< 0.0	< 0	1179
Other	5–10	2.0	657	55	2.0	292	80	0.4	694	1643
Total	100	20	6570	100	3.75	540	100	0.5	867	7977

Source: Gössling (2002)

Notes:

1. Industrialized countries: Australia, Canada, Europe, New Zealand, Japan and USA.
2. Reforming countries: Bulgaria, Chile, Croatia, Czech Republic, Estonia, Hungary, Korea, Lithuania, Latvia, Macedonia, Malaysia, Poland, Romania, Russia, Slovak Republic, Ukraine.
3. Developing countries: all other.
4. Percentage of distances that are covered with this means of transport.

Table 20.3 Global energy use and CO_2-e emissions in leisure-related transport, 2001

Means of transport	*Energy use*			*CO_2-e emissions*		
	pkm (billion)	*MJ/ pkm*	*PJ*	*pkm (billion)*	*g CO_2-e/ pkm*	*Mt CO_2-e*
Car	5155	1.8	9279	5155	132	680
Air travel[1]	1179	2.0	2358	1179	396	467
Other	1643	0.9	1479	1643	66	108
Water-borne	?	?	107	?	?	8
Total	7977	–	13,223	7977	–	1263

Source: Gössling (2002)

Note: 1. A factor of 2.7 was applied to account for the additional warming effect of air traffic (cf. IPCC, 1999).

2002), global energy use for 'activities' carried out by international tourist may have been in the order of 175PJ in 2000. This calculation excludes leisure-related activities at home, which seem at least of the same order as activities abroad. It can thus be assumed that the total energy use for activities is at least twice as high (350PJ) as that of international tourism alone.

Table 20.4 summarizes the worldwide energy use for tourism-related transport, accommodation and activities, showing that global leisure activities may have consumed approximately 14,080PJ of energy in 2001. Transport is responsible for almost 94% of this total.

Table 20.4 Global tourism-related energy use and resulting CO_2-e emissions (2001)

Category	*Energy use (PJ)*	*CO_2-e emissions (Mt)*
Transport (incl. ship, etc.)	13,223	1263
Accommodation	508	81
Activities	350	55
Total	14,081	1399

Source: Gössling (2002)

Overall, leisure-related energy use accounts for 3.2% (14,080PJ) of global energy use and 5.3% of CO_2-emissions (1,400Mt CO_2-e), even thought these estimates need to be seen as conservative because they do not consider life cycle energy requirements. Lenzen (1999) reports that such energy requirements are, for example, in the order of 25% to 65% of the direct energy use for passenger transport.

Exchange and Dispersion of Diseases

The spread of infectious diseases through human mobility is no new phenomenon. As Crosby (2003) points out, diseases have already spread from Europe to other continents with the rise of the colonial empires. For example, the transfer of the measles to Central and South America may have caused some 30 million casualties among the Indian native population (Wolf, 1982). Steamships in the 19th century globalized disease, with cholera and influenza pandemics spreading globally, often with devastating effects. The 1918–19 influenza pandemic, for example, took a toll of at least 25 million dead (Crosby, 2003). Today, a great number of diseases seem to be imported to Europe, with a speed that has entered a new dimension through the global network of air connections and the growing number of people using these connections. The most recent example is the Severe Acute Respiratory Syndrome (SARS). In the period from 16 November 2002 to 7 August 2003, the disease spread to 30 countries, infecting a reported number of 8422 humans, and causing the death of 916 people (WHO, 2003a).

Tourists are both at risk of acquiring new diseases and they also aid as vectors in the global dispersal of microbes, a process that has been reduced to a matter of hours with the rise of air travel (Wilson, 1997; Rodriguez-Garcia, 2001). In particular, tourists travelling to developing countries are at risk of being infected with formerly unknown diseases. For example, the statistical risk of infection may be as high as 30–80% for diarrhea, 0.3% for Hepatitis A, 0.25% for malaria, and 0.01% for HIV for travellers from Europe and North America to developing countries (Clift, 2000). HIV is of particular concern, because it can only move from one place to another as a consequence of human movement, and the speed with which the virus has moved geographically since the late 1980s is seen as a reflection of the volume and extent of human mobility, particularly tourism (Thirumoorthy, 1990; Quinn, 1994). Other diseases that can be exchanged and spread through travel to tropical countries include hepatitis A, B and C, yellow fever, cholera, dengue fever, filariasis, giardiasis, haemophilus meningitis, haemorrhagic fevers, hantavirus diseases, legionellosis, leptospirosis, listeriosis, lyme borreliosis, meningococcal disease, plague, rabies, schistosomiasis, tick-borne encephalitis, trypanosomiasis, tuberculosis, typhoid fever and typhus fever (WHO, 2003b). Again, there is little information about the global dispersion of these diseases by tourism, but incidences among travellers from North America and Europe to developing countries were 1.25% in the case of acute febrile respiratory tract infection, 0.3% for hepatitis A, and 0.03% for typhoid (India, north / north west Africa, Peru) (Clift, 2000).

Overall, travel to tropical and subtropical countries has been made responsible for a substantial increase in incidences of tropical diseases in industrialized countries (Degremont & Lorenz, 1990; Ostroff & Kozarsky, 1998; Loscher *et al.*, 1999). The relative importance of tourism in this process can be expected to become even more important with increasing tourist numbers and travel distances. Furthermore, a growing proportion of tourist activities is nature-based, adventure oriented and bound to remote areas (Ahlm *et al.*, 1994). All these trends ultimately mean an increasing exposure to a larger variety of (partly unknown) species and pathogens. Tourism may also indirectly increase health problems because reports of epidemics may be suppressed or played down in developing countries to avoid the negative effects of bad health news on the tourist industry (Goldsmith, 1998).

Use of water

Water is one of the most essential resources to humanity. In many countries, water availability has decreased, with an estimated 450 million people living under severe water stress, and an additional 1.3 billion people living under a high degree of water stress in 1995 (Vörösmarty *et al.*, 2000). Tourism can exacerbate water problems in countries facing water scarcity because it is often taking place in regions such as islands and coastal zones, with limited fossil water resources, low renewal rates of aquifers and few surface water sources. From a global point of view, tourism contributes in two ways to water stress: (1) causing a shift in water consumption from regions of relative water abundance to water scarcity, and (2) increasing global water demand because of altered water consumption patterns during vacation. Related to these aspects, water quality may often decrease through tourism, as a result of the discharge of untreated sewage, increasing nutrient loads and toxic substances in adjacent water bodies.

Water consumption by tourism

Water consumption by tourists has been poorly researched. Up to now, there are few detailed studies of tourism-related water use in different geographical settings and accommodation establishments. Water use per tourist varies widely, as exemplified by low consumption rates in city hotels in the high latitudes in contrast to those in large resort hotels in the tropics. The existing literature suggests that water consumption rates may vary between 100 to 2000 litres per tourist (t^{-1}) per day (d^{-1}) (GFANC, 1997; Lüthje & Lindstädt, 1994; UK CEED, 1994; Gössling, 2001). The World Wide Fund for Nature (2001) reports, for example, that the average tourist in Spain consumes 440 litres d^{-1}, a value that increases to 880 litres d^{-1} if swimming pools and golf courses exist. A survey of water consumption in a tropical island (Zanzibar) revealed that water use was lowest in small, locally owned guesthouses (100 litres $t^{-1}d^{-1}$), and highest in luxury resort hotels (up to 2,000 litres $t^{-1}d^{-1}$). The weighted water consumption of tourists was found to be 685 litres $t^{-}d^{-1}$ (Gössling, 2001).

In order to calculate the global water demand of tourists, consumption patterns in different forms of accommodation establishments need to be considered. For example, vacation homes account for 2% of beds worldwide, holiday villages for 2%, self-catering accommodation for 11%, pensions for 12%, campsites for 27% and hotels for 47% (Gössling, 2002). Assuming that occupancy rates in these categories are similar and not further distinguishing between locations in the high, medium or low latitudes, a weighted water use figure can be calculated. Vacation homes are not considered, as they will mostly be used by domestic tourists. Holiday villages (2% of beds) usually have indoor swimming pools, and they may also irrigate gardens (Lüthje & Lindstädt, 1994), two important water consuming factors (Gössling, 2001). Their water use is assumed to be rather high with 250 litres per tourist per day. In self-catering accommodation (11%), pensions (12%) and campsites (27%), water use will be lower. An average water use of 150 litres per tourist per day is assumed for these. In the remaining category hotels (47%), water use varies considerably (UK CEED, 1994; GFANC, 1997; Gössling, 2001; WWF, 2001). For the purpose of the calculation, the average water use per tourist is assumed to be 300 litres d^{-1}. Based on these figures, weighted global water consumption in accommodation establishments is about 222 litres $t^{-1}d^{-1}$, excluding, for example, water used for tourism-related food production. This can be compared to the global average per capita water use (domestic), which is 160 litres d^{-1} (database 1987–99, WRI, 2003). Domestic water use includes drinking water plus water drawn for homes, municipalities, commercial establishments and public services such as hospitals. However, there are great differences between countries. For example, daily per capita use for domestic purposes is 12 litres in Bhutan and 1661 litres in Australia (WRI, 2003).

The total water consumption by international tourists can be calculated only on basis of the available statistics. The World Tourism Organization (WTO) provides data on the average length of stay of tourists (business and leisure) in 97 countries, accounting for about 330 of the 692 million international tourist arrivals in 2001 (WTO, 2003a). Based on the data provided by WTO, the weighted average length of stay of international tourists is 8.1 days (available data for 1997–2001). It should be noted that this calculation excludes several important tourist countries such as the United States, Italy, China, United Kingdom, Russian Federation, Germany, Austria, Hungary, Hong Kong and Greece. Given an average water consumption of 222 litres per day and an average length of stay of 8.1 days, the 715 million international visitors in 2001 may have, for a rough estimate, used 1.3km^3 of water (excluding domestic tourism). International tourism may thus account for 0.04% of aggregate water withdrawal of 3100km^3 per year (data for 1985; Vörösmarty *et al.*, 2000). This estimate includes water used for irrigation, swimming pools, taking showers, flushing toilets, cooking, laundry and cleaning. The figure excludes water used for other tourism-related purposes (particularly building tourist infrastructure, food production, etc.), as well as domestic tourism, the latter being of significant importance (cf. Ghimire, 2001). It seems not possible to calculate the net increase in global water demand caused by international tourism because of the huge differences in domestic water use.

Global shifts in water consumption

The major proportion of tourist flows occurs between six regions: North America, the Caribbean, Northern and Southern Europe, North East Asia and South East Asia (WTO, 2003b). Of the 715 million international tourist arrivals in 2002, 58% took place within Europe, 16% in North and South East Asia and 12% in North America. Together they represent 86% of all international tourist arrivals. Within subregions, about 87% of all international arrivals in Europe are from Europe itself (some 350 million arrivals), while in the Americas, 71% of international arrivals are regionally (92 million) and in the Asia Pacific region 77% (88 million). Six major tourist flows characterize international travel, accounting for around a quarter of total arrivals, including those from Northern Europe to the Mediterranean (116 million), North America to Europe (23 million), Europe to North America (15 million), North East Asia to South East Asia (10 million), North East Asia to North America (8 million) and North America to the Caribbean (8 million).

Based on the global average length of stay of 8.1 days, each tourist travelling to another region may increase the water use in this region by 1800 litres. However, there might be great differences between the regions. For example, tourists to the Caribbean might use far more water as a result of the resort character of many hotels in this region (irrigated gardens, large swimming pools). Travelling abroad also means that water consumption in the source regions is reduced, which needs to be considered calculating the net water increase/decrease of the regions. As consumption patterns vary widely, even within industrialized countries, this is a difficult task. For example, while the global average consumption of water for domestic purposes may be in the order of 160 litres per capita per day, a Dutch citizen uses 71 litres, an Irish citizen 102 litres, a German 174 litres, a US citizen 653 litres, a Canadian 792 litres and an Australian 1661 litres (WRI, 2003). The calculation is further complicated by the fact that domestic uses include municipalities, commercial establishments and public services. The amount of water used for personal purposes may thus be substantially lower. For example, in the United States, per capita use for personal purposes is in the order of 380 litres per day (Solley *et al.*, 1998) as compared to 653 litres for domestic purposes (WRI, 2003).

Basically, any calculation of global shifts in water use through tourism should take into consideration the proportion of leisure versus business tourists, their average length of stay, the composition and character of different accommodation establishments, as well as water use patterns at home and in destination countries. As stated earlier, the available database does not allow for such sophisticated calculations, and a number of assumptions thus need to be made. Table 20.5 shows tourist flows, and the respective water use at home (hypothetical, consumption if tourists had stayed at home) and at the destination. The table illustrates that global water consumption increases by roughly 70 million m^3, if travel from Northern to Southern Europe is included. Looking at regions, it becomes clear that tourism shifts water use from one region to the other.

North America and North East Asia are the regions experiencing a net decrease in water consumption, while Europe, the Caribbean and South East Asia experience a net increase in water consumption (Figure 20.1). The major shift in water use occurs within Europe, with 116 million tourists travelling to southern Europe, particularly the Mediterranean. These movements account for a transfer of about 68 million m^3.

Temporal and spatial aspects of tourism-related water use

Several temporal and spatial aspects of tourism-related water use deserve to be mentioned. First, tourism shifts water demand from water-rich to water-poor areas at different scales, as transfers in water demand may occur both at large regional or continental scales (shifts from Northern Europe to Southern Europe, shifts from Europe and North America to the Caribbean), and at regional or local scales (shifts from the supplying centre to coastal zones). Second, tourists may need water during periods when its natural availability is restricted. As shown in Figure 20.2 for Zanzibar, arrival numbers may be highest when rainfall drops to a minimum. Consequently, this is the period when most water is needed by the tourist industry and recharge of the aquifers through rain is lowest. Temporal and spatial aspects of tourism-related water use thus need to be incorporated into considerations about its contribution to global water use.

Table 20.5 Global flows of tourists between regions and corresponding water use (2000)

Travel flows between regions	*Internat. tourists (million)*	*Water use home (l/cap/day)*	*Total home (million l)*	*Water use destination (l/cap/day)*	*Total destination (million l)*	*Increase/decrease by region (net)*	
North America–Europe	23	300	55,890	222	41,359	N. America	–43,691
North America–Caribbean	8	300	19,440	222	14,386	Europe	20,510
North America–North East Asia	4	300	9720	222	7193	Caribbean	21,579
Europe–North America	15	150	18,225	222	26,973	N. E. Asia	–16,945
Europe–Caribbean	4	150	4860	222	7193	S. E. Asia	20,898
Europe–North East Asia	5	150	6075	222	8991		
Europe–South East Asia	5	150	6075	222	8991		
North East Asia–North America	8	200	12,960	222	14,386		
North East Asia–Europe	8	200	12,960	222	14,386		
North East Asia–South East Asia	10	200	16,200	222	17,982		
South East Asia–North East Asia	5	150	6075	222	8991		
Total	95	–	168,480	–	170,831	–	2351
Europe N. to Europe S.	116	150	140,940	222	208,591	N. Europe S. Europe	–140,940 208,591

Source: Derived from WRI (2003); WTO (2003b)

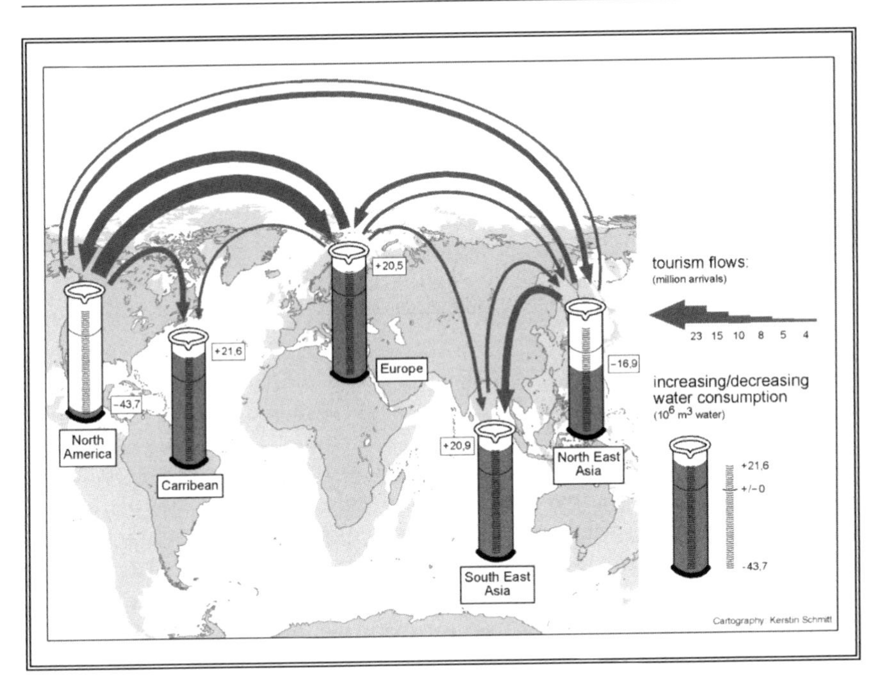

Figure 20.1 Tourism-related shifts in global water use

Source: WRI (2003); WTO (2003b); own calculations

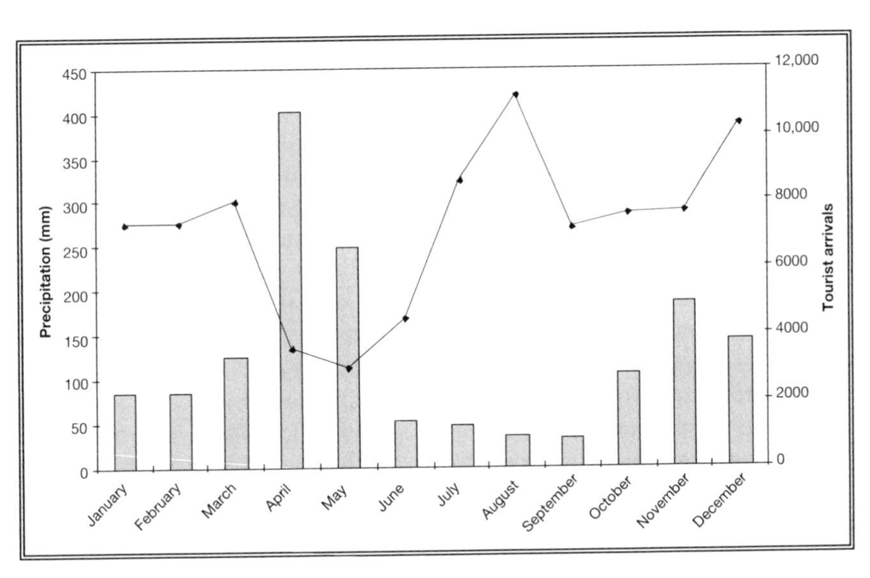

Figure 20.2 Water availability and tourist arrivals in Zanzibar, Tanzania

Source: Gössling (2001)

Water quality

It should be noted that sewage and effluent treatment are greatly neglected in many countries, and it is still a common practice to discharge the sewage from hotels directly into the sea. In the European Mediterranean, only 30% of municipal wastewater from coastal towns receives any treatment before discharge (Smith, 1997 cited in Kent *et al.*, 2002). As tourism substantially increases water use in the Mediterranean and other destinations, this sector is also responsible for global changes in water quality through the discharge of nutrients and toxic substances into adjacent water bodies.

Concluding Remarks

Tourism has been shown to contribute to four major areas of global environmental change, even though the calculations seem to indicate that this sector plays only a marginal role in global environmental change. It should thus be considered that the effects of international travel are related to activities in which human beings only participate during short periods (calculations of water use, for example, are based on an average length of stay of 8.1 days). Furthermore, only one-tenth of the world population participates in international travel. Looking at air travel, the environmentally most harmful form of travel, this proportion is even smaller. Environmental impacts may also be concentrated in areas that are particularly sensitive to disturbance, such as coastal zones. Finally, it should be noted that all aspects of global environmental change are interrelated and interacting. For example, global warming, to which tourism substantially contributes in per capita terms, will also lead to the spread of old, new and re-emerging infectious diseases (Kumate, 1997). Malaria, as one of the most important diseases in terms of casualties, is predicted to proceed into formerly unaffected areas with increasing temperatures – mostly those areas where the people with the least financial means to cope with such diseases live. Climate change may also entail changing rainfall patterns, which in turn will put additional stress on water resources (cf. Kent *et al.*, 2002). Overall, it seems difficult to assess the temporal and spatial scales over which ecosystems will be affected, and what this will mean for these complex, interacting and interdependent systems (Costanza, 2000). All stakeholders in tourism development should thus be aware of the two-way relationship of tourism and global environmental change. Tourism will be affected by global environmental change – but it is also an important driver of these processes.

References

Ahlm, C., Lundberg, S. Fesse, S. and Wistroem, J. (1994) Health problems and self-medication among Swedish travellers. *Scandinavian Journal of Infectious Diseases* 26 (6), 711–17.

Becken, S. and Simmons, D. (2002) Understanding energy consumption patterns of tourist attractions and activities in New Zealand. *Tourism Management* 23 (4), 343–54.

Clift, S. (2000) Tourism and health: Current issues and future concerns. *Tourism Recreation Research* 25 (3), 55–61.

Costanza, R. (2000) Social goals and the valuation of ecosystem services. *Ecosystems* 3, 4–10.

Crosby, A.W. (2003) Infectious diseases as ecological and historical phenomena, with special reference to the influenza pandemic of 1981–1919. Paper presented during the Conference World System History and Global Environmental Change, 19–22 September 2003, Lund, Sweden.

Degremont, A. and Lorenz, N. (1990) Importierte Krankheiten in der Schweiz: Entwicklung und Perspektiven. *Therapeutische Umschau* 47 (10), 772–9.

German Federal Agency for Nature Conservation (GFANC) (ed.) (1997) *Biodiversity and Tourism. Conflicts on the World's Seacoasts and Strategies for Their Solution.* Berlin: Springer.

Ghimire, K.B. (2001) *The Native Tourist: Mass Tourism Within Developing Countries.* London: Earthscan Publications.

Goldsmith, M.F. (1998) Health woes grow in a shrinking world. *Journal of the American Medical Association* 279, 569–71.

Gössling, S. (2001) The consequences of tourism for sustainable water use on a tropical island: Zanzibar, Tanzania. *Journal of Environmental Management* 61 (2), 179–91.

Gössling, S. (2002) Global environmental consequences of tourism. *Global Environmental Change* 12 (4), 283–302.

Gössling, S. (2003) The importance of aviation for tourism – status and trends. In R. Sausen, C. Fichter and G. Amanatidis (eds) *Proceedings of the European Conference on Aviation, Atmosphere and Climate* (pp. 156–61). Friedrichshafen, Germany, 30 June–3 July 2003. Air Pollution Research Report 83. Brussels: European Commission.

Intergovernmental Panel on Climate Change (IPCC) (1999) *Aviation and the Global Atmosphere.* A special report of IPCC Working Groups I and III. Cambridge and New York: Cambridge University Press.

Kent, M., Newnham, R. and Essex, S. (2002) Tourism and sustainable water supply in Mallorca: A geographical analysis. *Applied Geography* 22, 351–74.

Kumate, J. (1997) Infectious diseases in the 21st century. *Archives of Medical Research* 28 (2), 155–61.

Lenzen, M. (1999) Total requirements of energy and greenhouse gases for Australian transport. *Transportation Research Part D* 4, 265–90.

Loscher, T., Keystone, J.S. and Steffen, R. (1999) Vaccination of travelers against hepatitis A and B. *Journal of Travel Medicine* 6 (2), 107–14.

Lüthje, K. and Lindstädt, B. (1994) *Freizeit- und Ferienzentren. Umfang und regionale Verteilung.* Materialien zur Raumentwicklung, Heft 66. Bonn: Bundesforschungsanstalt für Landeskunde und Raumordnung.

Ostroff, S.M. and Kozarsky, P. (1998) Emerging infectious diseases and travel medicine. *Infectious Disease Clinics of North America* 12 (1), 231–41.

Quinn, T.C. (1994) Population migration and the spread of types 1 and 2 human immunodeficiency viruses. *Proceedings of the National Academy of Sciences* (USA) 91 (7), 2407–14.

Rodriguez-Garcia, R. (2001) The health-development link: Travel as a public health issue. *Journal of Community Health* 26 (2), 93–112.

Schafer, A. (2000) Regularities in travel demand: An international perspective. *Journal of Transportation and Statistics* December, 1–31.

Schafer, A. and Victor, D.G. (1999) Global passenger travel: Implications for carbon dioxide emissions. *Energy* 24, 657–79.

Smith, B. (1997) Water: A critical resource. In R. King, L. Proudfood and B. Smith (eds) *The Mediterranean: Environment and Society* (pp. 227–51). London: Edward Arnold.

Solley, W.B., Pierce, R.R. and Perlman, H.A. (1998) *Estimated Use of Water in the United States in 1995.* US Geological Survey Circular 1200. Denver: US Government Printing Office.

Thirumoorthy, T. (1990) The epidemiology of sexually transmitted diseases in Southeast Asia and the Western Pacific. *Seminars in Dermatology* 9 (2), 102–4.

UK Centre for Economic and Environmental Development (UK CEED) (1994) *A Life-Cycle Analysis of a Holiday Destination: Seychelles.* British Airways Environment Report No. 41/94. Cambridge: UK CEED.

Vitousek, P.M., Mooney, H.A., Lubchenco, J. and Melillo, J.M. (1997) Human domination of earth's ecosystems. *Science* 277, 494–9.

Vörösmarty, C.J., Green, P., Salisbury, J. and Lammers, R.B. (2000) Global water resources: Vulnerability from climate change and population growth. *Science* 289, 284–8.

Wilson, M.E. (1997) Population movements and emerging diseases. *Journal of Travel Medicine* 4, 183–6.

Wolf, E. (1982) *Europe and the People without History.* Berkeley: University of California Press.

World Health Organization (WHO) (2003a) *Summary Table of SARS Cases by Country, 1 November 2002–7 August 2003.* Available at: http://www.who.int/csr/sars/country/en/country2003_08_15.pdf.

World Health Organization (WHO) (2003b) *International Travel and Health.* Geneva: World Health Organization.

World Resources Institute (WRI) (2003) *World Resources 2002–2004, Data Tables.* Available at: http://www.wri.org (accessed 4 September 2003).

World Tourism Organization (WTO) (2003a) *Compendium of Tourism Statistics, 2003 Edition.* Madrid: World Tourism Organization.

World Tourism Organization (WTO) (2003b) *Climate Change and Tourism. Proceedings of the 1st International Conference on Climate Change and Tourism.* Djerba, Tunisia 9–11 April 2003. Madrid: World Tourism Organization.

World Wide Fund for Nature (WWF) (2001) *Tourism Threats in the Mediterranean.* WWF Background information. Switzerland: WWF.

Chapter 21

Making Tourism Sustainable: The Real Challenge of Climate Change?

JAMES HIGHAM AND C. MICHAEL HALL

Climate is an important and increasingly significant force shaping tourism and recreation. The preceeding chapters illustrate that climate change will inevitably alter the character of tourism resources, demands for tourism experiences, as well as the flow-on economic and social effects of the tourism industry. It will also significantly change the current competitive balance of seasonal leisure travel destinations. Although the consequences of climate change for tourism and recreation are likely to be profound, a high level of uncertainty surrounds future climate conditions and, therefore, precisely how climate change will bear upon tourism and recreation. What is clear, however, is that climate change represents a new and considerable challenge that demands the attention of, and informed strategic responses from, the tourism industry, government, non-government organisations and researchers within a range of academic disciplines.

The development of science and research agendas is a foremost challenge in responding to the inevitability of significant climate-induced change in the tourism industry. Science and research in the field of tourism climate change is in its infancy. There are encouraging signs that the development and adoption of scientific approaches to tourism climate change research is taking place. The refinement of research methodologies and measurement indices, development of cross-disciplinary research approaches, and emergence of academic discourses in this field have all occurred in recent years. The transition from research based on subjective value judgements to rigorous primary science demonstrates the considerable development of research in this field in recent years. However, the study of climate change, recreation and tourism is fraught with complexity. This is demonstrated, for example, by the uncertainty associated with future climate change scenarios, and the fact that standard meteorological data typically fails to capture the microclimatic characteristics of specific tourism destinations (e.g. coastal resorts) and recreational settings (e.g. alpine regions, littoral zones). High spatial and temporal variation in climate change models compounds the high degree of uncertainty in this field. Moreover, arguably, the relative lack of social science research on the relationship between tourism and climate change has not assisted a better understanding of the human dimensions of global climate change.

While the biophysical impacts of climate change, including extreme weather events, rising sea levels, altered ecosystems and changing patterns of snow cover, are recognised with increasing certainty the implications for the tourism industry, and response and adaptation options of government, industry and tourists remain unclear. It is noteworthy that the focus of research to date has been in the study of meteorology, environmental management and planning, less so in the field of recreation and tourism. The need for social scientists specialising in the study of recreation and tourism to address climate change research issues is evident given that the study of tourist responses to climatic conditions is associated with its own range of methodological issues. Foremost among them, it is necessary that research methods capture the visitor experience of actual weather conditions rather than through the use of hypothetical weather scenarios and/or seasonal means. As weather events become more extreme, mean weather data become increasingly meaningless for this purpose. It is necessary that research in this field is activity-specific, with the distinction between active and passive pursuits particularly relevant given the range of physical factors influencing the thermal state of the human body. Insights into how tourists and recreationists may respond and adapt to climate change scenarios provide a valuable avenue of research in this area. However, the study of tourist responses to climate change cannot take place in isolation from such considerations as economic fluctuations, technological advances, evolving demographic patterns in tourism generating countries, public health issues and environmental factors (e.g. water supply).

The lack of certainty that can be ascribed to future climate change scenarios has important implications for science in this field. First, standardised research methodologies are necessary to establish theory and to expand upon the existing body of rigorous scientific research. Second, alternative approaches to tourism climate change research are necessary. The assessment of sensitivity to climate change, for example, effectively circumnavigates the uncertainty associated with future climate change scenarios. Rigorous and scientific research can be directed towards understanding the sensitivity of tourism resources, or tourists themselves, to climate change. This approach allows the development of responses in anticipation of one of several climate change scenarios unfolding in the future.

It is necessary that the different information needs of tourism planners, government policy makers, tourism operators and tourists themselves are articulated as part of the process of developing a tourism climate change science agenda. Understanding the information requirements of these stakeholders is an important first step in this direction. The climate, tourism and recreation research needs of government and industry have been articulated by the International Society of Biometeorology, Commission on Climate, Tourism and Recreation. In hosting international conferences addressing this subject, the first in 2001, the Commission has highlighted a number of important research themes including:

- Climate as a natural resource for tourism.
- Implications of extreme atmospheric events for tourism and recreation.
- Methods for assessing relationships between climate and tourism.
- Economic significance of climate and climate variability for the tourism industry.
- Effects of weather and climate on tourism demand.
- Needs of the tourist and travel industries for climate and weather information.
- Advisory services for proper climatic adaptation of travellers.
- Development of Tourism Climate Index.
- Weather and climate as limiting factors in tourism.
- Contribution of tourism to climate change.

It is inevitable that as climate changes so too will tourism demand. Tourism is a discretionary activity. Levels and patterns of participation are immediately influenced in situations where the comfort of visitors is compromised. Tourist demand will also be moderated as and when visitors begin to make decisions influenced by travel energy budgets. Such decisions will affect changes in destination choice (distance of travel), transport modes and length of stay. This course of change will promote the importance of regional tourism product development, public transport services, the provision of 'slow' transport options (e.g. cycle networks) and energy efficient options of tourists (e.g. accommodation, attractions and activities). The energy efficiency of tourism businesses will be an increasingly influential element of competitive advantage in the future.

The implications of climate change for the tourism industry raises intriguing questions concerning changing visitor perceptions of tourism destinations, and changing patterns of tourist demand and behaviour. Little is known about precisely how climate influences tourism, or how climate change is likely to bear upon the tourism and recreation in the future. While climate is a resource that is exploited by the tourism industry, the suitability of different climates to different forms of tourism will be increasingly dynamic. Destinations once commonly associated with images of ideal climate are likely to find this status being eroded due to such things as increasing incidence of unacceptable weather, extreme weather events and changing visitor perceptions and preferences.

The study of changing preferences and demands for tourism and recreation can be meaningfully advanced at two levels. First, the macro level of analysis is important in that it affords insights into the implications of climate change for tourism at a general level. Broadly speaking, the development of tourism demand hypotheses and insights into climatic influences on patterns of mobility provide policy makers with the insights necessary to consider appropriate responses. This of course should not obscure important local and regional variations of climate change issues. A high degree of uncertainty surrounds regional changes in climate. This highlights the importance of micro level climate change analyses to the tourism industry. It is noteworthy that tourist responses to various aspects of climate will vary significantly between local/regional destinations particularly in countries such as France that demonstrate broad geographical and climatic diversity. The specific venues and settings at which visitor experiences are achieved, and indeed the activities pursued by tourists and recreationists, will also influence visitor responses to climate change.

Hitherto, tourist responses to climate change have been expressed primarily through the modification of visitor behaviour. Concerns relating to levels of UVB radiation and ozone depletion have induced responses such as awareness of and actions towards sun safety (e.g. the use of sunscreen treatments) and modification of activities across the hours of the day, rather than macro-level changes in patterns of tourist mobility. This is unlikely to remain the case given the expanding range of weather conditions at tourist destinations, and more extreme deviations from average weather conditions. It is inevitable that travel to climate-based high season destinations such as the Mediterranean coastal resorts will be significantly modified as summer climates become more extreme and daily weather patterns prone to periods of volatile change. The more specialised coastal resort destinations are likely to be most vulnerable to climate-induced change in patterns of tourist mobility. This, in turn, generates opportunities for new destinations, and possibilities for domestic tourism in regions that formerly generated outbound travel flows motivated by push or pull factors relating to climate (e.g. the countries of northern Europe).

To date the tourism and hospitality industry response to climate change issues has largely been one of denial. The state of fatalism arises from a range of factors including a lack of resources to implement long-term responses, uncertainty surrounding the manifestations of climate change, and the ineffectiveness of short-term responses to climate change issues. The perception of powerlessness among small business operators represents a significance barrier given the prevalence of small–medium sized tourism enterprises (SMEs) in many part of the world. Destination managers and tourism industry businesses are situated at the bottom of a hierarchy of flexibility in terms of responses to extreme whether events. While tourists and, to a lesser degree, tour operators, may respond immediately to weather event, destination managers and local industry operators are less flexible, and therefore lie particularly at the mercy of the natural elements. However, the vulnerability of the tourism industry to climate-induced natural disasters is an inescapable reality that the tourism and recreation industries must confront with clear and appropriate long-term visions and strategies.

Success in the planning and development of tourism destinations will therefore be determined to an increasing degree by responses to climate change issues relating to water management, land-use zoning, infrastructure standards, location and design and visitor services. These considerations and others such as energy consumption and efficiency will be critical to the planning, development and operation of tourist resorts. The changing demands that tourists may direct towards the tourism industry need to be anticipated where possible. These may extend to the provision of air-conditioned facilities, climate information services, public health services and sun-safe visitor attractions. A competitive advantage will be achieved at destinations that respond efficiently and appropriately to changes in tourist demand.

It is certain that destinations that rely heavily upon climate as a tourism resource will be challenged to contend with changing perceptions of ideal times to visit, extreme weather events, technologies associated with weather hazards and issues such as rising sea levels and ozone depletion. These destinations will be challenged to respond to changing patterns of visitor demand, possibly requiring a long-term strategy to diversify the tourism product. Equally, destinations that have developed in response to the expanding nature-based tourism industry are vulnerable to the manifestations and implications of climate change. In these instances the impacts of climate change may include the altered composition of natural resources, changing tourism and recreation opportunities, altered access to and seasonal availability of outdoor pursuits and compromised levels of comfort and enjoyment. Issues such as the redistribution or regional extinction of wildlife populations, and changing visitor management challenges (e.g. fire risk) highlight the relevance of climate change to nature-based forms of tourism. These require considered and strategic industry responses based on rigorous insights into changing demand and patterns of visitation. They will also bear upon changing infrastructure requirements, impact issues and assessment, visitor services and resource allocations at nature-based tourism destinations.

Responses to changing periods and conditions of snow cover will impact snow-based recreation and tourism operators in terms of operational viability, capital investment and operating costs, energy requirements and the consumption of resources, most particularly water. It will be necessary for operators to also plan and implement strategic responses to changing tourist perceptions of the quality of the tourism product and the impacts of operator responses to changing climatic conditions. This likelihood will in many cases require destination managers to work in closer collaboration with the tourism industry and policy makers to consider strategic responses. One such response may be the development of built attractions to complement or substitute natural attractions and activities, the appeal of which may be compromised by changing climate and extreme weather. Changing seasonal patterns of tourist mobility are likely to emerge as a significant challenge confronting destination managers. As some winter destinations face the marginal operating conditions in the winter season, coastal resort destinations may encounter evolving temporal patterns of tourist mobility across the tourism seasons, with shoulder seasons increasing in popularity at the expense of the traditional summer season.

Within the context of tourism destinations, global climate change will also demand the immediate attention of tourism businesses. Responses that minimise risk and safeguard the commercial viability of tourism operations in times of unpredictable weather have become a priority for tourism businesses in many parts of the world. Such responses have in the past included the design and development of infrastructure and facilities that offer visitors protection and comfort in times of extreme weather. It is increasingly important that such initiatives extend to contingency planning, adaptation of the tourism product, new avenues of promotion and marketing and, in some cases, the development of weather-specific forms of travel insurance.

The necessary response of the tourism industry to global climate change extends beyond the immediate commercial interests of the industry. The industry itself must demonstrate a commitment to assessing and responding to its own contributions to climate change. The tourism industry is characterised by its high level of fossil fuel dependency. As agriculture contributes significantly to methane emissions, so tourism transport contributes significantly to air pollutants such as carbon monoxide, carbon dioxide and nitrogen oxides. Furthermore, the prevalence of small-scale visitor operations is an important factor given the generally high level of energy investment per visitor when tourists are conducted in small groups. Industry initiatives and responses to energy use have been undertaken on the part of large tourism companies in pursuit of competitive advantage and energy efficiency leading to cost saving. The challenge remains for industry associations and governments to assist SMEs to follow suit.

Appropriate and effective responses on the part of the tourism industry and individual businesses need to receive high priority given the likelihood of energy taxes and transport (e.g. car use) restrictions in the future. The estimation of greenhouse gas emissions for specific transport sectors including private vehicles, aircraft and trains, provides compelling insights that are directing research, science and technology in this field now and in the future. Understanding the contribution of the tourism industry to fossil fuel consumption and energy use is an important contemporary research issue. It is evident that tourism and transport policy initiatives must be directed towards the reduction of greenhouse gas emissions in respect to the Kyoto Protocol.

Macro-level insights into the emerging realities of climate change and modified patterns of tourist mobility provide industry stakeholders and policy makers with new challenges. The need for closer collaboration and cooperation between governments is likely in a range of fields including the development and renewal of infrastructure, water regulation and management, and public health. The creation of legislation and planning policies including revised land-use zoning at tourist destinations is a likely consequence of climate-induced changes such as rising sea levels. The need for governments and industry to respond to the contribution that tourism and mobility make to climate change will required the reconsideration of future transport policies. Government policies directed at transport manufacturers, including reduced emission and energy efficiency targets (e.g. automobile and airline manufacturers), are now a reality. The pursuit of transport technologies that reduce fuel consumption via enhanced engine and transmission efficiency, and the exploration of new and renewable energy sources, is also a competitive goal of transport manufacturers. The importance and relevance of such initiatives to the tourism industry will become increasingly apparent in the future.

The issues that have been raised, explored and discussed in this book need to be addressed with priority given the economic importance of tourism in many parts of the world and, perhaps more importantly, the range of immediate and in some cases dire consequences of climate change. These issues present tourism sector stakeholders with the need to implement informed and strategic long-term responses to shifting and increasingly variable regional weather patterns, and dynamic global climates. This is a pressing challenge given the historical inability of the tourism industry to move beyond the short term, to medium- and long-term sustainability planning horizons. It is a challenge that will require willing dialogue and collaboration between industry, government and social and environmental scientists. Understanding and responding to climate change represents one of the more important, complex and challenging issues facing the contemporary tourism and recreation industries.

Index